SYSTEMS BIOLOGY - THEORY, TECHNIQUES AND APPLICATIONS

NETWORK STRUCTURES IN BIOLOGICAL SYSTEMS AND IN HUMAN SOCIETY

SYSTEMS BIOLOGY - THEORY, TECHNIQUES AND APPLICATIONS

Additional books in this series can be found on Nova's website under the Series tab.

Additional e-books in this series can be found on Nova's website under the e-book tab.

Systems Biology - Theory, Techniques and Applications

Network Structures in Biological Systems and in Human Society

Alexander V. Oleskin

New York

For permission to use material from this book please contact us:
Telephone 631-231-7269; Fax 631-231-8175
Web Site: http://www.novapublishers.com

Library of Congress Cataloging-in-Publication Data

ISBN: 978-1-62808-509-9

Published by Nova Science Publishers, Inc. † New York

Contents

Preface

This book focuses on *network structures* in biological systems and in human society. The term "network structure" is used in the literature in at least two different meanings. The broader meaning (denoted by this author as a network *sensu lato*) refers to any system composed of nodes (vertices) connected by links (edges). In terms of this interpretation, the analytical tools that deal with centrality measures, clustering- and community structure-related criteria, small-world behavior, and other network characteristics have provided important insights into the organization and functioning of various objects, including biological systems and human society.

However, there is a narrower interpretation of the term "network" that is predominantly used in the social sciences: a network structure is a decentralized, non-hierarchical system that is regulated by cooperative interactions among its nodes (a network *sensu stricto*). An example can be found in the Internet, which is largely based on this principle. In this work, the term "networks" is interpreted in the latter sense.

The characteristics of networks are considered in this work in comparison to other types of structures that are denoted as (1) *hierarchical (vertical, pyramidal) structures* characterized by a single dominant activity center (central leader, pacemaker); and (2) *(quasi-)market structures* dominated by competitive, rather than cooperative, interactions among the actors involved. This is an interdisciplinary work because the three organizational structures are considered with respect to biological systems and to human society, including its political system.

In the book, much emphasis is placed on interconversions and interactions between structures of different types. Disharmonious interactions between these structures pose the threat of the destruction of the system(s) involved. As far as human society is concerned, this issue is not merely of theoretical interest; recent history provides important examples that demonstrate the economic, social, and political consequences of the hierarchy–network–market imbalance.

The book also demonstrates that network structures, as well as hierarchies and quasi-markets, are widely spread in various forms of life, ranging from unicellular organisms to *Homo sapiens*. Decentralized network structures enable efficient behavioral coordination in the biosocial systems (groups, colonies, families, communities) of individuals belonging to diverse taxa. These network structures can be subdivided into several different organizational subtypes. In this book, they are exemplified by the cellular ("microbial"), modular ("cnidarian"), equipotential ("shoal"), eusocial ("ant"), neural, and egalitarian ("ape") paradigm. Different paradigms actually represent different vantage points from which researchers consider relevant biosocial systems of animals. All of these paradigms are necessarily anthropomorphic to some extent, i.e., they liken animals to humans.

Therefore, it is relatively easy to extrapolate all six of the aforementioned paradigms to human society. In terms of synergetics, they represent "attraction points," which are approached by human network structures under certain conditions. However, while nonhuman organisms simply *exist* in specific (sub)types of network structures, humans are capable of *creatively combining and modifying* them. Moreover, they can create uniquely human organizational scenarios in conformity with various social and political technologies.

Special emphasis is placed on the actual and potential applications of networks with respect to interdisciplinary scientific labs, creative teams of students in classrooms pertaining to innovative and interactive educational scenarios, business enterprises, groups of environmental and educational activists, and various kinds of political organizations, as exemplified by think-tanks and public policy centers, as well as revolutionary and protest organizations, e.g., the Zapatistas in Mexico and the worldwide alter-globalist movement.

The involvement of network structures in the development of civil society and the accumulation of social capital is underscored. Even though the book does not provide detailed recipes for the development of network structures, it is hoped that the information provided by this text will help innovative educators, scientific enthusiasts, environmental activists, political reformers, and all others interested in establishing decentralized, non-hierarchical, and cooperative structures to successfully carry out their creative organizational plans in different spheres of human society.

This work can also be used as a guidebook on network structures that is intended for high school, college, and university students specializing in the life sciences (including ecology, microbiology, ethology, and neurology), medicine, sociology, political science, management theory, psychology, and philosophy. For this purpose, the book offers a glossary; most of the sections in this book include *italicized* summaries for the students to use to recap the basic concepts that are discussed in this book.

In the final part of the *Preface,* the following points should be emphasized. Working on this interdisciplinary book was a challenging task. In particular, some of the fields of science that are addressed in this book are beyond the author's professional competence. However, the author has spent several years studying the specialized literature on sociology, political science, management science, and a number of other relevant fields of the social sciences and humanities. In addition, despite his microbiological background, the author has paid special attention to works that deal with the behavior of insects, fish, primates, and other biological species, as well as to the organization of eukaryotic cellular structures, including neural systems and their analogs.

The author is also aware of another challenge faced by this work. The term *network structures* in its narrow sense (non-hierarchical, distributed cooperative systems) was originally used, predominantly, in the literature on business management. In this work, this interpretation of network structures is extrapolated to other realms that extend far beyond the business world and even human society, i.e., to various biological systems. Generally, the network approach is still under development in the life sciences. Even the researchers that actually use this approach in their research predominantly deal with networks in a broad sense, as is preferred in the works of Steve P. Borgatti, Linton C. Freeman, Mark Granovetter,

Albert-László Barabási, Mark Newman, and other scholars and scientists that are mentioned in this work. Nonetheless, in spite of the indisputable and considerable potential of the concept of *sensu lato* networks and its related analytical tools used in all fields of modern science (including the life sciences and the social sciences), the more specific sense of *network structures*, which is preferred in this work, also seems to hold much potential value in terms of biological and social systems. Providing evidence in support of this statement is one of the main goals of this book. It is my aim that these concepts will be critically evaluated by audience members, including professionals that deal with these relevant fields. Hopefully, this book might also provide some food for thought for enthusiasts who aim to practically develop network structures in various spheres of human society, including within the political system.

Introduction

A major trend in modern science is to establish new links between formerly unconnected fields. Of special interest, in these terms, are interdisciplinary concepts that highlight general principles that may be applied to diverse systems in nature and human society. New fields of science developed in the 20th century such as systems theory, cybernetics, and synergetics, bring together data and concepts related to different areas of research, in order to unravel the in-depth organizational principles of the universe. The behavior of various kinds of complex systems has been the focus of attention of scientists/scholars specializing in a wide variety of disciplines (Wiener, 1948; von Bertallanfy, 1968; Setrov, 1971; Nicolis & Prigogine, 1977; Corning, 1983, 2003a, b, 2005, 2007; Moiseev, 1987; Jantsch, 1992; Camazine et al., 2001; Liseev, 2004; Sumpter, 2006; Knyazeva & Kurdyumov, 2007).

The quest for unitary principles which underlie the diversity of objects and phenomena in the world holds not only theoretical but also practical value. Revealing these general basic organizational principles should enable us not only to explain, but also to predict, and, in some cases, to control the behavior of a given system in order to attain our goals.

This work is concerned with comparative research on systems which belong to two specific realms in the universe: (1) biological systems (living organisms and their parts; various groups and communities formed by organisms, such as families, associations, etc.) and (2) systems in human society ranging from small groups of individuals to state-level political systems and transnational companies.

We will focus on one organizational feature that is characteristic of all systems, regardless of their nature. Specifically, this work deals with *network structures*.

The term *network structure* (or just *network*) has attracted much controversy; it is used in at least two different ways in scientific literature. In the wider sense, a network is defined as any "set of items, which we will call vertices or sometimes nodes, with connections between them, called edges" (Newman, 2003, p.2), and this definition has been adopted by a large number of scientists and scholars around the world. When defined this way, the term "networks" has already been applied both to human society and to the biological realm, and a large number of works comparing networks (in this broad sense) in these two types of systems has been published (see, e.g., Newman, 2003, 2012; Almaas et al., 2007).

However, since the 1980s, another interpretation of the term "network structures" has been used, particularly in the social sciences (Powell, 1990; Castells, 1996, 2004; Börzel, 1998; Meulemann, 2008; Kahler, 2009a, b). According to this more specific definition, not all sets of connected items may be called networks. A network lacks a central pacemaker (leader, dominant element), and its activities and collective behaviors result from cooperation among its members often involving a number of partial leaders with limited power and competence. This is exemplified by the World Wide Web, which is largely based upon this organizational principle.

Networks in the broad sense outlined above will be referred to in this work as networks *sensu lato;* whereas networks in the more specific meaning will be denoted as networks *sensu stricto*.

It is the latter meaning (a system lacking hierarchy and based upon cooperation) in which the term *network* will predominantly be used in this work (except subsection 1.1). Drawing on a number of classical and recent publications (Powell, 1990, 1996; Castells, 1996, 2000, 2001, 2004; van Alstyne, 1997; Börzel, 1998; Meulemann, 2008, Kahler, 2009a, b), network structures will be compared and contrasted with:

- *hierarchical (vertical) structures* with a single central leader (pacemaker) controlling the whole system (Figure 0.1).
- *(quasi-)market*[1] *structures* characterized by the prevalence of competitive interactions among the elements of a system over cooperation among them.

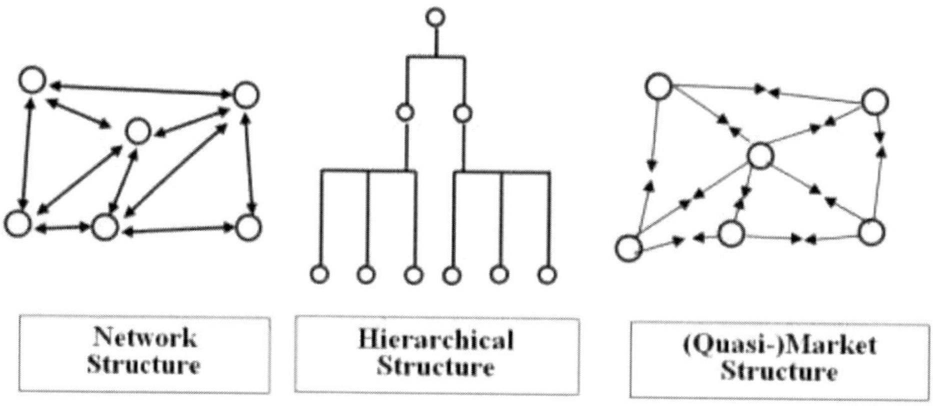

Figure 0.1. Comparing networks (*sensu stricto*), hierarchies, and (quasi-)market structures. ↔, cooperation; →← competition.

This comparative study is expected to provide two kinds of information:

First, data on network organization in biology (in ant societies, fish shoals, the brain, etc.) attracts our attention to important general systemic principles that are also valid for network structures in human society. Despite the self-evident differences between human society and biological systems, we can nevertheless take account of relevant biological facts in order to develop new potentially fruitful concepts in the social sciences.

For example, much attention in molecular biology is currently given to the *guided self-organization* of biomolecular networks. This process depends upon special biomolecules

[1] The prefix *quasi-* should be used for biological systems and the stages of evolution of human society during which there was no market as such, although competitive relations among individuals and groups were quite widespread. See also footnote 15 at the beginning of section 1.4.

(called *chaperones*) that enable other molecules to assemble into functionally competent, biologically useful supramolecular complexes while preventing the formation of dysfunctional, potentially harmful structures. In an analogous fashion, a political system can assume control of the self-assembly of networks in human society, by making good use of special mediator (adaptor) organizations that foster constructive (and prevent destructive) relationships between social/political actors.

Second, of considerable potential interest is the extrapolation of data obtained from human social structures for use in understanding biological systems. Importantly, the behavior of various animal species is often described by behavioral scientists (ethologists) in terms of concepts originally applied to humans. They are exemplified by such terms as altruism, aggression, and cooperation that are currently used concerning animal behavior. Not only people but also microorganisms can form a quorum, and research on microbial "quorum-sensing systems" is one of the most important topics in modern-day microbiology. Obviously, social knowledge should be applied to biological systems with a grain of salt, i.e., a critical analysis of the results obtained by extrapolating social data should be performed.

The comparability of biological and human social systems in organizational and behavioral terms does not diminish the importance of their differences. A significant difference is that only humans can consciously, purposefully establish social structures. Biological systems are simply the result of generations of evolutionary selection. Notwithstanding this difference, naturally existing organizational structures provide sufficiently interesting food for thought for those setting up structures in human society in terms of human/social/political engineering.

Comparative research on the common or analogous features of biological systems and systems in human society is within the competence of an interdisciplinary field called *biopolitics.*

Albert Somit, Steven A. Peterson, Roger D. Masters, Peter A. Corning, and other scholars (see Somit, 1968, 1972; Somit & Peterson, 1998, 2001a, 2011; Corning, 1983, 2003a; Carmen, 1997; Blank & Hines, 2001; Meyer-Emerick, 2007) who are members of the Association for Politics and the Life Sciences (APLS) believed that the term is to be used *"to describe the approach of those political scientists who use biological concepts, with neo-Darwinian evolutionary theory at the center, and biological research techniques to study, explain, predict, and sometimes even to prescribe political behavior"* (Somit & Peterson, 2011, p. 3). Ira Carmen (1997) claimed that "the biopolitical research agenda" is centered on "connections between our species' genetic constitution and our species' political behavior" (p.173).

In contrast, the French scholar Michel Foucault (1978, 2003 [1976]) and his followers defined the word biopolitics differently. According to Foucault's lectures, the political systems of Western Europe developed a system of *dispositifs,* i.e., practical measures to control society and the environment. These included tools to:

- monitor society (using censuses and other demographic methods)
- control human reproduction (using obstetricians and, more recently, family planning centers),
- manage health and morbidity (using health care institutions, sanitation, and hygiene),

- manage work capacity (education, safety regulations, and human engineering), and
- institutionalize social aspects of mortality (funeral institutions).

According to Foucault, *biopolitics* is concerned with the effects produced by the *political system on the biology of its citizens/subjects.* In Foucault's works, the term "biopolitics" is used as a synonym for "*biopower.*"[2] Biopower exercised by the political system of a state includes regulatory measures aimed at optimizing the biological characteristics and the work capacity of the population or, at least, maintaining them within the normal limits.

It should be noted that the term "biopolitics" is used in this book in the broad sense that encompasses most of the more specific meanings mentioned in the literature. Biopolitics is construed as *the whole totality of interactions between biology and politics, including both the impact of modern life-science data and concepts on politics and the influence of politics on human biology and the whole environment* (Oleskin, 2012).

An important subfield of biopolitics deals with the quest for biological driving forces and factors involved in the political behavior of human individuals and groups. Some scholars including Peter Corning (1983) considered human politics a more ancient phenomenon than human society itself (i.e., they admitted a certain primordial form of politics in the social systems of our immediate pre-human ancestors).

This subfield of biopolitics is closely linked to *ethology,* an actively developing branch of biology dealing with the behavior of living beings. In its classical interpretation, ethology focused on behavioral field research with emphasis upon innate, species-specific forms and mechanisms of behavior. However, more recently, ethologists turned to lab studies and paid tribute to acquired (learned) behavior patterns in addition to those directly controlled by genetic factors. Modern ethology is construed as "the study of behavior itself and in an evolutionary framework" (Barlow, 1989, p. 3). A subfield of ethology termed *social ethology* is concerned with interactions between individuals and groups in communities of animals or humans. The subfield of ethology dealing with human behavior, *human ethology,* assumes that "evolutionary theory after all is the basic theory of all manifestations of life and basic, therefore, for any understanding of human behavior, including those facets of human behavior which are the subjects of the various humanities" (Eibl-Eibesfeldt, 1997, p. 13).

In this context, of particular interest is comparative research on *biosocial systems,* i.e., collective systems formed by various uni- and multicellular organisms, including bacterial colonies, fish shoals, insect societies, monkey groups, etc. Of relevance to biopolitics is the issue of whether there are general principles governing the structure and the behavior of biosocial systems, regardless of the kind of living organisms that make them up.

This work concentrates on one of these universal principles–that of network organization. For comparison, we consider alternative organizational principles, such as the hierarchical and the quasi-market principle.

This work is also aimed at demonstrating that biologists and social scientists actually face the same challenge, despite their different professional interests. Any imbalance with respect to the network-hierarchy-(quasi-)market relationships endangers the whole biological or

[2] The Foucauldian followers Maurizio Lazzarato, Antonio Negri, and Michael Hardt did not regard these two terms as synonyms. In their opinion, the control exercised by the state political system over the biology of the citizens/subjects (a "form of power that regulates social life from its interior, following it, interpreting it, absorbing it, and rearticulating it") is to be denoted as *biopower* (Hardt & Negri, 2001). As for *biopolitics,* it primarily refers to organized resistance to biopower.

social system and threatens the system's existence. The problems recently faced by the European Union because of the financial meltdown seems to be an important example showing what economic, social, and political consequences can be caused by disharmony within the network-hierarchy-(quasi-)market system.

Research on network structures holds potentially practical value. Armed with knowledge about network organization and its general principles uniting biological systems and human society, activists from various social, cultural, and political movements can implement them in order to develop effective social/political applications.

This work is concerned with networks that lack a single activity center (leader, pacemaker), so that their behavior results from complex interactions among a number of nodes (partial leaders) that can both cooperate and compete with one another. The operation of decentralized network structures in various biological systems as well as in human society is considered. Networks are contrasted with other types of structures, i.e., hierarchical and (quasi-)market structures.

Chapter I

Comparing Networks with Other Types of Structures

1.1. Networks in the General Sense: Quantitative Characteristics

The interdisciplinary term *network structure* is used in a large number of areas of research dealing with various kinds of systems, ranging from star clusters to crystals to groups of elementary particles to technical devices. Actually, "the more we scrutinize the world surrounding us, the more we realize that we are hopelessly entangled in myriads of interacting webs" (Almaas et al., 2007, p.1-2).

1.1.1. Historical

The term *network* is based on a very ancient concept. Both the broader and the more specific meaning of this word were actually familiar to human society during the course of its history. Nevertheless, to see a network as a decentralized cooperative system would more closely reflect its original, literary, meaning. A network, when defined according to this earlier meaning, can be likened to a *fishing net.* By invoking this image, we can establish some important features that a network has, according to its modern scientific meaning:

1. *Decentralization*: A fishing net or a trawl obviously lacks a central node.
2. *Cooperation* among a network's elements, i.e., meshes that collectively perform their function (catching fish); the cooperation among the net's elements (meshes) is based on its *connectivity/cohesion*: the meshes of a fishing net are connected by threads or ropes

3. *Self-similar /Fractal*[3] *geometry* (any given part is a reduced copy of the whole structure). Part of a fishing net is a fishing net per se: it can be used for catching fish, albeit smaller ones. In other words, a net is a multilevel structure; it may be broken down into smaller nets and can form part of a larger network's structure.

A network's cardinal properties: *decentralization, cooperation, and self-similarity* have acquired culturally-determined meanings during the course of the history of human society. Their spectrum was extremely vast, and literary meanings were imperceptibly transformed into figurative or metaphorical ones by various cultures. Even the network in its original meaning, the fishing net, acquires metaphorical and allegorical overtones. "And Jesus, walking by the Sea of Galilee, saw two brethren, Simon called Peter, and Andrew his brother, casting a net into the sea: for they were fishers. And he saith unto them, Follow me, and I will make you fishers of men. And they straightway left their nets, and followed him" (Matthew 4:18-20, Bible, 2011).

In practice, irrigation systems with their canals and reservoirs actually implemented the fishing net model; taxes were levied and territories were defended using the structural pattern of a fishing net (Startsev, 2011).

Historically, the mathematician Leonhard Euler was one of the first researchers to use the network approach in science. He invoked graph theory, a subfield of mathematics that is very closely related to network analysis. Euler tried to solve a problem concerning seven bridges in Königsberg: to find a walk through the city that would cross each bridge once and only once. Euler revealed that the problem has no solution.

The term *network* was used in the humanities for describing relationships between individuals and groups in society, starting from the turn of the 20th century. Works by Georg Simmel (1908), the founder of the formal school in sociology, already contained the term "social network". The anthropologist Alfred Radcliffe Brown pointed out that "direct observation does reveal to us that… human beings are connected by a complex network of social relations. I use the term "social structure" to denote this network" (Radcliffe-Brown, 1931). He urged other anthropologists to conduct systematic studies using the network approach.

"Alfred Radcliffe Brown saw the social structure as a network of relations, while his student and colleague Lloyd Warner explored network cliques in his community studies. Some of Warner's ideas had been sharpened during his involvement in the famous Hawthorne studies, where the researchers had been struck by the similarities between the electrical wiring diagrams that littered the Hawthorne factory and the patterns of connection among the workers who produced this wiring" (Crossley et al., 2009, p.2).

In the 1930s and the 1940s, the psychiatrist Jacob Moreno and the psychologist Helen Jennings depicted, in the form of network structures, patterns of inter-individual interactions in their research concerning informal leadership in the US prison Sing-Sing and the Hudson School for Girls (Moreno, 1932; Jennings, 1943). Also of interest was the work of Elton Mayo and his colleagues concerning social networks formed by factory workers in the 1930s in Chicago as well as the mathematical models of Anatol Rapoport, as emphasized by

[3] Generally, *fractal* structures are *self-similar*, which means that they are characterized by similar or even identical structural patterns at different levels/scales. A microbial colony typically represents a fractal structure. Each of its parts (microcolonies) organizationally is a reduced copy of the whole colony.

Newman (2003, p.5). The "Southern Women Study" conducted in 1939 and published in 1941 dealt with 18 women from the American south that attended 14 social events. In the form of a network, the authors showed which of the women attended which events (Davis et al., 1941; quoted according to: Newman, 2012, p.380). A rigorous meaning of the term network was used in the social sciences beginning with the work of J. A. Barnes (1954).

More recently, this area of research has received considerable attention, and there have been a large number of publications in areas ranging from theoretical physics to computer science, the social sciences and the humanities, as well as the life sciences. "Increasing globalisation has seemingly interconnected everything, our communication networks (e.g., internet, mobile phone) and transport systems span the globe, the economic integration of nations and communities becomes ever closer and more complex with a growing number of multinational organisations and environmental problems transcending national boundaries. The fact that our world is becoming more and more interconnected has led to an exponential interest in understanding networks" (Krause et al., 2009, p.967).

1.1.2. Broad Sense of the Term "Network": Networks *Sensu lato*

To reiterate, a *network structure*, according to the meaning predominantly used in this work represents a complex system of cooperating elements (nodes) that lacks a central activity-controlling unit.

However, in this subsection, the alternative interpretation of the word "network" that is widely used in the literature on "network science" should be considered. A network, according to this meaning, is "a collection of points joined together in pairs by lines. In the jargon of the field, the points are referred to as *vertices* or *nodes* and the lines are referred to as *edges*", or links[4] (Newman, 2012, p.1). As for human society, "networks are a way of thinking about social systems that focus our attention on the relationships among the entities that make up the system, which we call actors or nodes" (Borgatti et al., 2013, pp.1-2). In many network structures, links between nodes differ with respect to their strength, the efficiency of the flow of information, and other features; this criterion is called link weight (w). If we adopt this more inclusive interpretation, network structures can be classified into "centralized networks" (essentially hierarchies) and "decentralized networks", i.e., *sensu stricto* networks aswell as (quasi-)market structures..

The characteristics of networks in the broad sense have been explored in mathematical terms. Many of the results of this mathematical analysis can be applied to networks *sensu stricto,* and used to distinguish them from both centralized hierarchies and competition-dominated (quasi-) market structures.

According to the broader meaning, the term "network structure" has been used by a large number of scholars in the social sciences, including Stephen D. Berkowitz, Steve P. Borgatti, Linton C. Freeman, Mark Granovetter, and many others. Currently, this more general sense of the word is quite popular in sociology, social psychology, and a number of other fields including the *en vogue* discipline of "network science" (see, e.g., Granovetter, 1973, 1985; Freeman, 1979; Scott, 2000; Segel & Cohen, 2001; Barabási, 2002; Newman, 2003, 2012; Watts, 2003; Evans, 2004; Newman et al., 2006; Borgatti et al., 2013). In human society, this

[4] In the following, the terms "nodes" and "links" will predominantly be used instead of "vertices" and "edges".

interpretation of the term *network* can be illustrated using the following examples: (1) a family, whose members are nodes (vertices) connected by their ties (links, edges); (2) an organization considered in terms of the relationships (links) between its members or their groups (nodes); (3) a number of commercial enterprises bound together by contract-based ties; and (4) a social movement whose activists are united by a specific ideological platform.

Networks are also exemplified by transportation systems, financial organizations with numerous branch offices, interwoven city utility grids, the World Wide Web, and the Internet. In general, the network approach has been successfully applied "to an extraordinarily variety of different communities, issues, and problems, including friendship and acquaintance patterns in local communities and in the population at large..., and among students... and schoolchildren..., contacts between business people and other professionals..., boards of directors of companies..., collaboration of scientists, movie actors... and musicians..., sexual contact networks... and dating patterns..., covert and criminal networks such as networks of drug users or terrorists..., historical networks..., online communities such as Usenet...or Facebook..., and social networks of animals" (Newman, 2012, pp.38-39).

In terms of the broad interpretation of networks discussed in this subsection, they can be classified using a variety of criteria including the following:

- Networks can contain identical or different nodes/vertices. "Taking the example of a social network of people, the vertices may represent men or women, people of different nationalities, locations, ages, incomes, or many other things." (Newman, 2003, p.3).
- Network links/edges can be identical or they can differ with respect to their importance (weight) or other features. For instance, relationships between individuals or groups can be formal or informal, friendly or hostile, and cooperative or competitive.
- Network links are considered directed if they run in only one direction (such as a one-way road between two points or connections in the World Wide Web), and undirected if they run in both directions (Newman, 2003, 2012). A whole network is directed if all of its links are directed.

Bipartite networks (graphs) contain nodes of two distinct types, with links running only between unlike types. "Affiliation networks in which people are joined together by common membership of groups take this form, the two types of vertices representing the people and the groups" (Newman, 2003, pp.3-4).

In addition, network structures can be quantitatively characterized in terms of a number of variables that are currently widely used in the literature on "network science". These network properties include node degree, clustering, path length, and others; they will be discussed in the following subsection.

In biology, molecular networks formed by functional proteins (enzymes, receptors, etc.) and genes connected by regulatory links have recently received a relatively high degree of attention (see, e.g., Barabási & Oltvai, 2004; Hahn & Kern, 2005; Stumpf et al., 2007; Almaas et al., 2007; Newman, 2012). In fact, "most biological characteristics" of living organisms such as heritable features including biopolitically important behavioral traits, arise at the cellular level "from complex interactions between the cell's numerous constituents, such as proteins, DNA, RNA, and small molecules". As a result, "interaction webs, or networks (including protein-protein interactions, metabolic, signaling, and transcription regulatory pathways) emerge from the sum of these interactions" (Barabási & Oltvai, 2004, p.101).

Network science has also been applied to animal groups and communities (Croft et al., 2008; Hill et al., 2008; Wey et al., 2008; Whitehead, 2008; Krause et al., 2009, and other recent publications). The network paradigm (in the broader sense) was used, for instance, in studies with insect societies (e.g., honeybee families, Naug, 2009), fish (Croft et al., 2004, 2005, 2008) and dolphin (Lusseau, 2003; Lusseau & Newman, 2004) schools, as well as groups of bats (Patriquin et al., 2010), meerkats (Madden et al., 2009), spider monkeys (Ramos-Fernández et al., 2009), and chimpanzees (Le Hellaye et al., 2010). Network analysis was applied, e.g., to dominance relationships in centralized network structures (hereinafter refered to as hierarchies) formed by primates (Appleby, 1983; Archie et al., 2006; Flack et al., 2006; de Vries, 2006). In terms of the life sciences, "one of the most attractive features of the network paradigm is that it provides a single conceptual framework with which we can study the social organisation of animals at all levels… and for all types of interaction (aggressive, cooperative, sexual etc.)" (Krause et al., 2009, p.687).

Mathematical analytic tools can be widely used to describe interactions among the parts (nodes) of various biological systems. "Modern social network analysis can contribute novel insights to animal behavior and how these findings may influence applications of behavioural studies. A common theme is the importance of relational data for understanding overall structure and the roles of individuals within that structure." (Wey et al., 2008, p.337). Representing an animal community as a network (or set of networks) helps us study, for example, the spread of contagious diseases or information within it.

Of particular note is research on the nervous system, including the brain, in which the concept of neural, or neuronal, networks was widely used. The work of McCulloch & Pitts (1943) was the first to construct a model of neural networks consisting of interconnected groups of nervous cells. Networks of blood or lymph vessels are studied in animal physiology. In a similar fashion, networks of vessels (tracheids, tracheas) in plants are dealt with in botany.

The network approach (*sensu lato*) is particularly efficient with respect to networks that are composed of a very large number of nodes; this is the case, e.g., with "webs of interaction between genes, proteins, and other molecules involved in the regulation of cell activity" (Crofts et al., 2008, p.3). However, network structures formed by animals or humans may consist of a small number of individuals, and then the network approach may yield less precise results, becoming too "coarse-grained". For example, it may be impossible to distinguish between random and scale-free networks (see below) if we deal with groups of only four or five chimpanzees.

Networks in the broader sense can also be applied to ecology. "Species in an ecosystem can interact in a number of different ways. They can eat one another, they can parasitize one another, they can compete for resources, or they can have a variety of mutual interactions, such as pollination or seed dispersal" (Newman, 2012, p.99). All these kinds of ecological interactions are represented by networks exemplified by food webs (predator-prey interactions). It should be noted that such *sensu lato* networks actually turn out to be a complicated mixture of hierarchical, network-type (*sensu stricto*), and quasi-market (competitive) relationships among the species involved, so that singling out structures based upon network principles (in the narrow sense) presents serious difficulties. This is one of the reasons why the nexus of ecological interactions is beyond the scope of this work, except for several short comments.

1.1.3. Quantitative Properties of Networks

An important parameter related to network nodes and links is *path length* (or *geodesic distance*), i.e., the minimal number of links separating two nodes in a network.

1.1.3.1. Small-world Phenomenon

According to Stanly Milgram (1967), maximally six links separate two individuals in the human population considered as a network, based on his studies in which the goal was to deliver a letter to a specific individual.

> "Milgram's initial experiment involved volunteers living in Kansas and Nebraska. Each was given a letter, and the goal was for the letters to reach a particular target person in Boston who was not known by any of the Midwestern volunteers. Each volunteer was told only the name and profession of the target, and was instructed to send the letter to someone whom they knew on a first name basis and whom they thought would be more likely than themselves to know the target person. Each recipient in the chain was given the same instructions. Each letter had a record of the path of people through which it passed. In several different experiments with varying protocols, Milgram found that between about 5 and 40 percent of the letters did get through to the targets. He found that the average path length (number of people passing on a letter) of these successful letters was six" (quoted from: Mitchell, 2006, p.1200).

Later research using e-mail demonstrated that contact between two arbitrarily chosen people can be made via a maximum of 5—7 links. In general, the small world phenomenon implies "that most pairs of vertices in most networks seem to be connected by a short path through the network" (Newman, 2003, p.9). Nonetheless, not all networks display the phenomenon.

A completely *regular* network, a lattice where all links between neighbors have an equal length (see Figure 1.5, B in subsection 1.2.4.1 below) does not exhibit small-world behavior. The average length of the *shortest path* (ℓ), or the *geodesic* line, between two arbitrarily chosen nodes in it, is proportional to $n^{1/d}$, where n is the number of nodes in the regular network, and d is the lattice's dimensionality (e.g., $\ell \sim n$ with an one-dimensional, i.e., linear, sequence of equidistant nodes; $\ell \sim \sqrt{n}$ with a two-dimensional, i.e., flat, lattice).

The addition of a few randomly arranged links, even with small probability, drastically decreases the path length between the nodes of the network, resulting in its *small-world* behavior (Watts & Strogatz, 1998): even seemingly distant nodes become connected by a disproportionately short geodesic line (i.e., a small number of link hops). In a completely random network, $\ell \sim \ln n$, i.e., the path length only insiginificantly increases with an increase in the number of network nodes (see 1.1.3.2 for the definition of a random network). A logarithmic (or still slower) dependence of average path length on node number is characteristic of some networks with more complex structural patterns. They are exemplified by networks which include parts with dense links between nodes (called clusters); these parts are connected by relatively few links. Such clustered networks (see below) are also characterized by the small-world phenomenon. They have very short geodesic paths despite their numerous nodes.

For instance, scale-free networks with relatively few nodes having numerous connections (hubs) and a majority of other nodes having few neighbors, which are considered below (1.1.3.2), demonstrate *ultrasmall-world* behavior: the path length is proportional to the logarithm of the logarithm of the number of nodes in them (Barabási & Oltvai, 2004).

The longest geodesic line (path) linking two nodes in a network is called the network's *diameter* (*D*); the longest path running between a given node and another node belonging to the same network is referred to as the node's *radius* (*R*).

Averaging l_i for all nodes in a given network yields its global mean path length *L*. This global variable can be used to characterize the properties of the whole network, e.g., to find out "how quickly a piece of information, starting from an arbitrarily chosen node," can "spread through" the whole network (Croft et al., 2008, p.68). According to Milgram's data cited above, the population of the USA, considered as one network, has an average path length (*L*) of approximately six, called the "six degrees of separation" (Milgram, 1967; Croft et al., 2008).

1.1.3.2. Node Degree: Random and Scale-Free Networks

The node degree *K* of a node in a network (called the *focal* individual) is the number of its neighbors with whom it is directly linked (hence $K = 5$ for node *1* in Figure 1.1). Averaging node degree *K* for all network nodes yields K_{mean}, the mean node degree of a network.

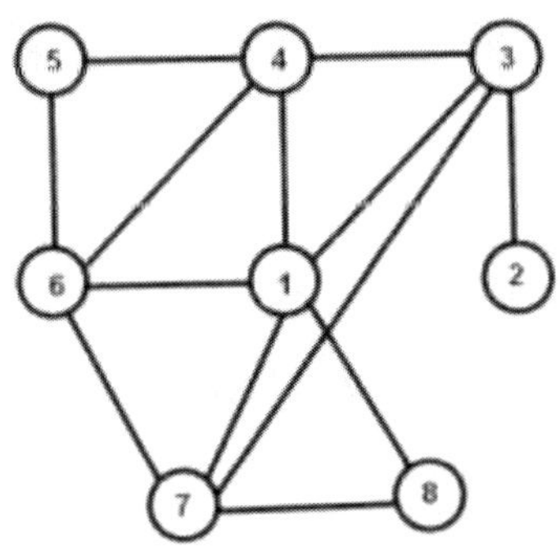

Figure 1.1. A simple network where nodes differ in terms of their degree (see text).

Of particular interest with respect to a network's organizational pattern is $<K>$, the sum of differences between the maximum node degree in a given network and the degrees of other nodes in the same network. This is one of the main criteria of a network's *hierarchization (centralization)* (see 1.2.4.1).

A quantitative property of whole networks enabling us to discriminate between different network types is *p*, the probability that network nodes have a given node degree *K*. What is the $p = f(K)$ curve like?

Let us first consider *random networks* (Figure 1.2, A) that were briefly mentioned in the preceding subsection. They are defined as follows: "we start with *n* nodes and connect every pair of nodes with probability *p*" (Almaas et al., 2007, p.9). Random networks investigated in the work of Erdös & Rényi (1959, 1960) are characterized by a normal (Poisson) node degree distribution. Hence Erdös-Rényi (ER) networks include typical nodes whose node degrees have the maximum probability. The node degree distribution is approximately that of a Poisson (Gaussian) curve with short tails (Figure 1.2, A).

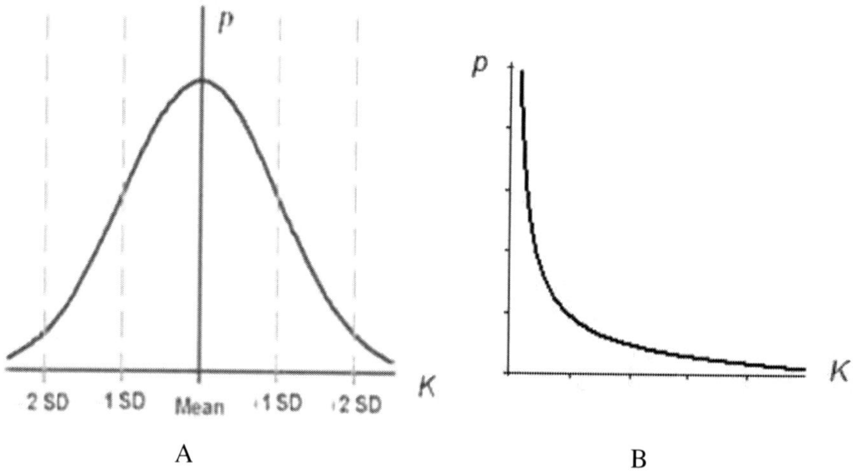

Figure 1.2. An idealized Gaussian curve for a random network (A) (SD is the standard deviation), and a power-law hyperbolic curve for a scale-free network (B).

However, a large number of networks do not belong to the ER type. It is emphasized in the literature that "real-world networks are mostly found to be very unlike the random graph in their degree distributions. Far from having a Poisson distribution, the degrees of the vertices in most networks are highly right skewed, meaning that their distribution has a long right tail of values that are far above the mean" (Newman, 2003, p. 12).

Some networks follow—or at least their node degree distribution is approximated by—the *power law*: $p(K) = a \times K^{\gamma}$, where a is a constant coefficient and $\gamma > 0$. These are *scale-free networks* (Figure 1.2, B); their properties were investigated by Barabási & Albert (1999). The growth of a scale-free network complies with the following two principles: "First, networks grow through the addition of new nodes linking to nodes already present in the system. Second, there is a higher probability to link to a node with a larger number of connections, a property called preferential attachment" (Almaas et al., 2007, p.9). Such Barabási-Albert (BA) networks are exemplified by the Internet and the World Wide Web, where new users preferentially link to the nodes that already have a larger number of ties. Similar behavior is displayed by networks composed of proteins or genes that interact or are connected by regulatory ties. Additionally, "dolphin social groups showed a 'scale-free' pattern (meaning that the degree distribution of the network follows a power law)" (Wey et al., 2008, p. 336).

Scale-free networks contain a small number of *hubs*, i.e., nodes having numerous links to other nodes plus a much larger number of nodes with few links to others. In terms of resilience, scale-free networks having few nodes with many connections and many other nodes with few connections generally perform better than random networks. Scale-free networks are, nevertheless, vulnerable to attacks targeting the hubs. It has been suggested that the measures aimed at stopping a dangerous infectious disease from spreading can be more effective if the hubs (the individuals linked to a maximum number of others) are preferentially dealt with (e.g., vaccinated, Mitchell, 2006).

Real-world networks are often too complex in terms of their behavior to fit the simplistic power distribution law: the probability of creating a link to a node is not only determined by its degree because nodes differ in their properties ("some Web pages are linked to because they are high quality, not just because they already have many links", Mitchell, 2006, p.1199). More sophisticated models, therefore, are required in order to take account of the complexity of real-world networks' behavior. "But the surprising fact remains that" simple models like the scale-free network "seem to capture some essential aspects of real-world networks" (Mitchell, 2006, p.1199).

1.1.3.3. Network Density. Clustering and Community Structure

Network density is defined as the proportion of actual links between nodes out of the total number of possible links (Croft et al., 2008, p.173). If, in a network with the three nodes 1, 2, and 3 (Figure 1.3), only the 1-2 and 2-3 links (but not the 1-3 link) exist, then the density is 2/3 = 0.666…

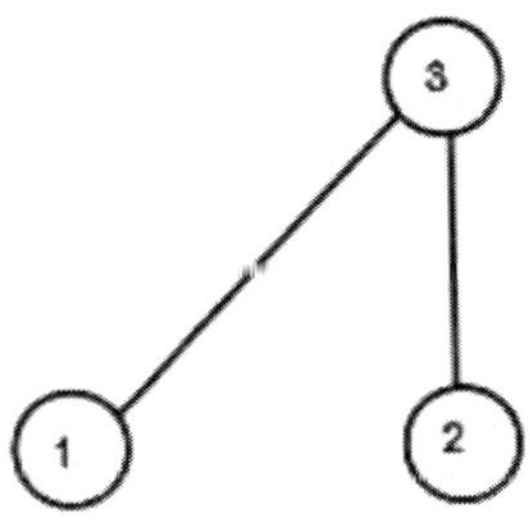

Figure 1.3. The network density is the proportion of actual links between nodes out of the total number of possible links (Croft et al., 2008)

A related variable is *clustering coefficient* (*C*). The clustering coefficient of a single network node corresponds to the ratio between actual links among its neighbors and those potentially possible (which would make the part of the network around the node in question a complete graph)[5]. Hence, for a node *i*, $C_i = E_i/[K_i(K_i-1)/2]$, where E_i is the actual link number, K_i the node degree (i.e., the number of its neighbors), and the denominator gives the total number of possible links between the chosen node's neighbors.

If a network structure includes parts with dense ties between nodes (i.e., a high *C* value) that are sparsely connected to one another, the network is characterized by *community structure*, or *cliquishness*. The term *clique*, originally used in the social sciences, is redefined in network theory as "a set of nodes where each node is directly tied to each other" (Wey et al., 2008, p.338)[6]. Similarly, the partly synonymous notion of *community structure* is used if

[5] In other words, "the number of existing ties between neighbors is divided by the maximal possible number of such ties (e.g., if an individual has four neighbors, these neighbors may have (4 x 3)/2 = 6 relationships among themselves). Clustering coefficient describes how densely (or sparsely) the network is clustered around the focal individual" (Wey et al., 2008, p.338).

[6] Strictly speaking, a *clique* is defined as a maximal subset of nodes such that every member of the set is directly connected by a link to every other (Newman, 2012).

there are groups of nodes ("communities") that have a high density of links, with a lower density of nodes between the "communities".

Areas with sparsely connected nodes between dense clusters in a network contain *structural holes* described by Burt (1992) with regard to networks in human society (see 3.7). In general terms, a network contains a structural hole if "the neighbors of a node are not connected to one another" (Newman, 2012, p.202).

Both of the two nodes in the middle of Figure 2.4 presented in Chapter two (section 2.3) which connect a dense cluster in the top part of the picture to the rest of the network are called *cut points* (Croft et al., 2008, p.53). Removing a cut point results in separating a network into several unconnected parts (called *components*). A more general term than a cut point is a *cut set* defined as a group of nodes whose removal disconnects a given pair of nodes (Newman, 2012, p.147). In Figure 1.1 above, nodes 4 and 6 represent a cut set because removing them disconnects node 5 from nodes 3 and 7 as well as, indirectly, from the rest of the network.

The clustering coefficient can be averaged for any part of a network or for the network as a whole, becoming its integral feature: $C = 1/n \sum C_i$ (Watts & Strogatz, 1998). Interestingly, for random (ER) networks with a finite average node degree K_{mean}, the average clustering coefficient is inversely proportional to the node number n in these networks (Barthelemy, 2008).

The term "clustering coefficient" at the level of the whole network or subnetwork is alternately defined by Mark Newman (2003, 2012) as the probability that two neighbors of a given network node are also directly linked to one another, thereby forming a triangle (a closed path). In other words, C in this meaning measures *triangulation*, i.e., the ratio between the number of triangles and that of three connected nodes in a given network, i.e., paths of length two (Newman, 2012). In these terms, a formula for calculating C is as follows:

C = (number of closed paths of length two)/(total number of paths of length two) in a given (sub)network (Newman, 2012).

Since the network paradigm has been applied to research on the spread of pathogens (agents causing infections) in an animal or human population, the notions of "path length" and "clustering" are used to estimate "the likelihood that a pathogen will remain a local outbreak or become global in a population" (Krause et al., 2009, p.968).

Apart from the clustering coefficient, a network node's *reach* is also relevant in this context; it is defined as the number of nodes connected to this node via two steps (e.g., the number of friends of friends, Sih et al., 2009). The notion of "reach" is of relevance to the biological applications of network science, particularly because it emphasizes that "indirect social connections matter"[7] (Krause et al., 2009).

An additional useful feature is denoted as *assortativity*. A network is considered asortative if nodes with similar properties (e.g., similar K values) are preferentially linked to one another; if, conversely, dissimilar nodes prefer to attach, a network is called *disassortative*. White American high-school students prefer to associate with other white students; in a similar fashion, black students tend to deal with other black students in the same school (Newman, 2003). Some biological networks, including those existing at the molecular level, are disassortative: biological molecules often tend to interact with molecules belonging

[7] In human society, of relevance in this context is *gossip;* individuals can be rewarded or punished not only by those directly affected by their behavior but also by third persons thanks to public opinion; apart from human society, this exists in nonhuman primates such as chimpanzee groups (de Waal, 1996, 2009).

to a different type or characterized by different node degree values: "hubs avoid each other" (Barabási & Oltvai, 2004).

A network structure sensu lato *is any set of nodes (vertices) connected by links (edges). Its functionally important properties include node degree, clustering (community structure), and assortativity. A remarkable feature of many networks is the small-world phenomenon. Among subtypes of network structures, random and scale-free networks are of special interest.*

1.2. Specific Sense of the Term "Network"

The word *network* has been used in a narrower sense in a number of works on management theory and other fields of the social sciences. Several notable articles were published in the 1980s and the early 1990s (Thorelli, 1986; Powell, 1990; Baker, 1994) and the term was widely used in the work of Manuel Castells (1996, 2000, 2001, 2004). A network structure is *a coherent decentralized (horizontal, non-hierarchical) system of cooperating elements (nodes).* Castells (2004) construed a network as a set of interconnected nodes lacking any kind of central controling unit. This definition of networks also applies to various biological systems that lack a central pacemaker and are composed of *cooperating* nodes. To reiterate, networks in the narrow sense (*sensu stricto*) are characterized by cooperation and horizontal relationships among their elements (nodes): they are contrasted with centralized structures (hierarchies) as well as structures predominantly based on competition between autonomous units (markets and their analogs).

Since the term "cooperation" is of paramount importance in terms of *sensu stricto* networks, the following subsection will deal with its definition and relevant implications as well as with other important terms and concepts that have been developed in ethology, a subfield of biology that is concerned with animal behavior.

1.2.1. Relevant Ethological and Sociobiological Concepts. Synergies

The word cooperation is familiar to most readers; its definition in terms of ethology is as follows: *Cooperation* denotes *interactions between two or more individuals or groups for the purpose of solving a problem or carrying out a task* (Dewsbury, 1978). An alternative, although in principle similar, approach to defining cooperation involves considering it from the viewpoint of a whole group (community). In these terms, *cooperators* contribute to the collective good within a distinct group at an individual cost (Hochberg et al., 2008, p. 3, modified). In human society, "a collective good is any good or service provided to the members of some collective (coalition, village, organization, nation, etc.) through the efforts of some or all of its members… Examples include large game that is widely shared, a community irrigation system, defense against enemy attack, a wide variety of rituals and ceremonies, and scholarly journals. Other social species produce collective goods as well" (Smith, 2010, p. 232). A classic example of cooperation is provided by social insects such as ants, whose societies are characterized by advanced cooperation and "labor distribution" patterns. All worker ants carry out "their duties assigned to them depending on the nature of work and the worker's skill" (Zakharov, 1987, p. 122).

Of relevance to networks in general—both in the biological sphere and in human society—is the issue concerning the reasons why independent individuals form more or less integrated networks that are based on cooperation among them. One approach invokes the originally sociobiological concepts of kin altruism and reciprocal altruism.

The term *kin altruism* refers to sacrificial behavior for the benefit of a close relative. The development of such behavior in the course of biological evolution promotes *kin selection* as suggested by W. D. Hamilton, D. S. Haldane, and J. Maynard Smith in the 1960's. Close relatives such as siblings, parents, or offspring have a large number of common genes. Hence helping a relative at the expense of one's own reproduction or survival chances actually promotes the spreading of the helper's genes within the population.

A classic example discussed by Hamilton (1964) is related to the behavior of social insects including bees, wasps, ants, and termites.[8] In terms of natural selection, worker individuals in these species are at an advantage if they take care of the reproducing queen instead of reproducing themselves. Females in social insects are diploid – their cells contain two copies of each chromosome while males (drones) are haploid, i.e., have only one chromosome copy. All working females have the same mother as the queen female. Calculations reveal that ¾ of the worker's and the queen's genes are identical (Maynard Smith, 1996). Helping the queen reproduce results in transmitting to the progeny more of the worker's genes, therefore, than if the worker reproduced herself because her own offspring would have fewer genes in common with her (1/2).

The term *reciprocal altruism* was coined by R. Trivers (1972). Unlike kin altruism, reciprocal altruism is self-sacrificial behavior that may benefit an unrelated individual, provided that the beneficiary is ready to perform an analogous act of altruism benefitting the altruist. "Rescue the drowning that will rescue you if you drown". Sociobiologists applied the reciprocal altruism concept to various forms of loyal behavior including the formation of coalitions by primates and dolphins. Coalition members typically assist one another, e.g., in courting females (de Waal, 1996). The reciprocal altruism concept is vulnerable to the criticism that altruists can be exploited by free riders that benefit from their help without helping others.

The reciprocal altruism strategy is less vulnerable to free riding if the animals involved are characterized by a sufficiently advanced nervous system and efficient sense organs. This enables the animals to individually recognize each individual with whom they have dealt in the past. Primates satisfy these criteria and, therefore, reciprocal altruism is widespread among them. Baboons and chimpanzees tend to help the individuals who have helped them before, e.g., give food to those that had shared food with them earlier. Chimpanzees even punish those who were reluctant to share food by behaving aggressively toward them or concealing information about the location of food (Cummins, 2001).

Kin and reciprocal altruism is related to the sociobiological *zero—sum* model. The altruist gets *zero*: it loses important resources or even dies while the beneficiary gets something ("The winner takes it all"), i.e., uses all resources made available by the act of

---

[8] Some biologists have recently expressed their doubts about the validity of Hamilton's mathematical calculations in this example. However, changing these details does not alter the main message.

altruism. However, several interacting individuals can implement a substantially different model termed the *win—win* model.

It is possible that all partners benefit from the interaction which can be interpreted in terms of Peter Corning's *synergy* model. Corning (1983, 2000, 2003a, b, 2005, 2007) widely uses the term *synergy,* which refers to "the combined (interdependent) effects produced by two or more parts, elements, or individuals… and is a ubiquitous phenomenon in nature and human society alike" (Corning, 2000, p.133). The following are important kinds of synergies:

- *Synergies of Scale.* "The term refers to the well-established fact that the sheer number of participants may produce combined effects that could not be achieved by any individual, or even a smaller group" (Corning, 2007, p.116). A larger coalition of male lions is more successful in commandeering a group of females; a lager chimpanzee group hunts more efficiently.
- *Modification of the Environment.* For instance, penguins in Antarctica cuddle together, maintaining a sufficiently high temperature despite the frost.
- *Cost and/or Risk Sharing.* Vampire bats share part of the blood swallowed by them with their hungry group mates, thereby reducing the risk of their dying of starvation.
- *Information Sharing and Collective 'Intelligence':* Animals frequently communicate information to one another, whether intentionally or not (Corning, 2007, p.117).
- *Labor (Function) Division, or Combination.* One of biology's numerous examples deals with cyanobacteria whose heterocysts specialize in fixing vital nitrogen from the air, whereas the rest of the cells in a cyanobacterial filamentous body carry out photosynthesis.

Although these features are characteristic of all social systems, decentralized networks enable all their members to equally enjoy the advantages of the social lifestyle whereas many hierarchies tend to preferentially benefit high-ranking members.

Cooperation is one of the forms of friendly (loyal) behavior, which also include *affiliation.* From the ethological and biopolitical viewpoint, affiliation can be defined as *an individual's tending to approach and remain near other individuals, particularly those belonging to the same family or social group, or* (in the animal kingdom) *to the same species* (Dewsbury, 1978). Experiments with animals, e.g., dogs, indicate that they are ready to do a difficult job if their only reward is meeting another individual of the same group. In humans, affiliation is associated with feelings of physical comfort, safety, and belonging.

Fish that form horizontal decentralized social structures (equipotential schools) exhibit the "tendency … to associate with other fish of the same species or those similar to them with respect to their form, color, and locomotive behavior" (Pavlov & Kasumyan, 2003, p.53). This example demonstrates that affiliation is selective. "Birds of a feather flock together": both humans and animals tend to deal with those who are similar to them, i.e., affiliation-based networks tend to be assortative (see 1.1.3.3). Animals as well as, in all likelihood, human individuals possess innate mechanisms enabling them to estimate the degree of gene-determined similarity between them and others based upon appearance, body proportions, and odors. Chimpanzees and, according to recent data, wasps can recognize related individuals by examining their "facial features." Wood lice recognize their mates and offspring using the sense of touch (Reznikova, 2005). Affiliation is closely related to kin altruism; affilative behavior towards kin and, more generally, "us" is typically accompanied by isolation from, and often aggressive behavior towards, "foreign" out-groups.

1.2.2. Network Structures in Human Society: An Introductory Discussion. The Hirama. Heterarchy

Network structures represent a form of spontaneous order that is established by decentralized agents and not central power (Fukuyama, 2002). Prerequisites for setting up network structures in society include the establishment of reputations among their potential members, their interpedendence, and often their seemingly unpragmatic altruistic behavior that consolidates the network. This enables its members to benefit from the combined assets of the other members, and, therefore, is profitable in the long term (Powell, 1990; Chuchkevich, 1999; Bezrukova, 2004). Importantly, interpersonal and intergroup relations can be based not on formal regulations but on trust and durability, as well as the "strategic dependence" of network members on one another (Meulemann, 2008), e.g., with respect to resources or technological know-how. Such relations are denoted as *social capital* in sociology and management theory.

Network structures in society are often not completely *flat*: they are characterized by the coexistence of a number of *partial* leaders with limited power and competence areas. This distinguishes them from *hierarchical* structures, which have a *single* center (leader or boss). In addition, networks are characterized by the prevalence of cooperation among their nodes over competition between them. This sets networks apart from the mainly competition-based *market structures* (to be considered in section 1.3). Cooperation in networks is promoted by

- *unifying goals, projects, behavioral norms and attitudes*; network-consolidating attitudes are strengthened and group identity development is stimulated by the mutual dependence of actors. This leads to sustained relationships between them (Meulemann, 2008)
- *efforts made by moderators (psychological leaders)* whouse psychological techniques to *encourage creative work and cooperation* in terms of carrying out the tasks of the whole network structure. Moderators are also responsible for process and conflict management, as well as risk reduction.

Networked organizations "provide a third alternative between top-down planning" typical of a centralized hierarchy "and the anarchy of the market" (Meulemann, 2008, p.31).

Network principles can be implemented by business entities, scientific research teams, social movements, and political organizations. Like new technology in general—ranging from nuclear energy to genetic engineering—network structures can have both positive and negative effects. The modern world is threatened by terrorists who belong to networks, as typified by *al-Qaeda*. Although they lack a rigid centralized hierarchy, such "dark" networks typically contain highly influential and potentially vulnerable nodes (hubs), i.e., many of them represent the scale-free network structures described by Barabási & Albert (1999).

Irrespective of their goals and behavioral norms, networks allow for competition among their members, as it promotes creativity and devotion; nevertheless, cooperation should prevail over competition. Many networks prefer the *competition for the sake of successful cooperation* principle. For instance, a networked creative lab may aim to carry out an innovative scientific or technological project, and the network's moderators harness the natural competitive spirit of the people involved, motivating them to accelerate their work on attaining the common goal, e.g., on developing a drug preparation. In contrast, market structures or their analogs (quasi-market structures) favor the *cooperation for the sake of successful competition* principle[9].

Network structures in modern-day society include small- and medium-sized businesses as well as giant multinationals that lack a central headquarters; networks also include various clubs, charitable foundations, and local administration bodies. The organizational properties of many network structures conform to the abbreviation *SPIN (Segmentary, Polycentric, Integrated Networks),* which, apart from networks in human society, applies to a large number of network structures in biological systems, particularly to those conforming to the modular ("cnidarian") paradigm of organization (see 2.2).

In recent years, the spreading of network structures in human society has been promoted by (1) modern information technology developments that facilitate the coordination of the behavior of network nodes by speeding up their communication (with the aim of achieving consensus) and (2) innovative organizational techniques enabling all network members to be actively involved in decision-making (van Alstyne, 1997).

Social networks are exemplified by a structure known as a *hirama* (Figure 1.4). This abbreviation stands for a *High-Intensity Research and Management Association* (Oleskin, 1994a, 1996, 1999, 2007a, b, 2012, 2013, 2014; Oleskin & Masters, 1997; Oleskin et al., 2001). This is a creative team which has been set up to deal with an interdisciplinary problem or issue such as *Small-Quantity Environmental Pollution Generators* or *Organized Terrorism*. The problem/issue is subdivided into several subproblems. For example, *Organized Terrorism* can be broken down into:

- *Ethological Approach to Terrorism*: this subproblem is linked to the evolutionary background of human aggression
- *Military Approach to Terrorism:* internationally organized terrorism as envisaged as a novel military strategy
- *Religious Approach to Terrorism:* this is concerned both with the terrorism-promoting role of religious fanaticism, fundamentalism, and intolerance and with potential methods of discouraging people from terrorism using peace-promoting ideas and dogmas contained, e.g., in Christian, Judaic, or Islamic doctrine.

However, despite subdividing the problem/issue into subproblems, the network is not subdivided into parts. Its members work, in parallel, on several (ideally on all) subproblems. The subproblems should, therefore, overlap and provide for a broad interdisciplinary vision of the network's focus. Roles or functions in this network structure are not fixed or defined as with the "offices" in a Weberian bureaucracy. Often only one person, *the subproblem leader*, is explicitly attached to a particular subproblem. The person collects ideas on this subproblem, which are generated by other network members. A partial leader in a hirama is a

[9]Examples of *cooperation for the sake of competition* include cooperation between firms #1 and #2 on the basis of a contract offered by #1 and signed by #2. Firm #1 deliberately includes in its contract such conditions that disadvantage its partner (#2), and #2 eventually loses its competition with #1.

variation on the more general theme of a *network leader* in human society, i.e., a person with intellectual, financial, material, communicative, informational, or other assets that are important in terms of business but who is not the boss.

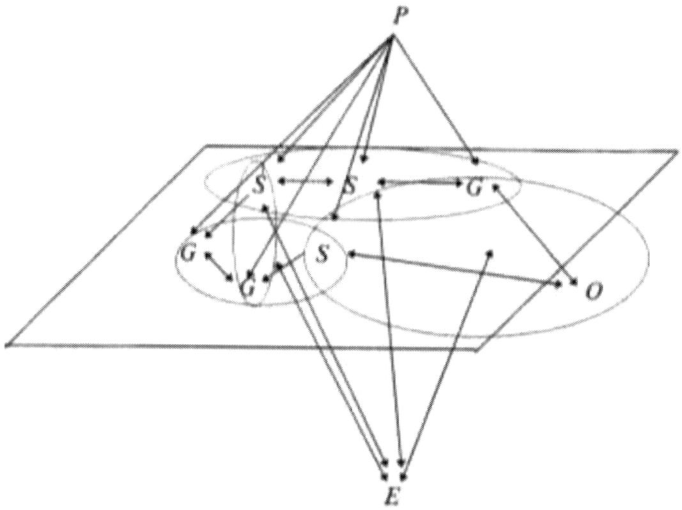

Figure 1.4. A hirama-type network structure. The structure is dynamic, and the creative subunits included within it are constantly in the process of formation/disintegration. Designations: *S*, subproblem leaders; *G*, ordinary network group members; *O*, an outsider collaborating with the network on one of the projects. Thin-line circles represent temporary creative subunits or discussion groups. These relationships all correspond to the "task fulfillment plane," as shown in the picture. The psychological and the "external affairs" leaders (*P* and *E*, respectively) are beyond this plane. Types of relations: → partial (task-limited) leadership; ↔ horizontal networking. From: Oleskin & Masters, 1997.

A partial leader responsible for coordinating work on a particular subproblem can be assisted by several experts on the same subproblem. Within the whole network, these relatively few people form a collective *hub* dealing with this subproblem. The subproblem hub interacts with unspecialized network members that are more numerous in many hiramas. Apart from hirama-like networks in human society, the *few specialists + a large pool of unspecialized individuals* pattern applies to a variety of decentralized network structures in the animal kingdom. Different hubs can alternately mobilize the same pool of unspecialized nodes for doing different tasks (e.g., for foraging for food in an ant family, see 2.4). A partial leader in a hub can spend part of his/her time on contributing to the work of other subproblem hubs, playing the role of an ordinary unspecialized pool member. Analogously, Israeli *kibbutzim* that largely comply with the network organization principle have elected members which perform socially important functions (an executive secretary, a treasurer, a household coordinator, etc.). In addition to these functions, they are often involved in the ordinary work done by *kibbutz* members, e.g., related to agriculture (see 3.2.4 for details).

Many hirama-type network structures also have a *psychological leader.* The psychological leader provides support, advice, and psychological help that is often sought by other members; he or she creates an atmosphere that promotes efficient work on all subproblems and helps other partial leaders interact with one another, mitigating or—still better—preventing internal conflict.

Otherwise, when partial "leaders do not recognize each other's leadership, the group can literally be torn apart". By contrast, if they "recognize one another as leaders, they should be better able to synchronize their leadership efforts so that decision-making and action are more effectively channeled within the group" (Mehra et al., 2006, p.235).

In addition, a hirama typically includes an *external leader,* also called an *"external affairs" leader.* The individual with this role is responsible for propagandizing hirama-promoted ideas, establishing contacts with other organizations, and shaping the group's pastime and leisure activities, thus contributing to the development of informal loyal relationships among members. Hiramas are free to make alterations in the network's organizational pattern. Additional leadership roles can be introduced such as:

- A commercial leader responsible for searching for sponsors and grant opportunities and for marketing and other profit-making activities
- An organizational leader who is particularly important while a hirama-type network is organizing its work and legalizing its status
- A spiritual leader (a "guru"). The operation of network structures depends on unitary spiritual values, often implying collective attempts at attaining certain ideals. This conceptual basis can be personified by the "guru" image. Importantly, the spiritual leadership should not develop into an authoritarian dictatorship, which would run contrary to the non-hierarchical principle underlying a network structure. To prevent this, many hiramas prefer a legendary guru or a long deceased person like Emiliano Zapata, a hero of the Mexican revolution in 1917, whose name is perpetuated by the networked Zapatista movement in present-day Mexico.

The hirama organization pattern is used, with certain modifications, for setting up interdisciplinary scientific research teams, commercial enterprises, and various communal structures. This subject will be discussed in Chapter 3.

A large number of network structures in human society possess fractal/self-similar properties. A part of a hirama can be regarded as a mini-hirama per se. Several hiramas may combine to form a higher-order network structure called a *hiramiad.* Some of these hiramas collectively perform the function of subproblem hubs in the hiramiad, while others may collectively behave as the unspecialized but mobilizable pool (Oleskin & Masters, 1997).

The term network (*sensu stricto*), as it is used in this work, is closely related to the term *heterarchy,* which is used, for example, in anthropological works. It means that individuals in a social system have no ranks or can be ranked in several different ways. A heterarchical system lacks a centralized hierarchy; a system which has one is a *homoarchy* (Crumley, 1995; Bondarenko, 2004). The term *heterarchy* is relatively widely used in studies on the structure of primitive hunter-gatherer bands. Many bands had partial, temporary leaders. Several leaders (e.g., the shaman, the war leader, and the leading hunter) could function in the same band, and the competence of each of them was limited. On the whole, such a band was characterized by *partial leadership*, which also occurs in the social groups of a number of animal species. A similar principle underlies modern decentralized networks including hiramas. With respect to the modern-day business world, heterarchy implies distributed authority which enables the parallel planning of all stages of projects carried out by a network of interdependent nodes (firms or semi-autonomous firm subdivisons) (Stark, 2002).

1.2.3. Networks in Biological Systems

This is an introductory subsection on biological networks, which will be followed by a more detailed discussion in Chapter 2.

As noted above, the term "network" in the broad sense has recently been applied to biological systems in a number of works. In this context, the question to raise is whether it makes sense to also extrapolate the narrower meaning to nonhuman life forms.

Evidently, a large number of biological systems are decentralized and favor cooperation over competition, at least within the boundaries of the systems, i.e., between individuals and subgroups inside it. For example, groups of apes (chimpanzees or bonobos) have hierarchies that are mitigated by decentralized interaction among individuals, based on mutual aid, food sharing, and grooming (de Waal, 1996, 2001). A cnidarian colony is a network without a central controlling unit, and the behavior of its nodes (zooids) is regulated in a decentralized fashion (Marfenin, 1993, 1999, 2002, 2009, 2011).

Decentralized cooperation-based networks are also widespread in the microbial world. Microbial colonies or biofilms consist of a multitude of cells, and a lack of a single central controlling unit does not prevent the effective coordination of social behavior (Shapiro, 1988, 1995, 1998; Ben-Jacob, 1998, 2003; Oleskin, 1994b, 2009, 2014). In an analogy to hiramas, microbial colonies may contain a numerically predominant pool of unspecialized cells together with a limited number of specialists. The cells of many tissues within the human organism form network-like structures. Stem cells form a pool of unspecialized or incompletely specialized cells that exist as reserve cells in the organism, against the background of specialists, e.g., epithelial or muscle cells[10].

Social insects combine hierarchies and networks. They have groups of specialists with leaders over them that perform situationally important functions such as foraging for food, cultivating aphids and collecting their nutritious excretions (Hölldobler & Wilson, 1990, 2009, 2010; Novgorodova, 2003). However, hierarchically organized specialists form part of a larger decentralized network structure. They coexist with an unspecialized pool of "generalists" that are typically inactive; they remain idle but can be mobilized in emergency (Schmidt-Hempel, 1990).

In a large number of biological systems, the term "network structure" can be interpreted not only in organizational, but also in geometrical terms. Predator dictyobacteria form nets that are composed of a large number of cells. Their prey (cells of other bacterial species) is trapped in their meshes.

Another example is provided by spiders that "live colonially, building separate nests in very close proximity to each other. One benefit to this might be that the between-web transmission of vibration caused by struggling prey may allow individuals to track spatial variation in prey availability and to make important judgments on relocation" (Krause & Ruxton, 2002, pp.24-25). In an analogy to commercial networked enterprises in human society, some other spiders team up to build a single web together and share the prey they capture. They "save on construction costs compared with the separate webs" discussed above (Ibid., p.25).

---

[10] There are tissues characterized by hierarchical organization. The nervous system combines hierarchies and networks (see 2.5 below).

1.2.4. Applying "Network Science" to Networks Sensu Stricto

Section 1.1. focused on the quantitative characteristics of networks in the broad sense (any system composed of nodes and links). In this subsection, the parameters considered in section 1.1. are applied to decentralized cooperative structures, drawing a distinction between them and other structures, i.e., hierarchies and (quasi-)markets, with particular emphasis on networks in biological systems.

In modern human society, there often are clear legal differences between the types of structures in question. The law regulates the functioning of official hierarchies, whether in the political systems or in business (bureaucratic enterprises); relations between agents in the "civilized" market are in conformity with legal norms as well. In some cases, the special status of decentralized non-hierarchical networks is officially acknowledged, and they have their legal rights. Even "informal" networks that lack a legally acknowledged status often emphasize their "informality" as a special distinctive criterion, setting them apart from the buraucratic world of legally acknowledged hierarchies and ruthless market structures that are based on competition and quasi-Darwinian "struggle for existence".

Legal criteria obviously do not apply to nonhuman systems. However, of relevance are some formal mathematical criteria developed for networks in the broader sense (*sensu lato*) as well as ecological characteristics; these criteria and characteristics be used for distinguishing networks *sensu stricto* from hierarchies and/or quasi-market structures.

1.2.4.1. Networks vs. Hierarchies

Several criteria can be used to distinguish hierarchies and networks in the narrow sense. These criteria obviously apply to both human society and biological systems.

1. Of special importance is *node degree (K)*, i.e., the number of links a given network node has with its immediate neighbors (see 1.1.3.2). Node degree is also called *degree centrality*. In the literature, the degree centrality of a node in a network is often normalized by dividing it by the total number of links in the network (i.e., the fraction of the all links connected to a given node is determined). The resulting parameter can vary between 0 and 1 (Freeman, 1979). Such a normalized degree centrality parameter was used, e.g., while disentangling the network structures formed by *al-Qaeda's* 9/11 hijackers in 2001 in the USA (Krebs, 2002, 2008; Miralles Canals, 2009). Normalized degree centrality values can also be used in formulas for calculating the hierarchization of a whole network.

   In these terms, the *hierarchization* of a network ($<K>$) is defined as the sum of the differences between the largest value of a node's degree centrality in the network (K_{max}) and those of all n nodes in this network: $<K> = \Sigma^{n}_{i=1}(K_{max} - K_i)/<K_{max}>$ where K_i is the degree centrality of any of the nodes of the network and $<K_{max}>$ is the maximum possible sum of differences in node degree for a network, which is calculated according to the formula $<K_{max}> = n^2 - 3n + 2$, where n is the number of nodes in the network (Freeman, 1979). Typically, such a maximally hierarchized network has a star-like shape (a star graph, Figure 1.5, A) with one central node. Accordingly, its hierarchization $<K> = <K_{max}>/<K_{max}> = 1$. In a centralized network (a hierarchy), most links are concentrated around a small number of nodes, whereas the difference, in terms of degree centrality, between nodes in a decentralized network (a network in the narrow sense preferred in this work) is insignificant, and degree centralization <*K*> tends to zero.

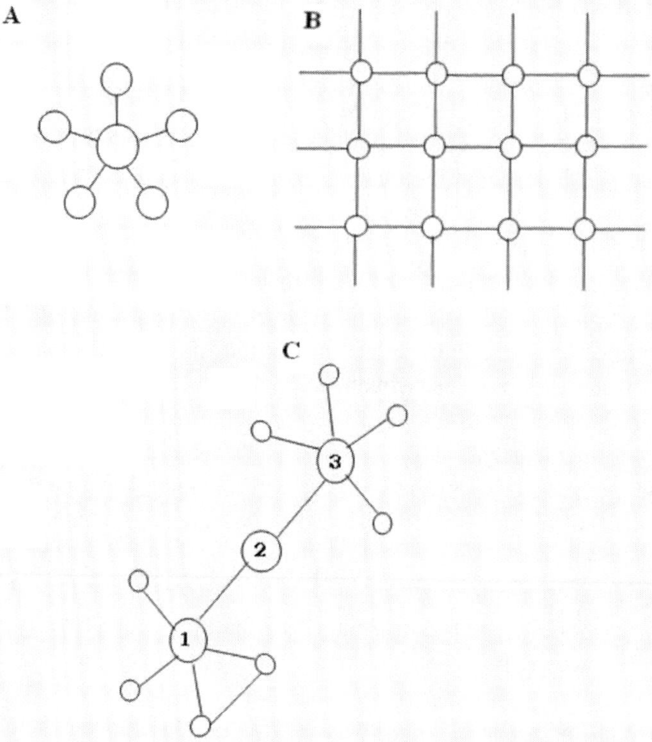

Figure 1.5. (A) A hierarchical structure (a star graph) where the central node has maximum centrality values on all the scales considered in this work; accordingly, the whole network exhibits a high degree, betweenness, and closeness centralization level; (B) a regular lattice where all the nodes have equal centrality values; hence, the whole network's centralization level is zero across all of the scales; (C) in a three-dimensional network, there are several hubs that have maximum scores on different centrality scales; nodes 1 and 3 are characterized by high degree values, whereas node 2 has maximum betweenness (and low degree) values.

2. Apart from node degree, other criteria are used to estimate the centrality of nodes in a network and to establish the degree of centralization (hierarchization) of a whole network. One of the alternative centrality criteria is called *betweenness, B*. It is based on the number of geodesic paths between every pair of other nodes on which the node in question lies (Freeman, 1979). In the network depicted in Figure 1.1 (subsection 1.1), the maximum number (9) of paths connecting other nodes run through node 3, and the minimum number (0) of paths runs through node 2. Some network nodes with a low degree value can nevertheless score high in terms of their betweenness. In particular, a cut-point node linking two otherwise unconnected parts of a network may have few direct neighbors (see Figure 2.4 in Chapter two where a node in the middle has only three neighbors but represents the only link between the upper and the lower part of the whole network of guppies). However, all paths

connecting nodes belonging to the two parts run via this node; hence, its betweenness may be very high.

For instance, a high betweenness value was characteristic of the group of bottlenose dolphins located at the border between two densely clustered populations in Doubtful Sound (New Zealand, Lusseau & Newman, 2004).

Network hierarchization in these terms (*betweenness centralization,* $<B>$) is determined by calculating the sum of differences between B_{max}, the maximum betweenness value among the nodes in a given network, and B_i, the betweenness of any node i in the network. The resulting sum can be normalized by dividing it by $<B_{max}>$, the maximum possible sum of differences in betweenness centrality for a network: $<B> = \Sigma^{n}_{i=1} (B_{max} - B_i)/<B_{max}>$. $<B_{max}>$ is calculated according to the formula $<B_{max}> = n^3 - 4n^2 + 5n - 2$, where n is the number of nodes in the network (Freeman, 1979).

3. Other criteria used in the literature to estimate network centralization include a node's *closeness* X to other nodes of the same network (Freeman, 1979 and subsequent publications). This parameter is calculated as the reciprocal of the averaged path length between a given node and all of the network's other nodes (see 1.1.3.1 above): $X = 1/(\Sigma l/n\text{-}1)$, where l is path length and n-1, the number of paths connecting two nodes in a network with n nodes. In a centralized network (a hierarchy), the central position of one or several node(s) results in comparatively short geodesic paths between them and the other elements, i.e., in a high X value. In a decentralized network, the difference between the length of paths leading to various network nodes is insignificant and, therefore, the closeness centralization of the network is close to zero[11].
4. An additional criterion called *eigenvector* value deals with the number of well-connected (i.e., having many links) nodes linked to a given node. It can either be large because a node has a large number of neighbors or because it has well-connected neighbors (or both). An individual in a social network, for instance, can be important either because he or she knows lots of people (even though those people may not be important themselves) or knows a few people in high places (Newman, 2012, p.170). By calculating the sum of the differences between the maximum eigenvector value and those of all nodes in a given network, its eigenvector centralization can be determined. This subject will be revisited in the subsection on "dark" networks in human society (see 3.2. below).

All the criteria can be used to discriminate between centralized hierarchies and decentralized networks, both in biological systems and human society. In his classical work on network centrality (hierarchization in our terms), Linton C. Freeman (1979) admitted that the three centrality criteria discussed by him (degree, betweenness, and closeness centrality) may correspond to different hierarchies in the same network (e.g., a node ranking high on the betweenness scale may rank relatively low on the degree scale).

Nonetheless, there may be some remarkable correlation between the three hierarchization estimates. This concordance is particularly significant within two extreme cases (Freeman,

[11] The hierarchization of a network ($<X>$) on the basis of closeness centrality of its nodes is defined in the literature, similar to degree and betweenness centrality, as the sum of the differences between the maximum closeness value and the closeness values of other network nodes, normalized by dividing it by $(n^2+ 3n -2)/(2n-3)$ (Freeman, 1979).

1979, see Figure 1.5): (1) maximally centralized networks exemplified by star graphs have high centrality (hierarchization) values with all these criteria (Figure 1.5, A); the same node obviously has maximum scores in terms of degree centrality, betweenness, and closeness and (2) entirely homogeneous networks (complete graphs). In the latter variety, all possible links between nodes are present. All nodes have the same betweenness, degree, and closeness values. Therefore, the whole structures are characterized by zero betweenness, degree, and closeness centrality values. This also applies to regular lattices (Figure 1.5, B).

The three criteria often are considerably discordant if we consider systems with intermediate properties.

These are the kind of systems to which most real-life network structures in the biological world and in human society belong. Their decentralized organizational pattern with split leadership is emphasized by the fact that *different* important nodes are characterized by high values in terms of *different* criteria. In the example of the hirama described above (1.2), it is evident that different partial leaders should have high ranks on different scales. The external leader who is expected to be an efficient broker should rank high on the betweenness scale. The psychological leader should be characterized by a high degree value, i.e., directly deal with a large number of hirama members. The organizational leader should have no trouble reaching other members of the structure she is to organize, i.e., her closeness centrality value should be sufficiently high.

In contrast, a very rigid hierarchy that is close to the maximally centralized structures envisaged by Freeman (1979) tends to have similar hierarchization/centrality values on all the different scales considered above. The node at their top (the dominant, the central leader) should be characterized by high values on all the scales, in contrast to the significantly lower values of the nodes located on the lower rungs of the hierarchical stepladder. A political example is provided by the Soviet dictatorial leader Joseph V. Stalin. He controlled the information flow through the whole political apparatus (i.e., had the maximum betweenness centrality value), personally gave commands to an enormous amount of Soviet people (i.e., exhibited the maximum degree centrality value), and had absolutely no trouble reaching even those who he did not deal with directly (i.e., had the highest closeness centrality value).

On a smaller scale, the team of 9/11 highjackers in 2001 in the USA was at least partly a hierarchy and not a decentralized cooperative network *sensu stricto.* Their "ring leader", Mohammad Atta, who personally participated in crashing a plane into the North tower of the World Trade Center, ranked very high on the three scales discussed in this subsection: he had the maximum scores on the degree and the closeness scale and the second highest betweenness value (Krebs, 2002). This example will be revisited in 3.2.8 below.

The convergence of the three centrality measures seems to be a significant distinctive feature of a true centralized hierarchy; it sets a hierarchy apart from a three-dimensional network where one of the partial leaders can temporarily come to the forefront.

However, using this criterion of hierarchization is problematic if a structure includes *more than one node* characterized by the highest values in terms of some or even all centrality scores. Is this still a three-dimensional network with a number of partial leaders or a subtype of hierarchy with a split pinnacle? Presumably, another criterion should be additionally used in this situation.

The question to raise is whether the nodes with concordant maximum centrality values on several (or all scales) are *functionally differentiated.* All creative subproblem leaders in a hirama (see Figure 1.4 above) may score very high on, e.g., the degree and closeness scales. However, since they specialize in different subproblems, each of them is a *partial leader.*

If the structure is more homogeneous, i.e., its nodes lack functional specialization, then the nodes with the highest centrality values try to attain dominance over the whole structure. This is a *hierarchy* with an "oligarchy" at its top level. It is exemplified by a group of spider monkeys (see the end of subsection 2.6.1 below) that is dominated by a small subgroup of high-ranking females that appears to represent a flat substructure *per se* (Ramoz-Fernández et al., 2009).

It should be noted that hierarchies with split pinnacles may lack stability and tend to evolve into (1) hierarchies with a single leader/dominant which prevails over its competitors or (2) typical decentralized network structures.

The difference between directed and undirected networks (see 1.1.2) is of relevance to the network-hierarchy issue. Individuals' behaviors are *asymmetric* and *complementary,* not congruent in hierarchies. In terms of networks *sensu lato* considered above, the links connecting them are *directed* (see 1.1.3). A high-ranking individual, e.g., in a chimpanzee group, *dominates* a lower-rank individual that *submits* to the "boss". For instance, the dominant individual *takes* the resources the subordinate *gives* him. This behavioral asymmetry, studied by animal ethologists, is obviously typical of human hierarchies as well, in sharp contrast to both network and (quasi-)market structures that are characterized by *symmetric, congruent* behaviors of their nodes (individuals) towards one another.

1.2.4.2. Networks vs. (Quasi-)Market Structures

Networks *sensu stricto* and quasi-markets differ in terms of their *ecological criteria.* In ecological terms, relationships between living organisms can be either positive (useful) or negative (harmful). This distinction is of relevance to the general concept of *signed* networks (or graphs) having either positive or negative links/edges between their nodes. Outside ecology, the concept of signed networks is applied to social structures. "For instance, in an acquaintance network we could denote friendship by a positive edge and animosity by a negative edge" (Newman, 2012, p.206).

In ecological terms, *plus*-relationships between organisms improve their survival chances and reproductive success (number of offspring). These relationships are considered cooperative (mutualistic) in ecological terms, and a system dominated by them, unless it is a hierarchy, is a network *sensu stricto.* A well-known mutualistic system such as a lichen, which is composed of cooperating algal or cyanobacterial cells and fungal hyphae, should be regarded as a network structure in these terms.

Minus-relationships decrease survival or reproduction rate. If they prevail over *plus*-relationships, whether in human society or in biological systems, this is a predominantly competitive *quasi-market* structure. It is of relevance that, in ecological terms, the word "competition" is often regarded as a synonym of the word "antagonism", although there are nuances of meaning that distinguish them.

It should be noted that antagonistic/competitive relationships may prove beneficial for the partners involved, even though they are minus-relationships *per se* (e.g., competition for oxygen between two fetuses in one womb limits the amount of oxygen available for each of them; competition between two firms for potential consumers limits their chances to market the goods produced by them). Such beneficial effects are compensatory, i.e., based upon efforts to prevail over the competitor(s). For instance, each of the competing firms attempts to improve its production, so that, in the long run, competition proves to be an effective vehicle of progress. This is related to Adam Smith's notion of the "invisible hand" operating in the market that can have biological analogs. A well known example is the stimulation of the development of the immune system of the animal or human organism by potentially virulent bacteria that release subtoxic doses of harmful substances.

However, the outcome of competition may not be beneficial for all the partners involved. Some of them, both in the human market and biological quasi-market systems, may eliminate. Alternatively, the winner and the loser may establish a hierarchical structure. In this case, competition is overridden by dominance-submission relations.

Networks and market-type structures—both in human society and the biological realm—differ in terms of the *clustering coefficient* C (defined in 1.1.3.3 above). Network organization typically implies a large number of dense clusters separated by areas with sparse links (this is exemplified, in the biological world, by leaderless fish shoals). Such cliquishness is less characteristic of, e.g., traditional markets in human society and similar predominantly competition-based structures in the animal kingdom. Accordingly, networks tend to be heterogeneous with respect to the clustering coefficient; C reaches high levels with nodes located in clusters (cliques, communities); it is quite low between the cluster areas within the network. Markets and their analogs typically have more evenly distributed C values, and this is due to the autonomous status of each of their nodes, i.e., agents that are involved in (quasi-) market interactivity (such as sale-purchase relations in human society and metabolic chains in the biological realm).

Naturally, a biological or human social system can change in time; it may become more hierarchical, more horizontal; more competitive or more cooperative. However, the fact that such structural alterations are possible does not imply that boundaries between networks, hierarchies, and markets are nonexistent. Transitions between these types involve serious organizational changes and give the agents involved new options for adapting to a changing environment. The fact that some primate species may have hierarchies during collective hunting and form more egalitarian network-type structures while relaxing after the hunt does not mean that there is no clear boundary between these two structures.

1.2.5. Network Structures in Information Processing Systems

Although this is not the main focus of this work, decentralized network structures are also exemplified by information processing networks. The Internet as well as its precursors, Usenet, LISTSERV, and particularly APRANET (used by the Pentagon), were based upon the network organizational principle from the very beginning. Each network cannot be controlled by any central unit and consists of thousands of autonomous computer networks, which can be connected by innumerable links that may bypass electronic obstacles (Castells, 1996).

More closely related to the biosocial focus of this work is not the Internet *per se,* "a physical network of computers linked together by optical fibre and other data connections" (Newman, 2003, p.7), but the networks made up of its users that form part of the *World Wide Web*. "The World Wide Web was invented in the 1980s by scientists at the CERN high-energy physics laboratory in Geneva as a means of exchanging information among themselves and among their coworkers, but it rapidly became clear that its potential was much greater" (Newman, 2012, p.64).

In the modern-day world, the term *social network* is widely used in a more specific sense than the meaning discussed in most sections of this work. It denotes communities of Internet users, including such global networks as *Facebook, Bebo, Twitter, MySpace, LinkedIn, Nexopia* (predominantly used in Canada), *Tuenti* (Spain), *Odnoklassniki* (Russia and Ukraine), and *QQ* (China). Indisputably, a majority of virtual social networks represent decentralized cooperative networks in organizational terms. Each user with his/her profile is an autonomous network node that can selectively connect to other users ("friends"). A prominent feature of many virtual networks is that they are based upon unifying ideas, norms, and customs, which is particularly characteristic of specialized networks such as *TheSocialGolfer* or *Great Cooks Community*.

However, despite its name, the World Wide Web has a mixed organizational pattern. Its networks often convert into hierarchies, partly due to external factors such as the influence of the political system, of the police, and of educational institutions. They also develop into market-type structures as the World Wibe Web[12] is largely commercial.

Obviously, the above discussion applies to other virtual social networks such as those consisting of mobile phone users, e.g., *Facebook Mobile*.

Importantly, "real-life" network structures composed of people and their groups (organizations, firms, etc.) increasingly use novel information technology based upon the Internet and become partially "virtualized". The information resources of network structures in society were developing during the last decade from presentation pages into more complex interactive web nodes which enable ongoing dialogues with various partners. Topical and functional online communities were set up. Information-exchanging virtual networks became an economical and efficient means of communication among the members of a social network *sensu stricto*.

Of still greater importance is the diametrically opposite process: the conversion of originally virtual networks into real-life network structures of people, which may have far-reaching social and political consequences. Internet-based networks can be "devirtualized" almost instantaneously during a flashmob, as was the case during revolutions in North Africa in 2011 and in Ukraine in 2013-2014.

In the literature on network structures, networks of all kinds are often compared to information networks, primarily those based on the Internet. It serves as a prototypical network structure. Special attention is typically given "to designing variable parameters for tasks, processors (or managers), their arrangement and communication between them" (van Alstyne, 1997).

The following is an attempt to subdivide network structures (*sensu stricto*, as adopted in this work) into several subtypes based on a number of classification criteria.

1.2.6. Flat and Three-Dimensional Networks

In structural terms, a network lacks a clear formal hierarchy and dominance-submission relationships are less important than horizontal interaction. Network members mutually

[12] Naturally, the Internet and the WWW have to a great extent been subject to commercialization; therefore, they were characterized by "marketization" from the very beginning of their existence.

influence each other's behavior and have approximately equal ranks. However, leadership exists in many network structures. This gives us grounds for classifying them into at least two subtypes.

1. A completely *flat* network is characterized by the equal importance of all its members (network nodes) and a lack of even partial leaders. Alternatively, each member can consider herself/himself apartial leader. One 1960's commune, *Twin Oaks Intentional Community* (Twin Oaks, 2014, http://www.twinoaks.org, revisited in 3.2 below) that was established in Washington, D.C., originally had 36 members and over 40 leaders (e.g., the kitchen leader, the hammock leader, etc.) so that some commune members combined several leader functions.

An almost ideally horizontal structure occurs in shoals of fish or seals. The individual that happens to be the first in a moving *leaderless (equipotential)* shoal migrates towards the back of the shoal after a short period of time and is replaced by another individuals who does not "lead the way" for a long time either. Most microbial colonies and many cell groups in the tissues of multicellular organisms also lack leader cells. In human society, attempts at establishing a completely non-hierarchical social system that were repeatedly made in various periods of history, maximally met with a partial, temporary success. Some hunter-gatherer bands were sufficiently close to this ideal system. Obviously, given the approximately equal importance of each node of such networks, they are likely to belong to the random network type. The node degree distribution within them tends to correspond to a Gaussian curve with very short tails; ideally, all nodes (with the possible exception of the marginal ones) can have exactly the same number of neighbors, making the whole structure a regular lattice.

2. *Three-dimensional networks* with *partial leaders* whose functions are limited to a particular situation/area of activity. (*Note:* The term "three-dimensional network" is not used in its *geometrical sense* in this work). Such a network may have a large number of leaders. Despite the partial leaders, the structure lacks the pyramidal shape that is characteristic of hierarchies. Nonetheless, the partial leaders typically represent nodes with a relatively high node degree (K), i.e., hubs in terms of network science (1.1.3). Therefore, the node degree distribution does not behave in conformity with the Poisson law, and the Gaussian curve is right-skewed, i.e., has a relatively long tail of high-degree hubs (Newman, 2012). It can be similar to a power law curve typical of scale-free networks.

In primitive society, one hunter-gatherer band could include several partial leaders with limited competencies (shamans, "headmen", skilfull hunters or fishers, and others, see 3.1.1, the passage concerning *split leadership).* For example, the shaman dealt only with spiritual matters. To reiterate, the split-leadership principle is being revitalized in present-day society in terms of various network structures including the hiramas described above.

One of the partial leaders can assume special power, at least during a certain period of time. The Internet-circulated text of the "Russian Doctrine" produced in 2005 by a group of scholars (A. Kobyakov, V. Averyanov, and V. Kucherenko) on behalf of the *Russian Entrepreneur Foundation* under the auspices of the Dynamic Conservatism Center assigns an important role to network structures in terms of Russia' development strategy. It singles out three-dimensional networks (with a vertical dimension): "Network structures can be flat… They are cemented by… a common worldview… At a more advanced stage of development, the network becomes three-dimensional and includes a stratum of goal-setting intellectuals and one of financial strategists alloting money for goal-directed projects" (Kobyakov et al., 2005, http//:www.rusdoctrina.ru).

The *flat network* → *three-dimensional network* transition was made by international network organizations dealing with human rights including those of "prisoners of conscience", such as Amnesty International. As the network size increased, connections between nodes (activists or their small groups) become increasingly distributed according to the power law that is typical of scale-free networks.

It should be reemphasized that the majority of nodes in scale-free networks have few links with other nodes, whereas a minor part of them is linked with a majority of other nodes from the same network. Accordingly, the network becomes increasingly centered on several nodes that concentrate *social power* (Kahler, 2009a). New members tend to join the nodes that were already linked to many other nodes.

In a similar fashion, collective work in ant societies, e.g., anthill construction, is initiated by several "project teams" working at different points. Thereafter, new worker ants entering the "construction site" join the most successfully working teams (Kipyatkov, 1991).

Obviously, one of the nodes with a large number of connections may eventually become the central leader. If this function is only transiently performed by a node that is bound to be easily replaced by another influential node, then the structure is still a three-dimensional network. However, the tendency towards the increasing centralization of power in three-dimensional networks can eventually enable the former temporary "project team leader" to assume the *position* of the permanent boss in the resulting rigid hierarchical structure.

Interestingly, the future network society described in the book *Netocracy...* by A. Bard and A. Söderqvist (2002) promotes the formation of three-dimensional networks with sufficiently strong leaders ("netocrats"). The netocrats use sophisticated password systems in order to hide in the network. This peculiarity of the *netocracy* reminds us of medieval Freemasons' lodges with clandestine leaders.

Analogs of human three-dimensional networks occur in biological systems including those composed of unicellular organisms. Cells of the slime mold *Dictyostelium discoideum* form a compact cell mass called the migrating slug, or pseudoplasmodium (Devreotes, 1989; Mutzel, 1995). This is a typical decentralized network structure. Under the influence of the regulatory signal cAMP, which is produced by a large number of cells, the slug forms a mushroom-shaped fruiting body with a stem and a cap that contains ripening spores. Invariably, the fruiting body appears on the upper side of the migrating slug which is characterized by the maximum concentration of cAMP and other signal molecules. The cells located in this part of the slug assume the role of temporary "project leaders" (to use a business term). The three-dimensional network becomes flattened if the migrating slug is mounted on the needlepoint, where it has no lower side with the cAMP pool diluted by the medium. As a result, several small fruiting bodies simultaneously form at different points of the slug surface, so that the needle resembles a medieval mace with several protrusions on it.

Despite all possible variations on the network theme, a lack of a single fixed activity center (a boss, a dominant individual, a central leader) distinguishes three-dimensional networks from hierarchical structures in biological systems and human society.

1.2.7. Dense and Sparse Networks: Modularity

The above discussion actually refocused our attention on the parameter called *clustering coefficient* that was defined in 1.1.3.3 as the ratio between actual links among its neighbors and those potentially possible. Based upon average clustering coefficient, network structures can be subdivided into *dense* and *sparse* networks (with high and low average C values, respectively). Decentralized network structures in human society include, for instance, (1) close-knit communes or brotherhoods where people are strongly motivated to cooperate and (2) loose associations of people, such as clubs, that are characterized by very weak ties between their members.

Networks in biological systems vary widely in terms of their clustering coefficient. In fungi, mycelium as a cell collective can be either dense or sparse, depending on the degree of aggregation of its hyphae (filaments consisting of cells or representing a single elongated cell with many nuclei). Some fungi can exist in the form of dense mycelium and unconnected solitary cells (yeast-like growth). In the latter case, cells act as autonomous individuals and compete rather than cooperate. The whole network structure, therefore, tends to convert into a biological quasi-market structure.

Of special interest are networks with compact subnetworks linked via a small number of nodes with high betweenness degrees (see 1.2.4.1), i.e., which include parts with a high clustering coefficient C (1.1.3.3), that are sparsely connected to one another. They are exemplified, in human society, by global terrorist networks such as *al-Qaeda* that includes local subnetworks. The nodes connecting local subnetworks are of paramount importance. They attain *bargaining powe*r (Kahler, 2009a) and a high betweenness centrality level even if they have a small number of links compared to nodes inside the dense subnetworks.

In a similar fashion, bacterial films including those formed by cyanobacteria are characterized by "cliquishness", or "community structure" as defined in 1.1.3.3: the central colony forms peripheral microcolonies connected with it by very narrow strands of cells that assume special importance in terms of maintaining the integrity of the whole structure (Sumina, 2006; Pavlova et al., 2007). An analogous community structure occurs in large fish shoals (Wey et al., 2008).

Such networks are characterized by *modularity*. "Modularity refers to a group of physically or functionally linked" nodes "that work together to achieve a (relatively) distinct function....Modules are seen in many systems, for example, circles of friends in social networks or websites that are devoted to similar topics on the World Wide Web. Similarly, in many complex engineered systems, from a modern aircraft to a computer chip, a highly modular structure is a fundamental design attribute" (Barabási & Oltvai, 2004, p.108). Typically, these "modules are not isolated from each other; they interact and frequently overlap... within a network" (Ibiḋ., p.112).

In biological systems, modules are exemplified by interconnected repetitive structural units (Marfenin, 1999) such as zooids (polyps and medusae) in colonial cnidarians (see 2.2). Modules form higher-order systems, often referred to as decentralized *modular organisms.* With respect to the neural networks of the brain (see 2.5), the term "module" denotes a network dealing with a specific function.

Modularity is also characteristic of some primitive and modern structures in human society. Recently, a number of social projects have been developed that envision establishing modular structures. For example, Dmitry A. Lebedev (2007, p.70) suggests setting up *Kinoviys*, or "social cells", i.e., groups of 350–450 people that should represent "full-fledged social organisms" within the higher-order network structure of the entire society.

Although a large number of modular structures possess manifest fractal properties, *modularity* is, obviously, not an exact synonym of *fractal/self-similar* structure, since a homogeneously dense network structure including no spatially segregated modules can, nevertheless, be self-similar. It can be cut into a large number of smaller networks, which are structurally similar to the entire structure.

1.2.8. Homogeneous and Heterogeneous Networks

Obviously, network structures can be subdivided into two types in the following way:

- *Heterogeneous networks.* Their parts (nodes, subnetworks) are characterized by different properties; accordingly, they can perform different functions, as is the case with partial leaders in hiramas (see 1.2.2); disassortive network structures (as typified by biomolecular networks where molecules tend to interact with molecules of a different type, 1.1.3.3) belong to the heterogeneous type.
- *Homogeneous networks.* Their parts possess similar or identical properties, which predisposes them to perform coinciding functions and may increase the stability and adaptability of the whole structure.

Some structures are characterized by intermediate properties compared to homo- and heterogeneous networks. The functions of their nodes (or subnetworks) significantly overlap, but they cannot completely replace one another in functional terms. Different brain layers contain functionally similar centers that, for example, are involved in regulating the sleeping-waking rhythm. Nonetheless, each center regulates the rhythm in a different fashion (Dubynin et al., 2003).

A network sensu stricto *is defined as a decentralized structure with cooperating elements. Networks are sufficiently widespread in biological systems and in human society. Mathematical analytical tools such as degree centrality, closeness centrality, and betweenness centrality can be used to distinguish networks sensu stricto from hierarchies, whereas ecological plus/minus criteria enable one to discriminate between network and (quasi-)market structures. Networks are also exemplified by information systems, particularly those based on the Internet. Networks can be subdivided into subtypes: they can be flat or three-dimensional, dense or sparse, homo- or heterogeneous.*

1.3. Hierarchies

Unlike the decentralized cooperative networks predominantly considered in this book, *hierarchies*, or *centralized networks,* have a fixed single center (leader, pacemaker, dominant individual, boss, president, etc.). The other elements in the structure are subordinate to the central element. These subordinate elements often differ among themselves in terms of rank, forming a typical hierarchical step-ladder. In other words, a hierarchical (pyramidal) structure can be defined as "any system of persons or things ranked one above another" (Collins English Dictionary, 2009, http://dictionary.reference.com/browse/hierarchy). The dominance-submission relationships and the rank order characteristic of hierarchies both in human society and in the animal realm set them apart from *sensu stricto* network structures that tend to equalize social ranks and establish cooperative interaction between their elements (individuals and their groups).

In human society, hierarchical structures can be defined as "a rank ordering of individuals along one or more socially important dimensions" (Anderson & Brown, 2010, p.57) such as power, status, or leadership. Many hierarchies combine several dimensions.

Hierarchical structures were characteristic of many traditional societies. They were exemplified by typical feudal-vassal relationships in medieval Europe. Pyramidal structures are also formed by bureaucracies including a wide variety of modern political, cultural, scientific, and business institutions. For instance, a business firm is headed by a President or Chief Executive Office (CEO) and their Vice Presidents and Directors, while the heads of its branches (middle managers) are their subordinates and dominate, in their turn, over those having still lower ranks (frontline managers). A quite similar pattern is often used by other kinds of organizations, e.g., an architectural firm.

A large number of nonhuman biosocial systems are also characterized by some degree of individual inequality. Individuals in such systems have clearly different ranks. "Applied to social organization, hierarchy is synonymous with rank order and involves the concept of social dominance. Individuals in a group yield to others in contention for something, such as food and mates, according to a more or less linear order" (Immelmann & Beer, 1989, p.131).

Hierarchical structures in the animal kingdom are exemplified by the communities of a large number of primate species. Most Old World monkeys form sufficiently rigid hierarchies, while they are superseded by partly non-hierarchical structures in the groups of closer evolutionary relatives of *Homo sapiens,* such as chimpanzees and particularly bonobos. Hierarchies are characterized by specific quantitative properties including the following (Anderson & Brown, 2010, with my modifications):

- *Tallness*, i.e., the number of levels in a hierarchy
- *Span of control*, number of subordinates of a particular boss (in human society) or of a dominant individual (in the animal kingdom)
- *Centralization*, which superficially seems to have a different meaning than in network science (Chapter one): "degree to which decisions are made by fewer individuals who are higher in the hierarchy…" (Ibid., p.62). Nonetheless, these high-ranking decision-makers should be able to contact a sufficiently large number of subordinates who should be informed of their decision; therefore, centralization in a hierarchy is directly related to the network science concepts of *degree, closeness, and betweenness centrality* because they are correlated with the availability of a relatively large number of other nodes (subordinates) that can receive the superior's instructions
- *Directionality of links* (see 1.2.4.1) between the central node and the subordinate nodes, i.e., the predominant control of the boss over the subordinates. Originally used with respect to human society, this term is also applicable to social animal species because their groups make collective decisions (e.g., concerning the direction of a group's movement) that can be influenced by high-ranking individuals.

1.3.1. Dominance and Leadership

The individual at the top of the hierarchy can use (and often combine) two different behavioral strategies, performing the role of the *dominant individual* or that of the *leader*. According to the literature concerning the behavior of animals (ethology) and humans (sociology, social psychology), the difference between dominance and leadership in a hierarchically organized group is as follows:

- *Dominance:* the individual at the top exploits the group's resources and limits their use by other group members. A dominant individual enjoys the right to behave as the individual chooses (Maslow, 1954) and to use food, shelter, sexual partners, and other resources without paying any attention to others' behavior. In ethological terms, dominance can be defined as "a social relationship when one individual can monopolize resources at the other's expense or usurp them from the others by using force or threatening to do so, even though not all aggression concerns immediate access to resources or contests over status, and some contests are decided by unprovoked, unilateral submission rather than by aggression" (Watts, 2010, p. 110). This statement highlights the fact that dominance is not necessarily linked to aggression. It may involve submissive behavior of others. "By persistently showing their submission, some primates secure a happy life for themselves," including the dominant's benevolence, protection from predators and strong second- and third-ranking (subdominant) individuals, assistance in obtaining food, and other advantages of the social lifestyle (Deryagina & Butovskaya, 2004, p. 137).
- *Leadership:* the highest-ranking individual guides the group's movements—leads the group literarily—and controls other collective behaviors (Lamprecht, 1996). Managing the group's movements may involve intention movements performed by the leader. The leader in a bird flock flaps its wings in readiness to take off. These movements are understood by other individuals, and they behave accordingly. In human society, leadership is largely a cultural phenomenon. Nevertheless, the term "leader" still refers to someone who leads; who guides other's behavior. Leaders are in charge of various human alliances, from small groups to whole nations.

These two aspects refer to the two edges of one continuum (Figure 1.6). The continuum can be subdivided into three main zones (cf. Eibl-Eibesfeldt, 1998):

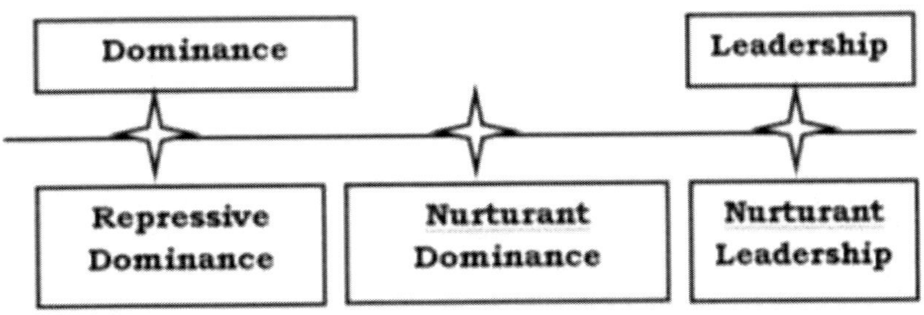

Figure 1.6. Dominance and leadership refer to different, but partially overlapping, zones within one continuum.

1. *Repressive dominance.* The dominant individual's behavior is predominantly aimed at gaining control over group resources, exploiting other group members, and limiting their freedom and independence.
2. *Nurturant dominance.* Dominance and leadership are at an equilibrium, which implies that the dominant individual takes care of the subordinates apart from exploiting them.
3. *Nurturant leadership.* Dominance is overpowered by careful leadership that involves guiding the behavior of all group members and providing support to them.

Primatologists argue that hierarchies in a number of primate species (including chimpanzees) have nurturant leaders at their pinnacles. Such a leader manages food distribution within the whole group. Other group members are encouraged to approach the leader to claim their share of food.

In a number of biosocial systems, e.g., in those of ungulates and birds, the dominant male and the leader are two different individuals. The individual who *claims* the first tidbit when food is distributed does not *lead* the group while it is moving. The dominant individual may prefer to stay in the center or even the rear of the moving herd/flock. The dominance-leadership continuum is disrupted, resulting in a split hierarchy with two distinct pinnacles, which is of some political interest in terms of power distribution within a political hierarchy. Some biosocial systems have several codominant and/or partial leader individuals.

1.3.2. Agonistic and Hedonic Hierarchies

In a hierarchical structure, the ratio between dominance and leadership varies depending on the strategy used to reach the hierarchy's top. The two main strategies used by humans and other primates to attain dominance/leadership are as follows (Masters, 1989):

- *Agonistic hierarchy.* The highest rank in such a hierarchy is acquired via agonistic interactions involving aggression or threatening displays demonstrating forcefulness and assertiveness. Many agonistic hierarchies are based on the might-is-right principle. Therefore, the individuals at their tops are likely to exercise repressive dominance, which is exemplified in human society by spontaneous hierarchies in prisons, barracks, kindergartens, and psychiatric hospitals. Chimpanzees are characterized by a partly egalitarian society in the wild but captive chimpanzees may establish agonistic hierarchies.
- *Hedonic hierarchy.* The hierarchy top is occupied by the most interesting, capable, intelligent, sociable, or knowledgeable individual to whom the whole group pays special attention (Chance, 1967). Even weak individuals can be in the group's focus of attention and, therefore, at the pinnacle of a hedonic hierarchy. A prerequisite is that they generate considerable interest with their group mates or supply them with important information (Barner-Barry, 1977, 1983). A male chimpanzee can attain a high social rank by securing the support of influential group members. Data on presidential elections in the USA have revealed that the most successful candidate typically displays non-aggressive nonverbal cues signaling a supportive and

reassuring attitude towards the voters. Hence the successful candidate for presidency establishes a hedonic hierarchy (Masters, 1989). In contrast to an agonistic hierarchy, a hedonic hierarchy is likely to involve nurturant leadership. Interestingly, in a primitive hunter-gatherer society, the leader was expected to resolve conflicts among members of his band/tribe and to make clever decisions in difficult social situations, i.e., he was to exercise nurturant leadership in a hedonic hierarchical structure.

Apart from the two aforementioned strategies for establishing hierarchies, there is an intermediate, or mixed, strategy in which physical superiority is supplemented by preferential attention given by the group involved. This mixed strategy apparently occurs in some lower primates such as vervets, resulting in the middleground hierarchy type characterized by *nurturant dominance.* The dominant individual claims valuable resources (food, mates, or shelter) but also performs some leader-specific functions including territory patrolling, the mitigation of conflict, and the coordination of group defense.

Agonistic hierarchies are often characterized by a high degree of centralization because they are based upon the unlimited repressive dominance of the individual at their top. The obvious reason is that, in this system, a successful aggressor dominates individuals that have acknowledged their defeat. To reiterate, I suggest that such rigid hierarchies should have central nodes that score high on all centrality scales mentioned in 1.2.4.1. Overall, agonistic hierarchies are expected to display high degree, eiegenvector, betweenness, and closeness centrality-based hierarchization levels.

By contrast, *hedonic* hierarchies based on preferential attention to particular individuals are less robust and more likely to split and convert into a structure with several leaders. A typical hedonic leader enjoys respect and popularity, i.e., predominantly scores high on the degree scale[13]; another individual may have more control over information flow and, therefore, have higher scores on the betweenness scale (see 1.2 above). In fact, the attention structure of ape or human groups is changeable. A hedonic leader's authority may be lost or at least limited as soon as a new emergent leader takes over. The new leader is likely to be temporary as well. At a given moment, the structure can include more than one leader. Such a system is actually in the process of transitioning from a hierarchy to a three-dimensional network.

A hedonic hierarchy may be turned upside-down by a sufficiently strong coalition of subordinates who put the former leader under their control and make him take the decisions that are preferred by them. This situation was described by ethnographers in primitive societies; it is possible in modern-day society as well. For example, a weak-willed school-class monitor merely does what other students require of him. Obviously, in this case the formal hierarchy is overpowered by an informal network structure that can be informally hierarchized if one of the students becomes the "dominant male".

Flat hierarchies with very few hierarchical ladder levels are similar to three-dimensional networks. In human society, they are exemplified by business firms in which the CEOs directly interact with frontline managers in the absence of middle managers. The border line between flat hierarchies and three-dimensional networks can be drawn somewhat arbitrarily

[13] In a hedonic hierarchy, the alpha individual can also score high on the eigenvector scale (mentioned in 1.2.4.1) if he/she belongs to an important coalition including influential individuals.

depending on how much power[14] (centrality) the preponderant leader has and whether this role is temporary (situation-dependent) or not.

On the economic macro-level, the transition from a hierarchy to a three-dimensional networks takes place whenever a firm's leaders delegate certain functions to its permanent partners instead of letting that firm's departments perform them; alternatively, some former departments may be given organizational autonomy, and a hierarchical firm converts into a network with the former boss becoming just an influential partner inside the developing internal network (see 3.7.2.2). This system is currently used by some formerly hierarchical businesses in China, including the Tianjin Automotive Group (Wang & Tanaka, 2011).

Transitional structures also exist in biological systems. In the animal kingdom, a relevant biosocial phenomenon is called *absolute dominance.* Such a system represents a two-level hierarchy with a single dominant individual and a number of subordinate individuals having approximately equal ranks. This hierarchical structure often results from overcrowding, e.g., when dense schools of stickleback males or groups of mice are confined to a limited amount of space (Reznikova, 2001, 2005). Subordinates per se form a flat network. Potentially, they may collectively restrict the dominant individual's power or even make the former dominant individual submit to their coalition. This can be called "hierarchy reversion" (this concept is revisited in 3.1.1. with respect to primitive hunter-gatherer societies).

1.3.3. Comparing Hierarchies and Networks with Respect to Their Performance

It is the type of the task/problem to be dealt with that determines the optimum organizational pattern of the structure to be used. Networks *sensu stricto* with decentralized leadership are more adaptive and flexible than hierarchies because of their loose coupling and openness to information. Studies on collective problem-solving by creative teams have demonstrated that "more hierarchical groups tended to perform better on simple tasks" requiring straightforward decision-making strategies. "However, flatter groups <i.e., networks – A.O.> tended to perform better on more complex, more ambiguous tasks that benefitted from the input of many members and that often required creativity" (Anderson & Brown, 2010, p.67). Since a disproportionately high amount of influence on decision-making is exerted by the leader in a hierarchy, a hierarchical group's success often crucially depends on the leader's skills and knowledge. Unfortunately, many groups tend to choose dogmatic, assertive, selfish, authoritarian, risk-taking individuals as leaders. The power and status enjoyed by the leader can make his/her personality still worse.

The potential assets of networks were acknowledged in Europe in the public policy domain in terms of the *New Public Government* doctrine, which has been promoted since 1995. Decentralized networks also work well in knowledge-rich environments because they have superior information-processing capabilities. "The theoretical rationale behind the prediction that decentralized leadership structures are related to superior team performance is that when there are many leaders within a group this enhances participation and information sharing among team members, which, in turn, enhances team performance" (Mehra et al., 2006, p. 234). In the same work, it is emphasized that a multileader group (a network *sensu*

[14] Notably, in the flat hierarchy of a firm the CEO still has sufficient power and his position is often permanent.

stricto) can only outperform a centralized hierarchy if the several partial leaders successfully coordinate their efforts and acknowledge one another's leadership functions.

Despite their potential assets, networks should not be idealized because they have their own limitations. Networks are prone to fall apart if the relatively few links connecting highly clustered fragments are removed. Their tendency toward organizational openness often endangeres the atmosphere of mutual trust ("social capital") which is a prerequisite for efficient network organization. In networked organizations, responsibility and accountability issues often pose serious challenges to the network structures involved. The advantages and disadvantages of networks will be revisited in 1.8. where interaction between different structures will be discussed.

The hierarchy—network distinction has relevance to modern psychology. Even though the self seems to be an integral system, it actually includes several personas. This split inside our inner world manifests itself while we are making important decisions: the different selves talk with one another and often express different opinions. Predominantly, the personas form a hierarchy dominated by the main self (that often is considered by an individual to be *the only self*). With some patients with mental disorders, psychiatrists have to deal with a typical network structure where the different selves have equal ranks. The psychiatric treatment of such patients is aimed at re-establishing the normal hierarchy. However, the hierarchy should not be too rigid: non-dominant selves should retain a kind of "voting right" during decision-making. Overall, a psychiatrist seeks to find an optimum balance point between a hierarchy and a horizontal network. A psychiatrist's job is similar to that of a social/political engineer who attempts to solve the same problem with respect to human social/political structures and the whole society.

As mentioned above, an organizational scenario used by a large number of network structures is based upon the interaction between a relatively small number of specialized leaders (with experts assisting them) and a large pool of unspecialized network members.

In sharp contrast, a typical scenario employed by a hierarchy implies that most individuals involved are specialists; the degree of specialization decreases as one moves up the hierarchical ladder, and the top is occupied by very few unspecialized high-rank individuals. The *pool of specialists+few generalists* scenario is characteristic of the subtype of hierarchies typified by modern bureaucracies. In a bureaucracy, each individual performs a specialized function (e.g., screwing a bolt in an assembly line) and even the boss may be perceived not as a generalist but rather as a specialist whose function is "general management". In a similar fashion, the emergence of a leader (an alpha individual) in a group of animals frequently represents the formation of the first specialized role in the group.

The distinction between hierarchical and decentralized structures was highlighted in the works of a number of scholars including Herbert Spencer, Ludwig von Bertalanffy, and the founders of cybernetics (Wiener, Ashby, etc.). Of particular interest in the context of this work is Alexander Bogdanov's (1980 [1921]) book on "general organizational science", or "tectology/tektology". This scholar emphasized that biological systems (living organisms and their communities) represent important models for studying general laws of complex systems' behaviors. Bogdanov distinguished between two different kinds of systems, referred to as *centralist* and *skeletal,* which are predominantly characterized by *egression* and *degression,* respectively:

- *Egression* is a process resulting in the formation of an active center, e.g., the development of the brain in an animal embryo or the emergence of a leader in an ape group or another biosocial system
- *Degression* refers to the development of a sufficiently strong "skeleton" making a system more stable and resistant. This is exemplified by the formation of an external skeleton in the literal meaning (e.g., the carapace of a crustacean or of a turtle). In a more general sense, the term degression denotes a protective barrier between the system and the environment. For instance, naked mole-rats, mammals that form large colonies like social insects, vigorously defend the boundaries of their territory from intruders.

As Bogdanov's commentator Milan Zelený (1988, p.336) pointed out, "so called *centralist* and *skeletal* forms are described as either egression (from "stepping out of line"), based on authority-subordination principle, or degression (from "delegating to lower level"), based on decentralized-distributed principle".

It is clear from the above that egression in Bogdanov's terminology corresponds to the process of hierarchy formation. Bogdanov (1980 [1921]) stressed that revolutionary periods in the history of human society are often characterized by egression, starting from hardly noticeable authoritarian behavior (associated with an incipient hierarchy) and ultimately resulting in a rigorous dictatorship in which "strong power" is exercised. As for *degression,* it is closely related to *matrix formation,* an organizational feature of complex systems that is of particular importance for decentralized cooperative networks; it will be considered in detail in section 1.6.

In biological systems as well as in human society, hierarchical structures are based upon dominance-subordination relations between high- and low-rank elements, in contrast to networks (sensu stricto), which are characterized by social rank equalization and cooperation. Dominance *implies the right of certain individuals to behave without taking account of others' behavior and to have preferential access to a group's resources. Dominance is supplemented by* leadership, *i.e., guiding the movements and other forms of individual behaviors in an hierarchical structure. The ratio between dominance and leadership varies depending on the hierarchy subtype involved. In particular, a hierarchy may be* agonistic *(based upon agonistic interactions among individuals) and* hedonic *(based upon the group's attention structure).*

1.4. (Quasi-)Market Structures

In contrast to both networks (sensu stricto) and hierarchies, market structures in human society typically include autonomous agents that enter into competitive relations with one another, while cooperation is only limited to contracts between the independent agents. Quasi-markets[15] are predominantly competition-based structures which lack true market relationships; they include not only human social structures but also communities formed by various animal species (Oleskin, 2012).

[15] The term *quasi-market* is used in a different sense by some economists: it denotes interactions between former departments of a firm that have acquired some degree of autonomy and entered into contract-based relations. Such a system is very closely related to what is called "internal network" (see 3.7.2.2).

1.4.1. Defining the Market

A large number of definitions of the *market* are given in the literature. This comparative approach will make use of a classical interpretation: "economists understand by the term *Market* not any particular market place in which things are bought and sold, but the whole of any regime in which buyers and sellers are in such free intercourse with one another that the prices of the same goods tend to equality easily and quickly" (Marshall, 1920 [1890], quoting Cournot, 1838). Many contemporary economists maintain that the market functions on the basis of competition and equivalence; the agents involved are free to choose the partners with whom they enter into purchase-and-sale relations (Economic Theory. Microeconomics, 2005). An idealized market – one that is free of hierarchies and networks – implies perfect competition with a large number of buyers and sellers which lack direct control over price levels that are set by an "interaction of the decisions these market agents made in the light of the prevailing price but over which they individually have no direct control" (Frances et al., 1991, p.7).

Cooperation in terms of (quasi-)market structures is only limited to contracts between independent agents. These contracts are made on the basis of market prices and the supply—demand balance. "The market coordinates 'automatically', so to speak. The pursuit of self-interest by individually motivated and welfare-maximizing individuals leads to the best outcome not just for them but also for the society as a whole. Coordination takes place in an unseen manner – via the 'guiding hand' of market exchange and the price system it supports" (Frances et al., 1991, p.3).

A hierarchy is comparable to an *"iron fist"*, a network to *the brain* (because consensual decision-making in a network resembles thinking), and a market to a guiding *"invisible hand"* (the comparison was drawn by Adam Smith; see Table 1 below).

1.4.2. Market Analogs (Quasi-market Structures) in Biological Systems

Some biological systems, like markets in human society, are characterized by predominantly competitive relations between autonomous agents. For example, some insects, crustaceans, and other kinds of animals form large conglomerations. Locusts can form enormous swarms. Individuals in them behave independently and compete rather than cooperate with another, although they respond to socially produced stimuli such as pheromones by, for example, changing their color (green locusts turn brownish-red in large conglomerations). Similar conglomerations termed "asocial aggregations" occur in fish species that do not form "social aggregations" such as shoals or polarized schools (see 2.3 below). Such "fish are not attracted to one another to a significant extent, their movements have no common orientation, and their responses to external stimuli lack coordination" (Pavlov & Kasumyan, 2003, p.8).

Quasi-markets in biological systems include analogs of production chains in human society[16]. Biological systems form *metabolic chains*. Chemical substances are stepwise

[16]Such human production chains are exemplified by the refining and chemical modification of crude oil by a series of oil-processing firms, so that the product of each firm is then supplied (according to a purchase contract) to the firm responsible for the next stage.

transformed by a chain of organisms, and each of them carries out just one stage of the process. Intermediate products are used as raw material during subsequent stages.

As in the marketplace, a biological quasi-market structure implies the coexistence and competition of several candidates for performing each chemical transformation stage. For example, several different bacterial species can degrade organic substances to simple molecules, supplying methane-producing microbial associations with "raw material" for biogas production; therefore, the different bacterial species can be considered as competitors in terms of their organic substance-degrading role within the overarching ecological quasi-market structure that, in this case, includes methanogenic microbes.

Reptiles are of interest in this context. In contrast to social animals, e.g., shoal-forming fish species and many insects, a majority of reptiles are "real individualists" (Semyonov, 2001). Many reptiles group because they jointly use food resources or prolong their exposure to the sun; such conglomerations are not due to social interactivity. For instance, "overwintering aggregation occurs in many snakes and some lizards. The animals may be responding individually to some aspects of the physical environment rather than to each other, although it is likely that young snakes find dens by following the scent trails of adults" (Heatwole & Taylor, 1987, p.205). Individuals in reptile conglomerations compete for resources, defend their territories, and mark them with odorous excretions (pheromones). Aggressive behavior is displayed towards intruders, particularly if they are adult individuals of the same sex. This occasionally results in fighting. Missisipi alligators use their jaws, sometimes breaking those of their opponents. More frequently, aggression is limited to ritualized threatening displays such as "snake dances", i.e., bloodless contests in which the loser finally takes flight (Semyonov, 2001, p.119). Irenäus Eibl-Eibesfeldt (1998), one of the disciples of the famous animal behavior researcher Konrad Lorenz, was strongly impressed by the behavior of iguanas on the coast of New Guinea. These reptiles do not engage in loyal behavior towards one another; on the contrary, they constantly demonstrate aggressive displays while competing for the territory, like merchants that compete for a market area.

Such quasi-market interactivity may cause the formation of dominance hierarchies, particularly if the area is overcrowded. These *agonistic hierarchies* (see 1.3.2) result from competitive interactivity and represent, therefore, *degenerate* quasi-market (competitive) structures in which market-specific competition is eliminated because one of the competitors gains a monopoly over all the resources involved[17]. For instance, "a large and strong sand lizard male usually plays the dominant role. Other males have subordinate roles" (Brehm, 2001, p.311).

The term "quasi-market" can be used also with respect to human society in reference to structures that are not called markets but are based upon competitive relationships. In a classroom, students can work individually, trying to outperform one another (in terms of a task given by the teacher). This quasi-market structure can be replaced by a network if the students are encouraged to work together, and the reward of each varies depending on the collective result. Finally, a hierarchy would imply the dominance of one student making most decisions concerning the task of the whole student group.

[17] In contrast, *hedonic* hierarchies (see 1.3.2) should be considered *degenerate networks* in which one of the partial leaders attains dominance. However, a hedonic hierarchy can subsequently evolve into an agonistic hierarchy.

1.4.3. Real-life Market Structures: The Formation of Networks or Hierarchies

In human society, the ideal market, one based on perfect competition, never existed, even though it was approximated to an extent by the free market competition that prevailed in Victorian Britain. The real market in modern-day society includes typical *network structures* based on prolonged cooperative relationships between firms that (1) produce similar products and form cartels, trusts, etc. or (2) are connected by supplier-consumer ties that involve mutual trust and are not limited to purchase contracts. Presently, market structures are "social molecules" composed of firms rather than the maze of interfirm interactions characteristic of perfect competition (White, 2002, p.114).

"A modern economy involves... the totality of social ties, i.e., stable links between individuals or firms that cannot be accommodated by the traditional concept of the market—hierarchy dichotomy... These networks of formal and informal relationships enable people to find jobs, exchange information, resolve conflicts, and build up trust" (Radaev, 2002a, p.6). Such networked alliances are formed by several enterprises, making them more competitive in the market. They are exemplified by textile-producing factories in Emiglia-Romania (Italy), a classic case considered by W.W. Powell (1990).

In the preceding section, intermediate forms spanning the gap *between hierarchies and networks* were considered. In the commercial realm, some interfirm alliances are characterized by an equilibrium between internal cooperation and internal competition, forming intermediate structures *between markets and networks.*

Interestingly, a similar transition from competitive quasi-market to cooperative network interactions occurs in some animal groups. "Competition for territory during the autumn-winter period among resident Great Spotted Woodpeckers *Dendrocopos major* or among representatives of other species with individual territories" results in the system's self-organization. The biosocial system converts into a network structure with predominantly horizontal long-term interindividual relationships (Fridman, 2009, p.337); the woodpeckers (or other birds) start personally recognizing one another, and their competition for resources (e.g., mates) is minimized. In sum, market structures and their animal analogs can gradually transform into networks, and this transition manifests itself in a gradual increase in interindividual cooperation (as contrasted with competition) and the whole system's coherence and connectivity.

It was relatively recently that networks received sufficient attention as "a third option" in a world otherwise dominated by hierarchies and markets. In terms of the neoclassical approach to economics, networks were considered in some works "only as hybrids between the two ideal poles" (Elsner et al., 2008, p.7) that correspond to markets and hierarchies, respectively. However, while markets are based upon "the paradigm of individually self-interested, non-cooperative, unconstrained social interaction" (Powell, 1990, p.302), networks are predominantly characterized by "voluntary, reciprocal, and horizontal patterns of communication" (Jung & Lake, 2011). This subject will be discussed in more detail in 3.7.1.

There seems to be a much sharper boundary between *markets* and *hierarchies.* These two terms were traditionally contrasted in the literature on economics and sociology.

Nonetheless, a tendency towards the *hierarchization* of markets occurred in the late 20th and the early 21st century. This was due to the establishment of oligo- and monopolies under the influence of industrial giants, particularly multinational corporations. Similar to intermediate (hybrid) forms between markets and networks, there are a variety of transitional stages between ideal markets and a rigorous hierarchy implying a complete monopoly of dominant enterprises over relevant economy sectors. "In between the strict monopolist case and perfect competition there lies a range of conditions that economists call monopolistic or oligopolistic competition" (Frances et al., 1991, p.7).

The hierarchization of the market also results from its subjugation to a political system. The political system provides subsidies, influences market prices, levies protectionist taxes, and, in some cases, resorts to coercion. However, too much coercion can obviously result in eliminating the market itself and replacing it with a communist-style command-and-control system.

An analogous process of converting quasi-markets into hierarchies takes place in some biological systems: their resources are monopolized by dominant individuals. This is the case with some of the reptile groups described above, as well as with fish species that do not form network-type shoals (see 2.3).

1.4.4. Comparing Networks, Hierarchies, and (Quasi-) Markets

Table 1 given below sums up the main properties of the three types of structures compared in this work. The table also contains some features to be discussed in the following sections, including internal control and communication, the matrix, and synchronization.

The table's line concerning the specialization of a structure's elements (nodes) summarizes the relevant ideas discussed in several sections dealing with networks, hierarchies, and markets or their analogs. This table is chiefly based on the literature on human social structures, but the concepts presented in it also apply to biological systems. Among the structure types considered in this work, special emphasis is placed upon network structures. The reasons why networks are preferred to other structures in this work are as follows:

- Network structures are ubiquitous (in their almost pure form or in combination with other structures) *in biological systems* ranging from colonies of unicellular organisms to cnidarian or bryozoan cormuses, social insect families, leaderless fish shoals, and egalitarian groups of higher primates.
- *In human society*, the disadvantages of hierarchies in the commercial sphere and the political system have already received much attention in the early 21st century; hierarchies face the serious challenges posed by the new century and often fail to function properly. They require the support and collaboration of their alternatives: network structures. As for modern-day politics, of particular importance is civil society, which is largely based on network organization. It holds special value in countries where the dominance of hierarchies obviously slows down economic, social, and cultural progress.
- From *a systemic viewpoint*, hierarchies and (quasi-)markets should be regarded as "degenerate" networks. *Hierarchies* appear in network structures if a temporary partial leader attains permanent dominance[18]. As mentioned above, *(quasi-)market*

[18] This directly concerns hedonic hierarchies. Agonistic hierarchies represent degenerate (quasi-)market structures, as the above (subsection 1.4.2) example concerning reptiles demonstrated. Since (quasi-)markets can be regarded as degenerate networks (*sensu stricto*), agonistic hierarchies emerge, in this case, as a result of two consecutive stages of degeneration. Alternatively, agonistic hierarchies can develop from originally hedonic

structures are established if the inevitable competition among network members overrides the normally predominant cooperation among them.

Table 1. Characteristics of hierarchies, networks, and (quasi-)markets (Powell, 1990; Frances et al., 1991; Meulemann, 2008; Elsner et al., 2008; Kahler, 2009a, b)

Characteristics	Hierarchical Structure	Network Structure	Market Structure (or its Analog)
Centralization	High	Low	Low
Cohesion, based upon cooperation among nodes	High	High	Low
Flexibility	Low	Medium	High
The elements (nodes) are	Dependent	Interdependent	Independent
Predominant relations between the elements (nodes)	Dominance and submission	Cooperation	Competition (plus contract relations or metabolic chains in biological systems)
Matrix development level	Different in different hierarchy types	Moderate (loose networks) to high (dense networks, "community structure")	Low ("wild" marketing, biological systems with intense internal competition) to medium ("civilized" marketing, contracts)
Control level	High	Low (except for matrix-dominated networks, see below)	Low
Functions of communication	Providing higher-rank nodes with information Information exchange between equal-rank nodes (limited)[19]	Information exchange and discussion aimed at collective decision-making	Advertizing, making contracts, signal exchange aimed at creating metabolic chains
Synchronization degree	High (due to the functioning of the pacemaker/boss)	Low but promoted by matrix-dependent synchronization)	Low but promoted by market-specific synchronization mechanisms or their analogs
Metaphor	Iron fist	Thinking brain	Invisible hand
Decisions are made by	Controlling units (leaders)	Whole structure (group)	Individuals (agents)
Degree of specialization of the elements (nodes)	High, increases from the top to the bottom (the boss is a generalist, the sub-ordinates are specialists)	Low, in three-dimensional networks decreases from the top to the bottom (specialized partial leaders and a pool of generalists)	High, evenly distributed within the whole structure

hierarchies. This also involves two stages of degeneration, i.e. the formation of the hedonic hierarchy and its conversion into an agonistic one.

19 Increasing communication among individuals with equal ranks may convert the hierarchy with a predominantly vertical information flow into a network. In many hierarchies, particularly, of the hedonic subtype, the *leader → subordinate* relations involve not only top-down control but also top-down communication.

Normal networks structures are often more organizationally complex than their "degenerate" variants (hierarchies and market-type structures). The intricate pattern of interwoven links among network nodes underlies the *communication power* phenomenon discussed by Manuel Castells (2009): the nexus of communications among network nodes per se is often sufficiently complex to match the complexity of tasks to be completed by the network. To an extent, the personal qualities of network members may be less important than the problem solving-facilitating communication pattern determined by the network structure per se.

This structure is sufficiently flexible to adapt for solving a new problem, either spontaneously (a network's *adaptability*) or thanks to the efforts of moderators or psychological leaders.

Notwithstanding these advantages, the structural complexity of a network might also present difficulties for its members; they may become confused, which also necessitates the involvement of an efficient psychological leader.

In the preceding text, hybrid (or intermediate) structures were mentioned. They include

- combinations of *networks and hierarchies*, exemplified by "layer-cake" structures which have alternate hierarchical and networked levels (ant systems considered in 1.8 and, in more detail, in 2.4 below);
- *network-market* hybrids, e.g., clusters of firms or of former subdivisions of one firm that have acquired some degree of independence (internal networks)[20]. These hybrids are characterized "by repeated, relatively stable, lasting, and relatively price-independent exchange relations, in this way being some hybrid between ideal cooperation and spontaneous decentralized private interaction systems where prices play some role" (Elsner et al., 2008, p.15);
- intermediate structures combining the features of *markets and hierarchies* discussed above in terms of the oligo- and monopolization of the "real market" in the modern-day world.

1.5. Communication and Control: (Super)Organisms

The functioning of networks, as well as hierarchies and (quasi-)markets, depends upon interactions among their elements (nodes). These interactions are prerequisites for the coordination of the behavior of the elements (nodes) of these structures (see 1.6 below) and can be subdivided into exchanging messages (communication) and controlling.

[20] Internal networks combine three types of structures: a hierarchy inherited from the former bureaucratic structure, a network of informal connections that are enhanced by removing the bureaucratic pinnacle of the formerly centralized firm, and a market structure based on contract relations and some degree of competition between the autonomized "splinters" of the original firm.

1.5.1. Communication vs. Control

Communication is based on *information exchange between individuals and/or their groups.* The message can be verbal (in human society) or represent a nonverbal signal, e.g., an alarm call or a pheromone (an odorous substance). Communication is an essential component of any kind of social behavior because it is hard to imagine social behavior without the exchange of information. Importantly, communication implies that the individuals receiving the message can choose whether and how they will respond to the message. The recipient of the message can ignore it or there may be a delay in the response.

Controlling action means that those under control have no choice. Controlling implies the *stimulus → preprogrammed response* pattern of interactions. In contrast, communication is based upon the *stimulus → receiving individual → variable (chosen) response* pattern. Therefore, the relationship between the response and the stimulus is probabilistic (stochastic), not mechanistic. During communication, the sender and the receiver can repeatedly swap their roles and even perform these two roles simultaneously. Swapping the sender and receiver roles equalizes the status of all the individuals involved in communication. Communication does not involve coercion but it may enable *influence* that is often mutual. Influence based on communication implies that "one actor /e.g., an animal or human individual or a whole group – O.A./ must transmit a message to another, and the second actor must receive, decide, interpret, and react to that message" (Knoke, 1994, p.3). In business, partial leaders in networked enterprises cannot *control* their partners but they do *communicate* with them and try to *influence* their behavior (see 3.7.1 below).

In human society, controlling is exemplified by a categorical order whereas communication implies giving recommendations and suggestions, not orders. Controlling behavior is displayed by a policeman saying in a polite tone: "Sir, let's go! Take your passport, please!" (the policeman has hand-cuffs in case the person did not hear him). However, communication takes place if a young man (e.g., the same policeman) invites a girl to take a walk: "Darling, let's go! It's such a nice evening!" The girl can decide herself whether to accept the invitation or not (controlling behavior would be a crime in this situation).

The difference between controlling and communication was actually discussed by the prominent American philosopher and pedagogue John Dewey in his work *Democracy and Education* (1916). "The parts of a machine work with a maximum of cooperativeness for a common result, but they do not form a community. If, however, they were all cognizant of the common end and all interested in it so that they regulated their specific activity in view of it, then they would form a community. But this would involve communication. Each would have to know what the other was about and would have to have some way of keeping the other informed as to his own purpose and progress. Consensus demands communication.

We are thus compelled to recognize that within even the most social group, there are many relations which are not as yet 'social'. A large number of human relationships in any social group are still upon the machine-like plane /i.e., they involve *control* not *communication – A.O./*. Individuals use one another so as to get desired results, without reference to the emotional and intellectual disposition and consent of those used. Such uses express physical superiority or the superiority of position, skill, technical ability, and the command of tools, mechanical or fiscal. So far as the relations of parent and child, teacher and pupil, employer and employee, governor and governed, remain at this level, they form no true social group, no matter how closely their respective activities touch one another. The giving and taking of orders modifies action and results, but does not of itself effect a sharing of purposes, a communication of interests" (Dewey, 1916, p.6).

The difference between controlling behavior and communication is also valid for the animal kingdom, particularly as far as higher animals such as mammals and birds are concerned; their individuals can make their choice on the basis of the information they receive from others. *Controlling action* is exemplified by the behavior of a male blackcock that mounts a female at a lek site. In contrast, male sparrows *communicate* with females to advertize themselves as potential mates. It is a female that chooses the partner (Fridman, 2007).

An intermediate subtype of interactions is controlling action with a feed-back loop. The element under control can to some extent influence the control-exercising element of the system by sending messages that evaluate the results of the controlling action. Controlling behavior can be modified or even terminated depending on this message. Controlling with a feed-back loop can be referred to as *regulation.*

Both controlling and communication occur in hierarchies, networks, and (quasi-)markets, but their contributions are different with different system types. Since the pattern of interactions among the elements of a hierarchy is centralized, controlling prevails over communication. An example case is an army where, as German *Wehrmacht* soldiers used to say, *Befehl ist Befehl* ("an order is an order"). Nonetheless, communication is possible even in a rigid hierarchy:

- Individuals with roughly equal ranks can communicate, although this communication is limited by superiors because its intensification may result in a formation of an uncontrollable network structure inside the hierarchy
- Low-ranking individuals communicate messages to higher-ranking ones, informing them about the situation but not making any decisions themselves.

Communication plays a more important role in a hedonic than in an agonistic hierarchy (such as the hierarchy of an army that is characterized by coercion). Since leadership in a hedonic hierarchy depends on the preferential attention paid to the leader by the group, the leader tends to use persuasion rather than coercion, i.e., communication rather than controlling; if controlling power is exercised, it represents *regulation*, i.e., implies a feed-back loop. Those under control can significantly influence the controller.

If an army is a typical hierarchy, many *clubs* can be considered typical network structures characterized by the prevalence of communication over controlling. A decision is made during a club meeting on the basis of consensus. A consensus is often difficult to reach, and the club moderator (an analog of the partial psychological leader of a hirama) has to make good use of her organizational skills. During a club meeting, temporary partial leaders emerge that try to assume partial control over the decisions made; nevertheless, it is communication that plays the dominant role. Club members verbalize their suggestions/recommendations, which are subsequently discussed at the meeting.

Apes such as chimpanzees or bonobos often form coalitions consisting of individuals with approximately equal social ranks. Such coalitions are characterized by constant intense communication, including not only vocal signals but also drumming (Arcadi et al., 1998; Remedios et al., 2009) and clapping hands (Kalan & Rainey, 2009). Communication is also a prerequisite for the functioning of the decentralized network structures of fish and social insects.

Controlling action gains in importance if a network starts transforming into a hierarchy. For instance, if a club moderator, instead of trying to reach a consensus, were to cleverly manipulate other club members, trying to coax them into making the decision she is interested in. Subsequently, such a moderator could assume power over the formerly networked club. A similar situation was described in the book by Bard and Söderqvist (2002), which is concerned with a future "netocracy".

One of the temporary situational leaders in an ape coalition may become the permanently dominant individual. He hierarchizes the formerly egalitarian coalition.

Control may be exercised over a network group from outside because the network may be part of a larger hierarchy.

Finally, networks include a subtype in which controlling prevails over communicating, even though such a structure is decentralized and cooperative. This *matrix-dominated* subtype of networks will be considered below.

Similar to networks, markets and their analogs prefer communication to controlling. Controlling is often regarded as deforming a market by hierarchizing it. This can result from the oligo- or monopolization of the market or from the influence of extramarket (political or criminal) agents. To reiterate, an ideal market implies only communication; every commodity producer, merchant, etc. makes her own decision as far as making contracts is concerned. In purely market-type structures, such communication is not especially content-rich because it is often limited to negotiations concerning commercial transactions. Communication becomes more multifaceted and more informal if stable cooperative relationships between market agents are established and network structures are formed.

1.5.2. The (Super)Organism

Many coherent biological systems characterized by the prevalence of controlling behavior over communication are regarded in the literature as *organisms* or, if their components are organisms per se, as *superorganisms*. An organism can be defined as "a complex structure of interdependent and subordinate elements whose relations and properties are largely determined by their function in the whole" (Merriam-Webster Dictionary, 2014, http://www.merriam-webster.com/dictionary) and a superorganism, accordingly, as "a collection of single creatures that together possess the functional organization implicit in the formal definition of organism" (Wilson & Sober, 1989, p.339). The term "superorganism' was applied in the literature to closely-knit animal groups, including societies of social insects such as bee hives, termite colonies, and ant families (Wheeler, 1928; Chauvin, 1963; Hölldobler & Wilson, 1990, 2009, 2010; Kipyatkov, 1991).

> "Consider one of the most organism-like of all insect societies, the great colonies of the African driver ants. Viewed from afar, the huge raiding column of a driver ant colony seems like a single living entity. It spreads like the pseudopodium of a giant amoeba across 70 meters or so of ground. A closer look reveals it to comprise a mass of several million workers running in concert from the subterranean nest, an irregular network of tunnels and chambers dug into the soil. As the column emerges, it first resembles an

expanding sheet and then metamorphoses into a treelike formation, with the trunk growing from the nest, the crown an advancing front the width of a small house, and numerous branches connecting the two" (Hölldobler & Wilson, 2009, p.xx).

However, other scientists object that congregations of multicellular organisms that represent autonomous individuals differing in terms of their behavior and "personality" cannot be compared to real organisms that are made up of cells and tissues, particularly if interactions among the individuals involve competition, aggression, and conflict (Zakharov, 1991; Ratnieks & Reeve, 1992).

Plausibly, the "superorganism" should be considered in more loose terms than the "organism" per se, allowing for a greater degree of individual freedom among its structural units and some degree of competition between them. Nonetheless, competition is ultimately overridden by cooperation among the units, which enables the coherent behavior of the whole system. Admittedly, *communication* associated with the ability of these units to make their individual choices can play a sufficiently important role in a superorganism, in addition to the direct *control* of the whole system over them.

However, different authors prefer different interpretations of both terms (organism and superorganism).

A relevant issue concerning human society has been raised since ancient times: *Is human society a congregation of many individuals or a single superorganism with human individuals behaving like specialized parts of this supercreature?* Traditional caste societies were compared, in different historical periods, to integrated organism-like systems. Separate groups, classes, and castes are the cells, tissues, and organs of this system. Democratic regimes, on the contrary, are comparable to loose cell associations like colonial algae such as *Volvox;* some of these regimes emphasize individual independence ("My house is my castle"), which amounts to the solitary lifestyle which is characteristic of unicellular creatures. Totalitarian systems of the 20th century adhered to the principle that a state is a single organism which conforms with the totalitarian state *(Stato totalitare)* concept suggested and implemented in Italy by Benito Mussolini.

Irrespective of the exact meaning, the components of a (super)organism are typically considered to be so interdependent that they cannot exist in isolation. For instance, in many ant species, individuals cannot survive for more than several hours without the whole ant society. In addition, a biological system (or, somewhat metaphorically, a system in human society) considered a (super)organism is expected to exhibit the following features (Wheeler, 1911, 1928; Dewey, 1916; Beklemishev, 1950):

- *The components* (cells, zooids, nodes, etc.) *exhibit functional differentiation* into subgroups that actually represent *organs of the whole system.* Worker and reproductive individuals in an ant society represent two distinct castes compared to two different organ systems in the whole "superorganism" (Hölldobler & Wilson, 1990, 2009, 2010) that represent the "soma" (the life-sustaining organs) and the "germ plasm" (the reproductive organs), respectively (Wheeler, 1911).
- *A general structural pattern and a developmental program are carried out by the whole system.* A highly integrated fish school, which possesses certain quasi-organismic properties, creates a general geometrical image persuading a predator to consider the whole school a single big fish.

- *The capacity for reproduction and regeneration* that is exemplified by the division of an ant colony into two smaller "daughter" colonies as soon as the original colony becomes sufficiently large; it should be noted that it is still an open question whether colonies of social insects can be considered superorganisms; Anatoly A. Zakharov (1991) questions this idea.
- *Size reduction and the structural simplification* of the components, the feature emphasized by Beklemishev with regard to the colonies of cnidarians. This is also typified by the formation of aberrant cell wall-lacking L forms in many bacterial colonies or biofilms. Some organs (such as the ovaries and the spermatheca) degenerate in individuals which belong to the working caste in the colonies of some social insects (Hölldobler & Wilson, 2009).
- *Repulsion* of components of other systems, in an analogy to the behavior of immune cells towards foreign cells (Wheeler, 1911, 1928; Dlussky, 1984).

In evolutionary terms, a superorganism, like an ordinary biological individual, is considered by many proponents of this term as a selection unit. Since natural selection selects for advantageous genes but is based on competition among organisms carrying them, it is assumed by most adherents of the superorganism concept that fitter superorganisms, e.g., ant colonies, outcompete less fit ones, and, as a result, the genes of their components (individuals) are selected for. "The colony effectively becomes the major target of selection, that is, it is a coherent 'extended phenotype' of the genes within colony members" (Hölldobler & Wilson, 2009, p.44). Hence, in addition to the competition at the individual level, evolution is associated with competition between whole superorganisms (see, e.g., Wilson & Sober, 1989). The repeatedly raised objection is that noncooperating individuals (free riders) that contribute nothing to the welfare of the whole superorganism (e.g., an ant colony) may gain a selective advantage and ultimately replace cooperators, which would ruin the whole superorganism.

However, much attention has been recently paid in the literature, particularly by Corning (1983, 2003a, b) to factors that prevent the spread of free riders in superorganisms that are otherwise composed of "honest" cooperators working for the good of the whole system. Suffice to mention that free riders will not be selected for if their cheating endangers the survival of all group members including the free riders themselves. In primitive human society, a mammoth would not be caught and the whole hunter band would starve if some hunters shirked, allowing the mammoth to escape. Free riders are severely punished, not only by humans and other primates but also in a variety of animal species (Cummins, 2001). Punishing free riders/cheaters is denoted as *negative reciprocity*, i.e., doing harm (or at least withholding help) to those who have done harm to you.

Generally, organisms (including superorganisms) can be subdivided into (1) *unitary* that are typically characterized by centralized organization and a set of interdependent *organs*, i.e., of functionally specialized and predominantly non-repetitive areas of the body and (2) *modular* that include many repetitive units that may or may not be capable of existing independently if separated (the concept of *modularity* was briefly discussed in 1.2.7).

From all the above, it is evident that hierarchies, if they exhibit a tendency towards the development of organism-like features, can become unitary organisms. They are typified by the human organism as a hierarchy with the central nervous system at its pinnacle. In human society, analogous features are exhibited by an authoritarian political system which is characterized by "machine-type" interactions (Dewey, 1916) between the central government and the citizens.

(Quasi-)markets based upon predominantly competitive relations among autonomous agents are unlikely to form any kind of organism-like systems unless they convert into a different structural type.

Networks tend to evolve into modular decentralized organisms. The somewhat paradoxical feature of such organisms is that their functioning involves controlling behavior exhibited by the whole structure in the absence of a central leader. The controlling agent exists everywhere and nowhere in the system. This is characteristic of colonial cnidarians (2.2 below) whose nodes (zooids) behave in conformity with the dominant behavioral trend that is demonstrated by a majority of other nodes. In this system, the interaction between zooids is facilitated by the stalk (coenosarc) to which they are all connected. In the following subsection, I introduce the term "matrix" in the metaphorical sense to denote the material substratum of the whole network or the set of social norms limiting the nodes' freedom of behavior.

Therefore, the terms "organism" and "superorganism" (discussing the difference between them in detail is beyond the scope of this work) both correspond to an advanced stage of the integration of either hierarchies that give rise to unitary centralized organisms or decentralized networks that, when sufficiently integrated, can be referred to as decentralized modular organisms.

Opponents of the (super)organism concept use other terms to denote close-knit, highly integrated biological systems. For example, microbial colonies or biofilms, regarded as analogs of multicellular organisms by such microbiologists as James Shapiro (1988, 1995) are called "cities of microbes" by other scientists (Watnick & Kolter, 2000; Nikolaev & Plakunov, 2007). Taking account of conflicts between genetically different members of the societies of social insects, Ratnieks & Reeve (1992) prefer the term "community of interests" to the term "superorganism" used, with regard to insects, by Wheeler (1928) and Chauvin (1963) as well as, more recently, by Hölldobler & Wilson (1990, 2009, 2010) and Kipyatkov (1991).

The different models (paradigms) of the organization of biological networks that will be discussed in Chapter two are characterized by different ratios between communication and controlling behavior. Their organismic features also manifest themselves to a different extent.

1.6. Behavior Coordination: The Matrix

The main question to be raised in this section is: *How do network structures coordinate the activities/behaviors of their nodes?* The reason why this question is particularly important with regard to networks is that they lack a permanently fixed single activity center (leader, pacemaker).

Behavioral coordination among individuals/nodes, i.e., the "concerted behavior of individuals, e.g., animals in a biosocial system or human individuals in society, often involving differentiated social roles" (Oleskin, 2012, p.227), presents serious organizational difficulties for *decentralized network structures*. Such problems are less likely to arise in *hierarchies* where coordination is secured by the central leader. The leader can exercise direct *control* including coercion, which is typical of many agonistic hierarchies, or, in hedonic hierarchies, engage in centralized *communication,* persuading most individuals into copying the leader's behavior. Importantly, both mechanisms work in human society and in the biosocial systems of various nonhuman species. Copying the behaviors of leaders is possible not only in hierarchies per se but also in three-dimensional networks with temporary/situation-dependent partial leaders.

Classical scholars dealing with complex systems such as Herbert Spencer and Alexander Bogdanov emphasized that even a system with only two active centers may be unstable because the controlling action of one center on the periphery collides with the control exercised by the other (Bogdanov, 1980 [1921]). The example used by Bogdanov in the book cited is an imaginary planetary system with two suns; the planets' orbits are likely to be unstable (Figure 1.7). Nevertheless, human three-dimensional networks with partial/temporary leaders and their analogs in biological systems are often quite stable actually. Moreover, they are less vulnerable than hierarchical structures that are typically destroyed by eliminating the central unit. Networks cope with the stability problem in two ways:

- In many networks, it is the functional specialization of partial leaders that enables the structure's stability. In Bogdanov's planetary system, the orbits become stable if only one of the suns is dark and serves as the gravity center, whereas the other is only responsible for illuminating the planet(s) and rotates around the dark sun together with them. Partial leaders in a *hirama* (see 1.2.2) are specialized in functional terms: they are responsible for dealing with different subproblems within the general creative project of the hirama.
- If partial leaders are functionally indistinguishable, they often differ in terms of their tempo and rhythm. Therefore, they can perform their controlling functions at different points in time. As a result, the network may have only a single active leader at any given moment. Although the brain is a multicentral network structure, it is often only one center or subgroup of centers that dominates it at a given point in time. This center delegates its functions to another center (subgroup) once the brain switches from one priority task to another[21]. In a network, the active centers are not fixed; they are situation-dependent and constantly migrate from point to point, in sharp contrast to a fixed hierarchical structure that is exemplified by a bureaucracy in human society.

[21] However, the brain can solve several problems in parallel. In this situation, there are several "partial leaders" with different functions within the brain.

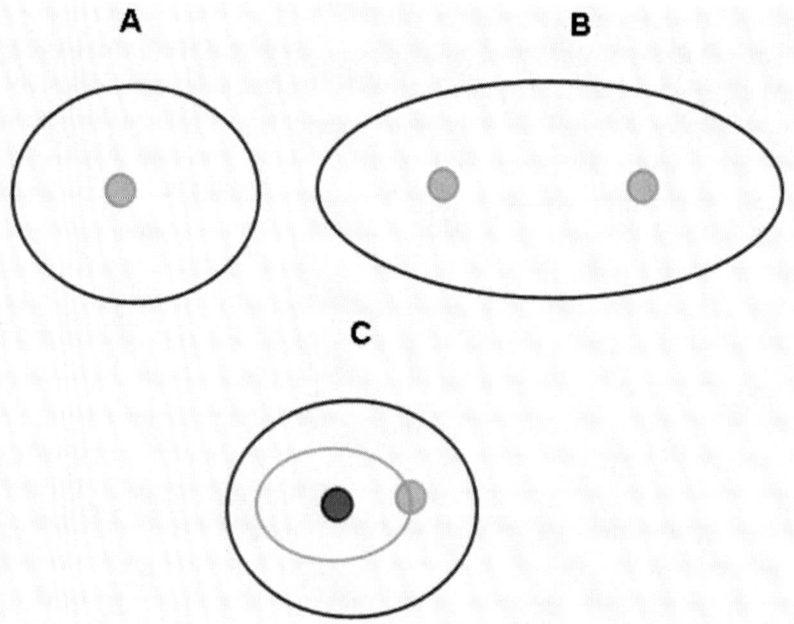

Figure 1.7. (A) A centralized system (a sun and a planet); (B) a system with two suns that compete for the role of the gravity center; the planet has an elongated and, in all likelihood, an unstable orbit; (C) one of the two suns shines without occupying the central position, whereas the other, dark sun, serves as the gravity center; this type of system with functionally differentiated centers can be stable. Based on: (Bogdanov, 1980 [1921]).

1.6.1. Heterochrony. Anticipatory Reflection

The preceding passage provides an example of *heterochrony*: different partial leaders are characterized by different rhythms. Generally, network structures tend to be *heterochronic,* i.e., each of their nodes or their subgroups is free to behave at its own rhythm, in the absence of a central pacemaker (leader, dominant individual, etc.).

In a heterochronic system such as a network structure, we can calculate the averaged rhythm of the nodes' development. However, some nodes lag behind the rhythm, the rhythm of some other nodes is close to the averaged rhythm, and there are nodes that are ahead of the averaged rhythm. The heterochrony of a network is linked to its remarkable property, which was called *anticipatory reflection* by Smirnov et al. (1982).

The term "anticipatory reflection" was originally used by Smirnov et al. (1982) with respect to the development of microbial populations (similar data were obtained by Gusev & Bobrova (1989)). It was revealed that the first developmental stage (the lag phase) in the life cycle of a bacterial population includes several "anticipatory" cell subgroups whose features are similar to those of the subsequent developmental stages of the same population. One cell subgroup in the lag phase population is analogous to the cells characteristic of the culture's *senescence* (i.e., stationary) stage. There are subgroups whose cells actively grow as if the population had already reached the *youth* stage (the logarithmic growth phase). It seems that they "'rehearse' the behavioral program carried out… in the main logarithmic growth phase" (Smirnov et al., 1982).

Hence, *anticipatory reflection* means that the initial stage of the development of a system includes second-order stages (substages) that resemble the later stages of its development; the system's development is a temporarily fractal (self-similar) process.

Imagine that a network and a hierarchy compete in carrying out a certain process. For example, a networked and a hierarchical research team are competitively working on similar scientific research projects. The hierarchy whose activities are centrally coordinated will probably be ahead of the more creatively chaotic network, as far as the averaged system rhythm is concerned (this rhythm is the dominant rhythm if the hierarchy is synchronized by the central pacemaker). The network will lag behind the hierarchy, but only in terms of the averaged rhythm, which is fictitious in a network. Its nodes have individual rhythms, some of which are ahead of the averaged rhythm. These "anticipatory" nodes may include those that *are also ahead of the competing hierarchy's rhythm*. For instance, we can predict that some of the members of the networked research team will come up with innovative ideas that predate those put forward by hierarchically organized research institutes. Indeed, this was the case with some of the achievements made by a network within the Russian microbiological community in the 20th century; this network focused on the population & communication paradigm in microbiology and generated innovative ideas (see 3.5).

Despite the heterochrony of networks, they make use of specific mechanisms that enable them to coordinate and at least partly synchronize the behavior of their nodes; such coordination is possible even in some completely flat network structures. They are exemplified by leaderless fish shoals or microbial biofilms. This non-hierarchical behavioral coordination has fascinated a large number of researchers. "How is it that thousands of neurons or fireflies or crickets can suddenly fall into step with one another, all firing or flashing or chirping at the same time, without any leader or signal from the environment?" (Wiener, 1948, quoted according to: Arenas et al., 2008, p.94).

The following is an overview of the mechanisms of coordination that have been discussed in the literature.

1.6.2. Local Interactions between Network Nodes (Short-Range Order)

Local interactions between network nodes (short-range order) enable behavioral coordination that depends on direct contact between neighboring nodes and involves communication between the nodes or controlling action often associated with a feed-back loop. In a neural network, a nervous cell transmits an impulse via the synaptic cleft to another nervous cell that becomes excited and can, in its turn, influence the state of the impulse-generating cell (this feed-back loop is emphasized in Hopfield's network models).

Local behavior coordination is characteristic of ants constructing or repairing an anthill, of lions collectively hunting large prey, and of some microorganisms whose cells are connected by nanotubes involved in communicating messages from cell to cell (Vysotsky et al., 1991; Dubey & Ben-Yehuda, 2011) or produce non-diffusing chemical signals that remain attached to the cell generating them. In humans and apes, local communication enables a network structure's members to copy their neighbors' behavior ("aping" or *Keeping up with the Joneses*). Interaction between neighbors results—both in biological systems and human society—in a relay transmission of regulatory messages. Gradually, the information spreads throughout the whole structure, so that originally *local* interactions become *global.*

An example case is a leaderless fish shoal or school. New important information spreads within it, resulting, if necessary, in a rapid reconfiguration of its spatial structure: "Stimulus situations that attract the fish are responded to by a whole shoal, even when only one individual has received the stimuli. The resolute swimming of this individual in one direction draws the other fish with it, ... and it is a question of quantity whether or not the whole shoal is pulled along" (Lorenz, 1966, p.140). The coordination principles of fish shoals/schools are considered in more detail in 2.3.6.

Information transmission is facilitated if the network displays small world behavior, i.e., seemingly distant nodes have only few links between them; this is the case with most random (ER) and scale-free (BA) networks[22] (see 1.1.3). Small world behavior is apparently characteristic of a variety of biological networks.

For instance, local interactions between neighboring polyps (zooids) in their decentralized colony enables the efficient global coordination of their rhythmic contactions. More specifically, the behavior of each individual node, e.g., of a liquid-pumping polyp, produces only an insignificant effect on the whole network. However, its influence is potentiated if the node's behavior is in unison with the dominant tendency of the whole network. For example, a sufficiently large group of polyps that contract synchronously to pump liquid overpower those with a different contraction rhythm; the latter slow down their contraction and change their own rhythm to match the network's dominant rhythm (Marfenin, 2002, 2009).

Yoshiki Kuramoto (1975, 1984) developed a mathematical model concerning a system of coupled oscillators whose natural rhythms are randomly distributed over a certain range. The model can be applied to the behavior of network nodes (Arenas et al., 2008). It predicts the *entrainment (synchronization)* of the nodes of one network, provided that their original rhythms are not too different. If the difference is below a certain threshold, the network is expected to contain clusters of interacting nodes with synchronized behavior. A further decrease in the initial difference between individual node rhythms makes all nodes of the network behave in step with one another.

Interestingly, Kuramoto (1984) admitted that even weakly coupled oscillators can become synchronized if certain conditions are satisfied (primarily, the difference between individual rhythms should be sufficiently low). Obviously, communication causing a probabilistic response is a weaker kind of interaction than controlling action resulting in preprogrammed behavior. However, according to Kuramoto's work, initially asynchronous nodes of a network may become at least partly synchronized as a result of communication between them. It is of relevance that the coordination of the behaviors of social insects (ants, wasps, bees, and termites) involves not only controlling action but also communication inside their societies (see 2.4.2 for details).

In human society, Granovetter (1973, 1985) emphasized the "strength of weak ties" between human individuals, in contrast to strong ties often involving direct controlling, as exemplified by the relationship between a boss and his (her) subordinates. Social ties, even

---

[22] To reiterate, random and scale-free networks mainly correspond to flat and three-dimensional networks respectively

relatively weak ones, obviously gain in importance if they become more regular and numerous, and involve a larger number of partners.

In social insects, an increase in the size of their colonies may result in a transition from disordered to ordered collective behavior. "For example, small colonies of the pharaoh ant (*Monomorium pharaonis*) forage in a disorganized manner (no trail) with a transition to organized pheromone-based foraging in larger colonies" (Detrain & Deneubourg, 2006, p.174).

1.6.3. The Effects of Distant Regulatory Agents Creating Long-Range Order

Global behavioral coordination within a whole network structure involves diffuse stimulators, i.e., chemical substances or physical fields. Such distant regulatory agents either work as *preprogrammed response-causing stimuli* (an example concerning microorganisms is given in this subsection) or as *information-carrying signals*. In other words, they can directly *control* the nodes' behavior or just *communicate* messages to them, so that their response varies depending on the nodes' state.

The macro- and microstructure of the colony of the bacterium *Escherichia coli* is generated under the influence of the attractant substance (aspartic acid) formed by all its cells. Complex patterns such as concentric circles or hexagonal lattices result from the superposition of two spatial aspartic acid gradients: (1) produced by the colony center and (2) generated by the cells on the periphery (Budrene & Berg, 1991, 2002; Mittal et al., 2003). The interaction between the two gradients results in an inhomogeneous pattern of aspartic acid distribution within the colony. The cells respond to the inhomogeneous pattern by forming conglomerations at concentration maxima.

The coordinating agent can be produced by the nodes of the network or, alternatively, generated outside the structure. This mechanism of coordination can operate in all structure types but its contribution is enhanced in decentralized cooperative systems.

An additional important factor facilitating network-wide behavioral coordination is a network's *matrix*. This term stands for the totality of consolidating factors and mechanisms, both material and immaterial (see below).

1.6.4. The Matrix

The word "matrix" has several different meanings. With respect to networks *sensu lato,* a widely used meaning of the term is a rectangular array of numbers. For instance, in an *adjacency matrix*, each node in a network corresponds to a particular row and a column in the matrix, depending on its position in the network. An adjacency matrix can be used to show the arrangement of nodes and links in a network, instead of a picture visualizing the nodes and links (see, e.g., Figures 1.1, 1.3, and 1.5).

In terms of the *sensu stricto* interpretation of the word "network" predominantly used in this work, two other meanings of the term are of primary relevance:

1. A material structure cementing the system: this meaning is widely used in microbiology, cytology, and histology. In a tissue or a microbial colony/biofilm, the matrix consists of extracellular biopolymers (polysaccharides, collagen and other proteins, or extracellular DNA strands) that bind cells together. The matrix performs *structural* and *protective* functions and promotes the *spread of signals*, i.e., intercellular communication. The matrix can be compared to a system of buildings in a town — a "city of microbes", a metaphorical term denoting biofilms (Watnick & Kolter, 2000). The term "matrix" can be extrapolated to non-microbial biological systems. Analogous functions are performed by the structure (coenosarc) that integrates individual cnidarians, bryozoans, or ascidians into one coherent system, a colony (cormus, see 2.2 below). In the world of insects, the functional analogs of a "city of microbes" are anthills, beehives, or termite mounds. Naked mole-rats that live in large colonies, like social insects, have systems of tunnels in their subterranean "cities". Various buildings, churches, and other cultural artifacts perform structural, protective, and communicative functions in human society, in an analogy to the microbial matrix. Irrespective of the kind of system involved, the matrix in the broad sense denotes *the interorganismic* (not belonging to any individual organism per se) *materials that are responsible for the system's structural integrity and promote long-range communication and regulation* of the processes carried out by the system.
2. A more popular meaning pertaining to the sphere of the humanities and social sciences. It deals with "a situation from which a person or society can grow and develop" (Longman Dictionary, 2003, p.1015-1016). In this situation, the regulatory factors forming part of the matrix are predominantly *immaterial.* They include social norms and restrictions that, in an analogy to extracellular biopolymers, guide and restrict the behavior of individuals and their interactions. For example, social insects, such as ants or bees, acknowledge individual and collective territories; they distinguish in-group from out-group members, and interaction between these two kinds of individuals is heavily influenced by a complex system of social norms and rituals (Zakharov, 1987,1991, 2005, 2009; Hölldobler & Wilson, 1990, 2009, 2010). Apes such as chimpanzees and bonobos live in a "democratic regime", i.e., a mitigated hierarchy having important network-like features. For example, their "public opinion" imposes restrictions upon the dominant individual's power (de Waal, 1996). He is expected to obey social norms, e.g., respect the rights of individuals in a social group and avoid using violence in conflicts with females and young males. Such norms and prohibitions constitute the immaterial matrix of the social group. Human social structures have statutes, programs, shared goals, values, norms, and views. They unite individuals and modify their behavior, promoting their identification with the structure involved. In particular, matrix-like factors that consolidate social structures and political movements lacking hierarchy include (1) *superideas*, such as the establishment of a Palestinian state or resistance to globalism and (2) rituals/ceremonies that often objectify these superideas. Group-specific norms and rituals also exist in primitive hunter-gatherer bands consolidating these largely non-hierarchical, egalitarian structures (see 3.1.1 for details); they occur in nonhuman primates in the form of behavior patterns that "vary locally in a way

that appears inconsistent with either genetic or straightforward environmental explanation" (Whiten, 2010, and see 2.6).

In sum, our interpretation of the term "matrix" combines two conventional meanings of this word that deal with its *material* (superorganismic-level structures) and *immaterial* (social norms, behavioral rules, and rituals) components, respectively. These two components are closely interrelated and interdependent; in many systems, the material component can be considered an objectification of the immaterial one. For instance, the statues, temples, and other cultural artifacts created in honor of a political leader actually symbolize the social conventions concerning his/her special privileged status.

The description of the matrix evokes the closely related concept of *degression* contained in Alexander Bogdanov's works, cited above. To reiterate, degression refers to the formation of a sufficiently strong protective structure, a "skeleton" in a system. This "skeleton' can be external (like a carapace) or internal (typified by the human skeleton), material and immaterial. A system of symbols representing an ideology can serve as a skeleton securing an organization's integrity and helping it protect its boundaries despite external influences.

1.6.4.1. The Matrix in a Hierarchy

Generally, a matrix is present in a structure of any type, including hierarchies. However, it is developed to a different extent in different subtypes of hierarchies. As far as the immaterial component is concerned, it plays a minor role in agonistic hierarchies topped by a dictatorial dominant individual ignoring most social norms, including the sense of justice that seems to be common to all primates (de Waal, 1996, 2001). Only one social norm is obeyed: the dominant individual is "legally" granted exclusive privileges with respect to food, shelter, mates, and other resources.

In contrast, hedonic hierarchies in which an individual's rank depends on social skills and other merits—an idealized image of Weberian bureaucracy[23]—are often characterized by a well-developed matrix. In particular, the matrix of a bureaucracy includes a system of regulations, statutes, and instructions that regulate the interactions among the people in it. As Weber (1956 [1907]) emphasized, these official regulations distinguish a bureaucracy from a more agonistic medieval hierarchy that was based on the dominance of the lord (king, chief), so that explicit instructions were not mandatory. In Bogdanov's words, egression, i.e., the hierarchization of a system, is additionally reinforced by degression, the formation of protective structures such as (in human society) palaces and fortresses for defending the political elite in wartime as well as the ideology that sanctifies the existence of the hierarchy and the special rights of the elite at its top.

1.6.4.2. The Matrix of a Network

The unifying matrix is still more important for decentralized networks that lack a "boss". Importantly, the matrix of a network structure contains *distributed implicit information* that no individual member possesses. It can only be used when all members of, for example, a

[23] Such an idealized Weberian system where ranks depend on socially important individual merits can be denoted as *bureaucratic meritocracy*. However, networks in human society can generate a different kind of meritocracy (*network meritocracy*) in which it is the network structure-based civil society that empowers socially recognized experts to contact the central government on its behalf with regard to political, economic, social, cultural, or environmental problems (to be discussed in subsection 3.8.6).

commercial network, collectively solve problems. In various network structures both in human society and biological systems, "individual units do not have a complete picture of their position in the overall structure and the structure they create has a form that extends well beyond that of the individual units" (Sumpter, 2006, p.5).

This implicit information provides a network structure with a program of its development that may contain, in a cryptic form, the final target state the structure is to achieve. This is what Alexander G. Gurwitsch (1944) called the "dynamically pre-existing form", e.g., an imaginary geometrical figure whose form a real structure such as a group of cells tends to approximate. For instance, the cells of a developing floral receptacle tend to form a parabolic structure that exists as implicit information from the beginning of the receptacle's development. Similar ideas were suggested in *On Growth and Form* by Wentworth Thompson d'Arcy (1945). The presence of a matrix preprogramming a structure's development is also related to the biological and social phenomenon referred to as *equifinality*. The elements of a developing structure, such as the migrating cells of a sea urchin embryo, can choose between several different pathways that all lead to the same target state for the fully developed structure, i.e., a mature sea urchin. In human networks members may choose different individual strategies for attaining the same goal, such as establishing a global Muslim fundamentalist state (one of the ultimate goals of the *al-Qaeda* network).

In a hierarchical structure, it is only the boss who is responsible for guiding the structure's development in accordance with its intended purpose. In a decentralized network, each node makes an effort to bring its own behavior into line with the guidelines provided by the network's matrix. Apart from behavior per se, the matrix controls each node's individual features, standardizing them to some extent. Recent data suggest (Croft et al., 2005) that a fish shoal predominantly consists of individuals with similar-sized bodies, food preferences, and behavioral traits (e.g., courage).

Even though dominance-submission relationships between individuals in a network are overriden by egalitarian cooperative interactions and communication prevails over controlling, the matrix can behave as an impersonal *quasi-leader*. Its instructions (e.g., "put on clothes worn by the members of our network") are carried out like the orders of an army commander. The situation is paradoxical because the non-hierarchical structure contains an *invisible, "imaginary"*, quasi-leader that may – in some networks – exersize considerable control over their nodes. In optics, the real focus where light rays converge in a system of mirrors and lenses is contrasted with its imaginary focus where the continuations of divergent rays intesect. Published works on international humanitarian network structures demonstrate that important decisions are often dictated by the network as a powerful agent endowed with its own will rather than by the opinions of individual members of such a network, however influential they might be. The logic of the development of Amnesty International (AI) has encouraged this networked organization to predominantly focus on prisoners of conscience (Kahler, 2009a). The choice was made by the matrix as the quasi-leader rather than by any of AI's members.

As mentioned above, cnidarian colonies contain the coenosarc, a scaffold to which individual polyps (zooids) attach; it plays a matrix-like role in the cnidarian system. In many species, the coenosarc is the dominant part of the system while the polyps are reduced to tentacles and mouths that supply food to the single colony-wide gastral cavity running through the coenosarc that exploits them.

In some human social structures, the matrix reprograms the behavior of each of their members until they actually lose their individuality. Such a matrix works like a "paralyzing poison" inhibiting individual initiative, e.g., in some primitive human societies, as emphasized by ethnographers such as Boris Porshnev and Claude Levy-Strauss.

A somewhat similar situation is characteristic of sectarian movements such as the *Amish* (USA): despite a lack of rigid hierarchies, there are detailed instructions based upon the social conventions under which the Amish have been indoctrinated (such as preferring a simple lifestyle, wearing plain dress, and refusing to adopt many conveniences of modern technology).

Apart from stringently controlling individual behavior, the matrix can also influence network nodes in a more subtle way. Spontaneously or intentionally established network structures in human society have common goals, adopted values, and shared behavioral strategies. This can account for the similar or even identical results of independent creative work done by network members. As already mentioned in this work, a network spontaneously established in the Russian microbiological community in the late 20th century included a large number of formally independent researchers. Surprisingly, they came up with very similar ideas concerning the subject they all were interested in (microbial communication and biofilm formation, Oleskin & Kirovskaya, 2006).

1.6.4.3. The Matrix of (Quasi-)Market Structures

Market structures and their analogs contain a matrix consisting of sets of norms and conventions that standardize price levels, the rights of agents, the conclusion of contracts, the rules of fair competition, etc. This "market matrix" is related to what Adam Smith called "the invisible hand". However, the matrix is typically less rich in content than that of many networks exemplified by hiramas which are characterized by intense cooperation and information exchange among their members. The matrix is so "thick" in networks partly because of the fact that their members share many basic values, ideas, and principles, whereas markets emphasize independence and often tolerate agents with completely different value systems as long as they do not violate the basic norms of commercial interaction.

Biological analogs of human market structures, i.e., diverse conglomerations of competing individuals, also demonstrate the relative paucity of their matrix. In locust conglomerations and iguana groups, only relatively few social norms exist: for example, iguanas have individual territories. Territorial claims and counterclaims are made according to specific social rituals including defensive postures.

Behavioral coordination among individuals/nodes presents serious organizational difficulties for decentralized network structures. Nevertheless, the following coordination-promoting factors function even in completely flat (leaderless) networks structures: (a) local interactions between network nodes (short-range order) and (b) distant regulatory agents (long-range order). In addition, biosocial systems and human society contain, apart from individuals, a specific element that can be denoted as their matrix. It includes both a material (collective constructions) and an immaterial (social norms, behavioral restrictions, etc.) component. The matrix is of particular importance in terms of the coordinate behavior of decentralized cooperative network structures.

1.7. Networks, Hierarchies, and (Quasi-)Markets: Transformation Scenarios

Although we predominantly considered pure organizational types (network, hierarchical, and quasi-market structures) in the preceding text, real-life systems often combine several types. Different structures are interwoven and interact in a complex fashion within them. A large number of systems are manifestly *polystructural,* i.e., consist of several different structures. A *matrix* organization[24] in business includes (1) a linear, i.e., hierarchical, structure headed by the boss, which is responsible for the general sales & distribution strategy, research & development activities, etc. and (2) a horizontal network structure with project leaders that interact with the pool of workers, reminiscent of the *hirama* pattern (see above). Additionally, the whole matrix organization behaves as an agent within the market structures of capitalist society.

A mitigated dominance-submission hierarchy coexists with a network of horizontal relationships in a chimpanzee or bonobo group, so that the whole system is polystructural (societies of many insect species, notably of ants, are also manifestly polystructural, Zakharov, 1991; Hölldobler & Wilson, 1990, 2009, 2010).

There are also systems that are apparently dominated by one structural type — a hierarchy, a network, or a (quasi-)market. However, if we change our perspective, we can often reveal elements of other structures inside them. Although the social structures of a large number of lower monkeys are rigorously hierarchical, they also include relatively egalitarian networks of young individuals with partial temporary leaders. As noted above, the Internet is actually a polystructural system, despite its name.

Moreover, structures can interconvert. If time is considered the fourth spatial coordinate, then transformations of structures over time and their interactions at a given moment are variants of the same phenomenon. Different structures are located at different spatial or temporal points of the space-time continuum. Various systems in the biological and the social sphere display a wide variety of transformation/interaction patterns. In human society, these transformations or interactions partly result from efforts that have been intentionally made by the people involved. "Agents choose to cooperate or not in a market, hierarchy, or network as a function of their individual attributes and their beliefs about the attributes of the other agents with whom they may interact" (Jung & Lake, 2011).

Louis Meulemann (2008) suggests that specialists (meta-governors) should "design and manage situationally optimal mixtures of governance styles", i.e., types of organizational structures. Meta-governors dealing, e.g., with public policy issues, are to "combine governance style elements", "switch between governance style elements", and "maintain the mixture". For example, "during a crisis, a hierarchical command and control style should be in place, because time is crucial and quick decisions are needed". After the crisis, market-type organizations are preferable, while in a crisis-free period, "parties cooperate in the form of a network and work on enhancing mutual trust and understanding" (Meulemann, 2008, p.vii).

[24] The meaning of the term "matrix" in this situation differs from that used throughout this work: it denotes a specific network-type organizational pattern (with one or several project-oriented partial leader(s) plus the department's head) rather than the backbone of any network structure.

Ethological data demonstrate that many biosocial systems, e.g., those of some primates, are flexible and adaptable. The same species can form different structures and mix them in varying proportions, depending on the situation. For example, a relatively rigid hierarchy formed by a group of primates during a hunting raid is replaced by a loose horizontal structure when the group decides to relax after a successful hunt.

1.7.1. Structural Transformations

The issue to raise is whether there are typical interconversion scenarios. One widespread scenario involves *network hierarchization.* To reiterate, Bogdanov (1980 [1921]) denoted this process as egression. In primitive human society, an increase in status differences within a primitive kin group can result in the formation of a permanent hub personified by a patriarch who manages the work and controlles the distribution process.

Hierarchization occurs in cell cultures. Cells communicate and cooperate, which results in forming a multicellular network. Subsequently, leader cells may emerge. They guide the further development and the behavior of the network[25].

While growing on a solid nutrient medium, skin epidermal cells form a monolayer with decentralized regulation of the behavior of each cell largely based on local cell-cell interactions. However, the structure becomes partly hierarchical over time because leader cells emerge in it. They move faster than their neighbors but stay in contact with them (Samoilov & Vasiliev, 2009).

Hierarchy formation in networks was documented and extensively studied with respect to human social structures. Such hierarchization is characteristic of a large number of revolutionary, sectarian, and other underground organizations. Initially, they represent amorphous networks lacking a hierarchy (brotherhoods). However, their clandestine activities necessitate the establishment of the elite that guides these activities and possesses secret information or esoteric knowledge that is not available to less reliable ordinary members of the same network. Rigid hierarchies with permanent leader positions were established in the originally non-hierarchical *Red Brigades* in Italy in the 1960-1970's (Alexeenkova, 2005). In a similar fashion, the rapid expansion of the terrorist network *al-Qaeda* was accompanied by its increasing centralization. However, its centralization degree decreased after the terrorist attacks on September 11, 2001 because the network faced the threat of the destruction of its central unit (Kahler, 2009a, b) during a retaliatory operation.

At this point, reference should be made to Michels' (1949) "iron law": even parties and other political movements that originally have no hierarchies and set themselves egalitarian (Early Christian, socialist, communist, etc.) goals establish increasingly centralized bureaucratic structures during the course of their development, forming political elites that are concerned about their own interests.

A continuation of the hierarchization trend is the transition from a hedonic to an agonistic hierarchy, from nutrurant leadership to repressive dominance. For example, a charismatic political leader is initially on the pinnacle of a *hedonic* hierarchy because the people pay

[25] However, some cell networks never convert into hierarchies; such network structures are exemplified by networks formed by predatory dictyobacteria for the purpose of trapping their prey (other bacteria).

special attention to him and he inspires them with his political ideas. At a later point, the leader becomes unwilling or unable to make efforts in order to inspire his people. He loses his charisma, and the hierarchy becomes *agonistic*. A formerly charismatic leader can retain his power only by *repressive dominance* involving the persecution of dissidents. A similar change from nurturant leadership to repressive dominance is also possible in the biosocial systems of various animal species, particularly in primates.

Taking account of the distinction between flat and three-dimensional networks drawn in 1.2.6, one of the pathways from flat networks to hierarchies can be represented as follows: *Flat Heterogeneous Network → Three-dimensional Network → Hedonic Hierarchy → Agonistic Hierarchy*. According to this scenario, a prerequisite for the hierarchization of a flat network is its heterogeneity that implies functional specialization of its nodes, i.e., potential partial leaders. The emergent goal-setting leader is initially perceived as one more specialized partial leader, even though (s)he has the potential to convert the network into a hierarchy. Completely homogeneous networks (see 1.2.8) have partial leaders whose functions broadly overlap or coincide like those of most body segments of an earthworm (except for the rostral and caudal segments). Their hierarchization can be caused by increasing competition among the functionally identical partial leaders that actually establish a transient *quasi-market structure* within the network. If the competition results in the victory of one of the nodes involved, the hierarchization pathway is as follows: *Flat Homogeneous Network → Quasi-market Structure → Hedonic Hierarchy → Agonistic Hierarchy*.

However, hierarchical structures are not necessarily long-lived. A widespread transformation scenario involves their conversion into networks. Such a scenario unfolded in the Soviet Union during the *perestroika* period and contributed to its demise. Obsolescent hierarchies were replaced by networks in other historical periods and other parts of the world as well. As for biological systems, we can return to cell cultures containing leader cells. Their "de-hierarchization" is caused by contact inhibition: cells stop growing, dividing, and moving once they come into contact with a sufficient number of other cells. Therefore, leader cells lose their ability to grow/move faster than other cells and become indistinguishable from them. The regulation of the cells' behavior becomes decentralized and involves diffuse chemical signals.

In forest ecosystems, dominant tree species perish after a drastic change in the environment that is caused by a forest fire or another catastrophe. In this situation, smaller plants such as shrubs are given a chance to spread within the ecosystem and to dominate it.

It is suggested in a number of works that ants and some other insects can form their biosocial systems in accordance with a *parasocial scenario*. It is assumed that the system is originally created by a number of individuals of the same generation that live in the same composite nest and cooperatively nurture the offspring (Hölldobler & Wilson, 1990). Initially, they establish a non-hierarchical network structure. At a later stage, one of the females dominates her contemporaries, becoming the *de facto* queen. At this stage, "group members are organized in dominance hierarchies, which, from time to time, are challenged and overturned by members of the society ready to take the top position" (Hölldobler & Wilson, 2009, p.xxi). The secondary hierarchy formation is still later displaced by a new trend towards decentralization, which is caused by the increasing functional differentiation of worker individuals; specialized leaders emerge in each of their functional groups such as foragers, brood-rearing workers, etc. (Zakharov, 2005, 2009), and the whole scenario is as follows: *Flat Network → Hierarchy → Three-dimensional Network*.

Secondary networks may subsequently be hierarchized. Networks may function as intermediates between the old and the new hierarchies. For instance, networked, initially informal, groups of supporters of young Peter the Great paved the way for the reformed hierarchical government in Russia at the turn of the 18th century (Sergeev & Sergeev, 2003).

Of theoretical and practical interest is the issue of whether or not *the interconversion of organizational structures is inevitable*? In particular, should network structures always develop into hierarchies, and vice versa?

1.7.2. Isolationist and Interactive Networks

Generally, network structures can choose between two different strategies vis-à-vis other structures, including hierarchies, (quasi-)market structures and other networks. These strategies, or network subtypes, are the *isolationist* (*degressive* in Bogdanov's terminology) strategy and the *interactive* (*egressive*) strategy.

The two network subtypes differ in terms of the role played by *communication* processes. Generally, network structures can use communication processes to achieve two different kinds of goals. According to Robert D. Putnam (2000), they can (1) establish links between formerly unconnected nodes (these are *bridging* networks) and (2) consolidate existing node clusters (*bonding* networks).

As pointed out in subsection 1.5.1, communication is essential for the functioning of any kind of networks. However, *isolationist* networks promote *internal* communication (and, therefore, represent bonding networks according to Putnam), i.e., they exchange messages between already interconnected network members/nodes and suppress communication with outsiders. In contrast, *interactive* network structures engage in both internal and external communication, performing both the bonding and the bridging function. The following is a brief discussion of the differences between the network subtypes; their ability to maintain their non-hierarchical cooperative organization despite external influences will be compared.

1.7.2.1. Isolationist Networks

Isolationist networks are characterized by strong "exoskeletons"—closed borders—and unlimited communication inside these borders. In human society, they are exemplified by various closed clandestine decentralized societies. In particular, some sectarian movements seek to distance themselves from the vanity of mundane life and to establish direct contact with a higher power.

Other isolationist networks represent communes with strictly controlled entrances and often a closed perimeter of walls and fences, i.e., *gated communities*. Such gated communities are exemplified by many of the Israeli *kibbutzim* (see 3.2.4 for more detail) that were originally based upon socialist principles (self-sacrificial work, free food, clothes, and accommodations) as well as Zionist ideas. In order to maintain their ideas and principles, kibbutzim tried to create a sufficiently strong barrier between their members and outsiders (despite using some of them as volunteers and, more recently, as hired workers). A potential new kibbutz member was to successfully pass entrance tests and complete the whole multistage ritual before he/she was actually enrolled.

It should be noted, however, that a large number of *kibbutzim* combine the features of isolationist (closed) and interactive (open) networks: despite forming gated communities, they actively engage in nation-wide political activities.

The brain, as a multilevel neural network (a network composed of neural networks), is characterized by efficient internal communication. However, the brain is covered by an external skeleton in the literal meaning, i.e., the skull.

During the course of its organizational development, an isolationist network has to cope with the problem of hierarchization in accordance with Michels' iron law. If the structure is interested in retaining its decentralized cooperative organizational pattern, it pays special attention to what Bogdanov called "degression". For instance, many non-hierarchical primitive societies in Papua-New Guinea, Australia, or Africa as well as sectarian societies attempt to shut out all kinds of factors that can potentially cause organizational changes. The number of members of a band, group or sect is kept constant, their functions, e.g., the rites they perform, remain unchanged for a long period of time, and their interactivity with the social environment is limited. The development of a degressive matrix including group-consolidating values and behavior norms and rituals is aimed at isolating the network structure from other social structures that are possibly based upon different organizational principles.

An analogous strategy is carried out by microorganisms whose populations remain viable for a long time without performing active metabolic processes. Such *persisters* cause chronic infections during which the few surviving microbial cells are difficult to detect because they exist in a dormant state, e.g., inside blood cells.

In the modern-day world, a similar "persister" strategy is used by network structures aiming to perpetuate their isolation and to exist in the "pores" of society where they can live in conformity with their utopian or religious ideas. An example case is the so-called DIY (do-it-yourself) strategy (Poldervaart, 2009), preferred by a number of youth movements including squatters which occupy vacant buildings.

Inevitably, such isolationist structures face the strong influence of the social environment in a shrinking world with its powerful communication channels. Besides, the "persister" strategy does not conform to the goals of many networks. They try to spread their ideology rather than to persist within the pre-existing boundaries.

1.7.2.2. Interactive Networks

These networks are interested in intense interaction with the social milieu. This is typical of a large number of global networked organizations, e.g., the International Eco-Ethics Union, which consists of interconnected networks called "local chapters", as well as of firms "without boundaries" that actively collaborate with, and include in the network, external agents such as suppliers, clients, and even competitors. This often results in setting up joint ventures or research and development commissions that exchange technological know-how (see item *Creating "organizations without boundaries"* in 3.7). Interactive networks aim to increase their size, to spread in the social environment, and to recruit new nodes (members, participants, or partners). Their boundaries are dynamic, fuzzy, and often unclear. Manuel Castells predominantly considered interactive networks when he emphasized their openness and the capacity to expand without limits by incorporating new nodes if they use similar communication codes (including values and tasks to be done). The outer "skeleton" is not as strong as with isolationist networks. However, interactive networks typically have a developed internal skeleton, i.e., a system of ideological principles and symbols that cement the whole structure[26].

Unlike isolationist networks, interactive networks assign an important role not only to matrix development (degression in Bogdanov's words) but also to hierarchization (egression) that, however, takes place outside the network per se. In their efforts to increase their size by enrolling new members, interactive networks attempt to establish their social dominance over other social structures and, in human society, to promote their ideological principles and goals. As a result, the entire interactive network occupies the top position in a hierarchy that is composed of whole organizations, i.e., becomes the collective leader of this interorganizational hierarchy.

Some ant families which are characterized by three-dimensional network organization assault families of a different ant species, whose workers are then enslaved by them. The victorious ant family, although a three-dimensional network per se, assumes the dominant role in the emergent interspecies hierarchy.

Isolationist networks may represent a stage in the development of interactive networks; during this stage, networks await opportunities for switching to the interactive strategy. The interactive strategy is preferred by international networked organizations that deal, e.g., with human rights (*Amnesty International*), debts to be paid by Third World countries (*Jubilee 2000*[27]), and poverty (*Make Poverty History* which is part of the international *Global Call to Action Against Poverty* campaign) as well as by criminal "dark networks" concerned with drug trafficking, e.g., in Columbia, or terror attacks (the *al-Qaeda* network). Such networks are often characterized by *scalability*, i.e., a capacity for rapid growth at low costs without altering the organization pattern (Kahler, 2009a, b).

1.7.2.3. Revolutionary and Symbiotic Strategies

Interactive networks can choose between two different strategies. For instance, network structures (biofilms) of microorganisms that spread inside the human organism can choose between:

1. Destroying host tissues and occupying their area; this is characteristic of infectious processes such as gangrene or phlegmon;
2. Establishing a symbiotic relationship with the host which may benefit from the symbiosis, as is the case with the normal microbiota of the gastro-intestinal tract.

In human society, the analog of strategy (1) is called the *revolutionary* strategy. An organization using it has to overcome the resistance of the political regime. Upon overthrowing it, the new rulers of the country can carry out their political program. In a number of works including the article on Italian Red Brigades cited above (Alexeenkova,

[26] To an extent, the development of isolationist and interactive networks is analogous, respectively, to *chronic* and *acute inflammation* processes. Both kinds of processes result in the formation of pathological network structures that consist of bacteria, host tissue cells, and a peculiar purulent matrix. With chronic infections, the inflammation focus is isolated by a barrier zone, as often is the case with lung tuberculosis. With acute infections, there may be no clear-cut boundaries of the inflammation area; the infection constantly spreads towards formerly intact tissues. Typical examples are gas gangrene and cornea ulcer.

[27] Jubilee 2000 was actively operating as a consolidated structure before the turn of the century. After 2001, Jubilee 2000 became a loose global network with regional or national hubs including Jubilee South, Jubilee Scotland and Jubilee Research in the UK, Jubilee USA Network and other organisations.

2005), it was demonstrated that a network choosing such a strategy is very likely to convert into a hierarchy before taking over politically. In accordance with Michels' iron law, the new political system often becomes rigorously hierarchical, despite the communist, socialist, anarchist, or Christian slogans originally promoted by the network of revolutionaries.

A large number of network structures in modern-day society do not envisage taking power and destroying the existing political system. Networks often set themselves purely religious, cultural, scientific, environmental, or educational goals. Even many politically networked movements avoid using violence and, instead of trying to monopolize power, confine themselves to furthering the interests of a certain social class/stratum/group and contributing, in collaboration with other political actors, to the improvement of the political system and to the welfare of the nation.

In this case, the only acceptable alternative is strategy (2). It involves negotiations with other social (and political) structures and is aimed at reaching a compromise and, optimally, establishing mutually beneficial relationships with them. In an analogy to the behavior of the normal microbiota of the human organism, I call this strategy *symbiotic*.

The transition from networks to other structural types is not always prevented by using the symbiotic strategy. Growing networks inevitably interact with other structures of all kinds, such as other networks, hierarchies, and (quasi-)markets. If direct conflict with these structures can result in destroying an interactive network or altering its organization (e.g., a network may develop into a hierarchy in order to more efficiently struggle with another hierarchy), peaceful interactions involving negotiations and compromises often pose their own dangers. For instance, an important issue is who should represent a network, e.g., an environmental association, during its negotiations with a hierarchy (an administrative body)? The hierarchy usually expects the network to elect an official representative authorized to talk on behalf of the whole network with the hierarchy's leader or his plenipotentiary. This poses the threat of *hierarchizing* the network because the hierarchy is likely to consider the authorized representative as the actual leader of the network. European colonizers established hierarchies in originally non-hierarchical primitive societies of Equatorial Africa, including those of Pygmies. The tribes appointed leaders that conducted negotiations with the colonial powers. They exploited the Pygmies with the help of these leaders, who organized road construction, ivory collection, and cocoa bean harvesting (Soengas, 2009, pp.102-103).

In such a situation, two scenarios are possible: (1) the leaders assume real power over the "former network", as was the case with the Pygmies and (2) they are prevented from taking power because the network explicitly acknowledges them as only *partial leaders* who verbalize, during negotiations, decisions collectively made by network members. To reemphasize, the second option is typical of a hirama-type network that includes a special external leader representing the hirama if it contacts other social structures[28]. North American Indians also had leaders with limited authority that only furthered the interests of their tribe during a pow-wow involving representatives of several tribes.

---

[28] In a similar fashion, a special *commercial leader* existing in some hiramas is responsible for their interaction with (quasi-)market structures while preventing the marketization of the network itself.

1.8. Constructive and Destructive Interactions between Networks, Hierarchies, and (Quasi-)Markets

The preceding section prepared the groundwork for a general, somewhat paradoxical, idea concerning the issue of preventing networks from being transformed into other structures, particularly hierarchies: *If the isolationist strategy does not work, a network can try to retain its organizational structure (and successfully develop it) if it establishes close contact with a hierarchy. This contact should not involve antagonism between the two structures; conversely, it should represent a mutually beneficial symbiosis enabling a reasonable distribution of functions between them.* In Alexander Bogdanov's terms, a network can avoid converting into a hierarchy by allowing egression to develop only outside the network's boundaries. This can be achieved in several ways:

- The network internally retains its non-hierarchical structure but aims to dominate other social structures, including a hierarchical partner, or at least to hold a sufficiently important position in a higher-order hierarchy, which enables the network to promote its ideas, values, etc.; in other words, the *internally* networked structure assumes the dominant role in a larger *external* hierarchy;
- Functions whose performance stimulates hierarchization are delegated to the hierarchical structure that interacts with the network;
- The network plays a subordinate role in the hierarchy-network tandem and exists in the hierarchy's shadow.

Irrespective of the scenario preferred by the partners, an efficient interaction between different structure types is a prerequisite for the optimum functioning of the whole polystructural system made up by them. Such polystructural systems are exemplified, apart from human society, by the human (or animal) organism. Regardless of the kind of system involved, hierarchies and networks constructively interact, i.e., support each other, within their framework. Their tandem is so important because both hierarchies and networks, on their own, possess serious disadvantages. It should be noted that, although this subsection predominantly deals with human social structures, many (dis)advantages of human networks or hierarchies are also characteristic of their biological analogs (e.g., the human organism discussed below).

Important *advantages* of network structures include their flexibility, adaptability, and creativity (see 1.3.3) as well as the capacity to deal with fuzzy issues and interdisciplinary tasks. However, these are situation-dependent advantages; there are also situations in which hierarchies perform better than networks.

Although one of the goals of this work is to promote network organization as a promising social technological project, the author is aware of the *disadvantages* of network structures (each of these serious disadvantages may limit their practical applications unless other structures such as hierarchies can provide their compensatory support):

- managing *networks* presents serious difficulties;

- they are slow in making decisions (by reaching a consensus);
- programming/planning their development is difficult or unfeasible.

There are data that a network is slower than a hierarchy in performing routine tasks according to a fixed plan.

However, *hierarchies* cause their own serious problems:

- they lack flexibility and adaptability;
- they tend to routinize their activities without adequately responding to challenges posed by a dynamic, changing environment;
- they stop functioning if the central control unit is removed, whereas networks are not so vulnerable.

The author's personal experience giving classes at high school showed that a creative task is completed more quickly by a hierarchically organized student group (with a leader reporting the results to the teacher and classmates) than by a network (see 3.6.1 for details). However, a networked team finds a more insightful solution to a problem formulated by the teacher (Oleskin, 2007a, b; Oleskin et al., 2001).

1.8.1. Constructive Network-Hierarchy Interactions

A combined system can benefit from using both its hierarchical and network components, provided they function in harmony and are not in conflict (such harmonious cooperation may require the involvement of a special mediating structure, see below). The hierarchical component of the polystructural system should be at the forefront whenever the functions to be performed demand a fixed plan and/or quick decision-making based thereupon. In contrast, roles delegated to the network component could include, in human society, developing long-term political, economic, social, scientific, technological, and cultural strategies and promoting their implementation in society (this subject is revisited in 3.8 in the context of the political functions of network structures). Real-life examples of hierarchy-network tandems include large companies, e.g., *Mitsubishi*, that use advisory or consultant-based network structures.

"Systemic contradictions" (in Bogdanov's words) characteristic of purely hierarchical structures, including insufficiently strong connections between the top and the bottom levels in a multilevel hierarchy and excessive power in the hands of uncontrollable bosses, can be mitigated by incorporating network components into these structures. This is *"counter-differentiation"*, or *"democratization"* (Bogdanov, 1980 [1921]): the subordinates become involved in regulating the structure's activities (cf. the Ostrom's polycentral political system discussed in 3.8 below); the bosses have to tolerate this.

Partly non-hierarchical groups of young monkeys that coexist with the hierarchical structures of the adults perform quasi-educational functions. While playing, young monkeys familiarize themselves with adult-specific activities such as hunting, food collecting, struggling for one's social status, mating, and rearing infants.

The presence of a network "in the shadow" of a hierarchy increases the stability of the whole system and enables it to remain viable if the hierarchical component becomes defunct. The network takes over. Apart from prolonging the existence of the polystructural system, the network can acquire enough power to enable the restoration of the hierarchical structure, perhaps in a changed form. The defeat of Saddam Hussein's hierarchically organized army did not put an end to organized resistance to American troops in Iraq. Conversely, the resistance increased because of the autonomization of local and global networks (including *al-Qaeda*) that had been suppressed by Saddam's hierarchy and became politically influential after its demise.

The network component of a combined system can orchestrate activities that are beyond the hierarchy's reach. The central political apparatus is often unable to assume control over political life at the local level, especially in a large country. Local networks can provide support to the central administrative body in such situations.

Of relevance both to biology and medicine is the human organism that functions under the combined influence of

- A hierarchical structure dominated by the central nervous system (CNS)[29] that controls the activities of various organs and tissues by generating impulses and chemical regulatory substances that can be produced by the CNS per se (neurohormones) or by CNS-regulated endocrine glands (hormones);
- A complex system composed of network structures that includes two components: (1) human cells that form local or global (distributed within the organism)[30] chemical signal-releasing networks; (2) symbiotic microorganisms inhabiting various niches in the organism, especially the gastrointestinal (GI) tract.

Constructive interactions between the hierarchy and the networks improve physical and mental health; they are a prerequisite for adequate social behavior and, ultimately, political activity.

In the following, I address the functions of the GI tract-inhabiting microbial association. Despite the absence of a central leader, microbial cells in the intestinal lumen or in biofilms on the mucous membrane effectively coordinate their behaviors. This behavior coordination depends on chemical mediators. Microbial networks are involved in food digestion, intestinal motility regulation, and the maintenance of the acidity (pH), temperature, and gas phase composition within normal limits, promote the maturation of the intestinal mucosa and contribute to its barrier function with respect to harmful substances and pathogenic microorganisms and viruses, stimulate the activity of the immune system, and provide the organism with useful compounds such as vitamins, short-chain fatty acids, and hormones (quoted from: Oleskin, 2012, p.173).

[29] Despite representing the top of the interorganismic hierarchy, the CNS combines hierarchical and network structures in its internal organization

[30] Such distributed and, nevertheless, coherent cell network structures are exemplified by the network of immune cells that release important regulatory substances

1.8.2. Destructive Interactions between Networks and Hierarchies

Despite the positive aspects of hierarchy-network interactions, their relationship is not necessarily harmonious, and the tandem can under certain circumstances be detrimental to each of the two partners and the whole polystructural system involved.

Destructive interactions of hierarchies and networks are caused by the typical properties of each of the structures. As for hierarchies, some of their potentially deleterious properties were listed above. Additional potentially negative features are exhibited by the subtype of hierarchies called *bureaucracies.* These currently widespread hierarchical structures, discussed in detail in classical works by Max Weber and other scholars of the early 20th century, will be considered in Chapter three (see 3.3). At this point, suffice it to say that they are characterized by the existence of communication barriers between their parts and between the whole bureaucracy and its environment as well as by the tendency towards formalizing and routinizing many aspects of the behavior and the lifestyle of people employed by the bureaucracies. This results in the suppression of their initiative and creativity.

These properties of hierarchies inevitably influence their relationships with networks that try to form alliances with the hierarchies. For example, hierarchies attempt to prevent networks from participating in making decisions and even from expressing their opinions concerning them. In human society, networks functioning as expert pools are often pressured by the hierarchy into approving of its decisions instead of criticizing them. Some politicians overtly express the concern that networks "are suspicious" and suggest using them as a kind of smokescreen for an actually hierarchical regime without granting these bogus networks any real political rights.

In biological systems, there are analogous situations in which networks are negatively influenced by hierarchies. Normally, many ecosystems represent networks that often include a large number of species that interact "on an equitable basis". However, if a single biological species assumes the dominant role in an ecosystem, the development of other species within its boundaries is limited or completely suppressed, as is the case if the spruce dominates the mixed forest ecosystems of the East European Plain. Gradually, the species diversity of such ecosystems is impoverished.

Nevertheless, interactive[31] network structures (the subtype of networks that actively interact with hierarchies) also possess a number of systemic properties that can cause serious problems in tandem systems.

1. *A large number of interactive networks tend to ignore territorial boundaries and actively communicate with "outsiders".* In contrast, many hierarchies both in biological systems and human society mark and defend the borders of their territory (in the literary or, in human society, also in the metaphoric sense). In many hierarchically organized monkey groups, the dominant male vigilantly controls the group area, whereas young individuals forming a parallel network structure tend to disregard the boundaries of this area and play with individuals from the area of another group. The hierarchy of the political system of a state makes efforts to safeguard the state's territorial integrity. The territory to be defended includes not

[31] The other subtype of network structures, i.e., isolationist networks, tend to avoid contacting hierarchies as well as other networks or (quasi-)market structures..

only the geographic area but also confidential information and commercial, military, and political secrets. Networks tend to pay less attention to territorial boundaries and to build bridges across them in order to communicate with interesting people on the other side of the boundaries (functioning as bridging networks in Putnam's terms). Important political developments were caused by links established by networks of Russian ecological activists with environmental movements outside Russia in the late 1980s. The Russians actively absorbed not only environmental information of their foreign colleagues but their political ideas as well, and proceeded to implement these ideas in their home country. Even if a network sets itself the goal of supporting a hierarchy and jointly carrying out a social/political/cultural project, the network tends, nevertheless, to establish links with "outsiders" that do not belong to the network-hierarchy tandem. Networks are predisposed to share important secret and confidential information with competitors and enemies. Online social networks such as *Facebook* or *MySpace* face serious privacy issues. Their users supply too much information about themselves, and online networks cannot guarantee the confidentiality of private information that is of interest to advertisers or even criminals.

2. *Networks tend to grow in such a way that they negatively influence structures interacting with them.* Even a benign tumor can deplete its host organism of nutrients by using them for its own development. In human society, network structures tend to spend a disproportionately large amount of money and expend other resources in order to attain the goals they officially set themselves. In the absence of a central controller, network members may be tempted to spend part of the resources on pursuing their personal goals. In various creative networks (networked research labs, political think tanks, etc.), a large number of network members tend to exaggerate their own contributions to projects developed by the networks and, accordingly, to claim a disproportionately large reward. Apart from squandering resources, this can also result in internal conflicts inside such network structures or between them and the hierarchical component of the tandem. Hierarchies including bureaucracies may be characterized by corruption and embezzlement. However, bureaucrats have to overcome a certain psychological and legal barrier because they should violate the rigorous behavioral rules that are normally obeyed by a bureaucratic structure (Weber, 1956 [1907]). In contrast, the relative freedom the members of a network structure enjoy is often considered one of their advantages, and the members may envisage the right to liberally use financial resources as one of their inalienable rights.
3. *Networks forming part of a network-hierarchy tandem do not follow the rhythm set by the hierarchical partner in their behavior, threatening to disrupt this rhythm*—to cause what cardiologists call *arrhythmia* (a condition in which the primary heart pacemaker is temporarily inactive and different parts of the heart beat with different rhythms). In the summer of 1917, after dethroning the Czar, the democratic provisional government of Russia still planned new military operations on World War One's Russo-German front but the soldiers disobeyed the generals, who had limited power. The soldiers listened only to the networks of political activists belonging to various factions (including the Bolshevists) that encouraged them to

stop fighting and even to fraternize with the enemy. Therefore, all the military operations failed, and the Germans made important advances.

4. *Networks attempt to perform regulatory functions instead of the hierarchy.* This tendency to replace the hierarchy in terms of its regulatory functions is useful if the hierarchy fails to function properly. However, if the hierarchy is still functional then the network component's attempt to take over can cause the system to behave chaotically. Such a scenario was depicted in the futuristic book by Bard & Söderqvist (2002).
5. *Networks are more complex in organizational terms than hierarchies*, and their complexity may be out of proportion with the advantages they have (Powell, 1990). During a series of classes conducted by the author of this work, students in an experimental classroom were subdivided into two teams organized along hierarchical and network principles, respectively. It was revealed that the teams of students found it much easier to adapt in a hierarchical system than in a network (Oleskin et al., 2001). The organizational complexity of networks may hamper the functionality of network-hierarchy tandems. In such situations, the successful operation of the tandem requires the efforts of a team of skillful mediators (see below).

The properties of networks that cause problems for network-hierarchy tandems can be regarded as advantages under specific circumstances. A network's propensity to spread information across boundaries is a prerequisite for the successful operation of cross-border networks, unfortunately including those of drug dealers or terrorists (Kahler, 2009a, b). The tendency of a large number of networks to ignore the rhythm set by hierarchies may be regarded as a positive feature as far as networks of scientific enthusiasts are concerned. The tempo of the research work done by these enthusiasts may be faster than that of hierarchical research institutes, enabling the enthusiasts to advance seminal ideas that are several decades ahead of their time (this was discussed in 1.6.1 in relation to the concept of "anticipatory reflection"). The capacity of some networks to take power, assigning subordinate roles to hierarchies, endangers the power of hierarchies, e.g., of the government with the president at its pinnacle. However, this capacity becomes a network's advantage if the goal is to overthrow the hierarchy of the political apparatus, as was the case with President Yanukovich's dismissal in Ukraine in 2014.

As for the human organism as a polystructural system, its health is endangered by disruptions in the tandem formed by the hierarchy of the central nervous system and the microbiota's networks. This can result in serious problems including not only intestinal diseases (irritated bowel syndrome, Crohn's disease, ulcerous colitis, and colon cancer), but also liver dystrophy, rheumatoid arthritis, spondylosis, multiple organ failure and mental disorders (autism, Tourette's syndrome, ADHD, etc.)" (Oleskin et al., 2011, p.172). The influence of the microbiota "on our physical and mental health is mediated by chemical agents including neuromediators". In such a pathological situation, symbiotic microorganisms do harm rather than good to the host organism. The above items (1–5) concerning network properties in general also apply to the host-microbiota tandem below:

1. The networks of GI microorganisms *tend to communicate with "outsiders"*, i.e., microbes that do not belong to normal symbiotic species, including virulent bacteria. The signal substances (*pheromones*) released by the normal microbiota of the mouth

and the pharynx stimulate the growth and virulence of the dangerous pathogenic species *Pseudomonas aeruginosa.* Limiting such "cross-border" communication by suppressing the normal microbiota with antibiotics results in mitigating *Pseudomonas aeruginosa*-caused problems, even though the antibiotics may not affect the pathogen per se. Recently, more subtle methods of cutting off pathogen-stimulating communication have been suggested, based on manipulating the mechanisms of microbial communication including quorum-sensing systems (see 2.1.5). They are expected to yield a new generation of antimicrobial drug preparations (Shpakov, 2009).

2. An important goal to attain, in medical terms, is *to keep the development of microbial network structures under control.* The excessive growth of microorganisms, even useful ones, may be harmful to the organism (although their insufficient growth may also cause problems because of the "vacancies" that can be used by unwanted intruders). The widespread intestinal symbiont *E. coli* releases chemical regulators including DOPA (Shishov et al., 2009) that converts into catecholamine neurotransmitters in the bloodstream and the brain. Changing the number of intestinal *E. coli* cells is likely to cause alterations in their concentrations. For instance, excessive growth of the *E. coli* population is likely to increase the activity of the dopamine-using (dopaminergic) systems of the brain. Excessive activity within the dopamine-using system manifests itself in hypersociability and euphoria and may promote the development of schizophrenia (Dubynin et al., 2003).
3. The symbiotic microbiota of the human organism normally functions in sync with the biorhythms of the host organism, and *any discrepancy between the rhythms/tempos of the two structures* in the tandem poses the threat of serious problems in terms of physical and mental health. Since networks generally tend to ignore the rhythm of their hierarchical partners, the human-microbiota tandem is extremely fragile and the concordance between the rhythms of its components can be easily disrupted by seemingly insignificant external factors such as the wrong diet, stress, or usually non-problematic low levels of toxic substances in the environment.
4. Despite the influence of microbial chemical regulators on the human organism, the dominant role in it is normally played by its own main controlling organ, the brain. In pathological situations, *an analog of "netocracy"* (Bard & Söderqvist, 2002) *can be established*, and the activities of the brain are heavily influenced by microbial networks with their chemicals. This may account for the fact that a number of nervous system problems and psychological disorders, including autism, Tourette syndrome, and ADHD, are associated with disruptions in the functioning of the GI microbiota (Bai & Ouyang, 2006). Even in healthy humans (or animals), their microbiota attempts to assume control over the functioning and behaviors of the host organism. According to a recent hypothesis (Norris et al., 2013), bacteria influence the host's appetite, inducing the host to prefer food that contains the nutrients which are required by the bacteria. Pregnancy-associated changes in eating habits may be directly caused by alterations in the microbiota composition that occur due to hormonal changes. Generally, host-microbiota interactions are bilateral: the CNS hierarchy produces substances, including neurotransmitters, that influence the microbiota. For example, catecholamines released into the bloodstream under stress stimulate the growth of potentially pathogenic microbiota. The microorganisms, in

their turn, may challenge the host hierarchy's dominance by releasing their own regulatory substances, which are aimed at putting the host under their control.

5. The *organizational complexity of networks* is also characteristic of the decentralized structures formed by human symbiotic microorganisms. This causes serious problems in terms of the prevention or therapy of infections. Most drugs produce multiple effects on the microbiota, which, to reiterate, includes not only symbionts but also (potential) pathogens; these two subtypes of microbiota constantly communicate and actually form part of one complex, coherent interorganismic network structure. An extremely difficult task, therefore, is to make the positive effects of a drug (such as an antibiotic) outweigh its negative effects. The antibiotic may kill not the pathogen but, instead, useful microbes that could otherwise restrict the pathogen's reproduction and migration inside the organism.

1.8.3. Role of Mediation Structures

All the potentially harmful properties of networks do not inevitably cause negative effects. In hierarchy-network tandems in human society, organizational measures can be taken in order to prevent the negative influence of network-specific problems and to enhance the positive effects of networks.Taking such measures is the job of *mediation structures* that facilitate network-hierarchy interactions (Rimsky & Sungurov, 2002; Sungurov et al., 2012).

These intervening structures can include three main parts:

- The network component of the network-hierarchy tandem can create a special *subnetwork* that is directly responsible for dealing with the hierarchical component (personified by the external leader in a hirama);
- The hierarchy can establish an *additional department* which bears the responsibility of contacting the network component;
- An intermediate structural part can be sandwiched between the subnetwork and the hierarchy's department. This is the *meta-governor* described by Louis Meulemann (2008). The meta-governor is tasked with creating an optimal "creative mix" composed of different types of structures. If the meta-governor is a special social group rather than a single individual, then the structure of choice is a network rather than a hierarchy. The flexible and dynamic character of a network enables it to efficiently respond to environmental changes and adequately restructure the system if necessary.

Analogous mediating structures are likely to exist in some biological systems. Presumably, mediating functions can be performed in the human/animal organism by the immune system. By limiting the growth of potentially harmful microorganisms and favoring beneficial microbial networks (at least "letting them live"), the immune system could help the host and the microbiota to establish a harmonious tandem system and stimulate the microbiota's activities that are useful for the hierachical component of the tandem (the organism with the CNS), such as releasing neuromediators, hormones, and other biologically active substances (Oleskin et al., 2011). Mediating structures in biological systems have not been sufficiently well understood and await further research in the future.

1.8.4. Interactions between Networks and (Quasi-)Market Structures

Apart from hierarchization, a network structure can also undergo marketization. One strategy of preventing marketization is establishing sufficiently close ties with an already existing market structure, as exemplified by Israeli *kibbutzim*. They let some of the workers that are not *kibbutz* members do some of the jobs (see 3.2.4 below). Therefore, contract relations are maintained not between *kibbutz* members but between the *kibbutz* as a single agent and the hired workers. This can keep marketization outside the boundaries of the network component of the network-market tandem, although hired workers, nevertheless, pose a serious challenge to a *kibbutz'* basic principles (see 3.2.4).

The network-market tandem can set up its own mediating structures, whose job should include reasonably distributing functions between tandem components and securing the dominance of non-market cooperative relationships inside the network component per se.

1.8.5. Network-Network Relations: Establishing Higher-Order Networks

Apart from hierarchies and (quasi-)markets, networks can interact with one another. Such interactions between networks that enable them to form alliances (higher-order networks) are characteristic of both human society and a variety of biological systems. The weakly hierarchical (three-dimensional) loose network groups of chimpanzees can form alliances, i.e., higher-order networks that can be called megacoalitions.

In human society, interactions and mergers between originally independent networks are a prerequisite for the development of civil society as a system of horizontal links and relationships that involve a significant part of the population (Mezhuev, 2008, p.6, civil society will be discussed in 3.8). Only sophisticated, large-scale networks within a civil society can be considered by the hierarchical structures of its political system as sufficiently important partners for establishing tandems. Small networks are ignored by the political hierarchy; they exert no influence on it unless they form strong alliances. However, the strategy of merging networks into a larger structure can result in the *hierarchization* of the entire structure: some networks may tend to dominate other networks. Nevertheless, if the goal is to keep the whole higher-order structure flat, we can make use of methods originally developed for hiramas as small-sized network structures. They can form alliances and further merge into larger flat network structures (*hiramiads*) if they comply with one of the two scenarios given below (Oleskin & Masters, 1997):

- *Creating temporary project-oriented associations of networks.* A hirama-type network (see 1.2.2) subdivides the problem(s) to be solved into several parts (subproblems); accordingly, there are several partial subproblem leaders in each network. If one of the subproblems is also dealt with by other networks, all the networks involved can form a temporary union that is focused on this common subproblem. The external leaders of each of the network conduct negotiations which also involve creative leaders that specialize in the relevant subproblem. As a result, loose networks composed of networks form (with the main decision-making power resting with the individual small-sized networks). An additional example of such relationships might be the interlocking memberships on the Boards of Directors of large industrial, commercial, and banking firms.

- *Establishing relatively rigid second-order structures for the purpose of long-term cooperation among a number of networks*, with each network specializing in a specific part of the overall task. Individual networks can join together according to the hirama pattern. In this case, each of the hirama functions (subproblem, psychology, "external affairs", organizational, and commercial leaders) corresponds to a specific network, which, in keeping with its organizational principles, breaks down its subproblem into "sub-subproblems". Thus, one of the networks deals with the external affairs of the whole collective, and this task is further subdivided inside it. These networks can be supplemented by a number of non-specialized groups, equivalents of members with no leadership duties in a first-order hirama. These "free lancers" can alternately generate ideas on different subjects, temporarily forming unions with some specialized network groups.

There is, in principle, no reason why the above pattern cannot be further applied, in order to form third-order, fourth-order, etc. networks. The resulting structures will represent horizontal, non-hierarchical, and non-bureaucratic networks. Coherent interest groups and pressure groups can be created on the basis of such higher-order network structures. They can make up the backbone of *civil society* (revisited in 3.8.5), enabling it to engage in dialogues and debates with the political system and businesses. Establishing relationships between networks can be facilitated by internetwork interaction-mediating structures. I reemphasize that mediator structures can also be used to facilitate constructive network-hierarchy and network-market interactions; more traditional market-hierarchy interactivity, which was discussed in a number of classical works (e.g., Coase, 1937), is not considered in this book.

1.8.6. Interactions among Structures in Multi-Order Systems. Homo- and Heteronomous Systems

The preceding subsection concerning higher-order networks has set the stage for a more general discussion on *multi-order (multilevel) systems*.

Multi-order networks, hierarchies, and (quasi-)markets are composed of groups of lower-order structures. Such structures are characteristic of both biological systems and human society. For instance, colonies of unicellular organisms and structures formed by nervous cells inside the brain can be considered multi-order networks (=networks composed of networks). Obviously, hierarchies and (quasi-)markets may also represent multi-order systems. The question to raise is whether networks always consist of smaller networks (that are composed of still smaller networks) and hierarchies of smaller hierarchies (as is the case with most bureaucratic structures)? Systems with structurally similar (isomorphic) levels will be referred to as *homonomous* in this work. Apart from pure structural types (networks of networks, hierarchies made up of hierarchies, or markets consisting of markets), homonomous systems include mixed structures, provided that the same ratio between different structure types and the same pattern of interactions between them is characteristic of all their structural levels.

Homonomous systems exhibit *fractal/self-similar* organizational patterns: parts of the system are reduced copies of the whole system in structural terms.

This feature is referred to as *topological self-similarity*, and one way to quantify it is based upon the *Horton-Strahler (HS) index*. Consider a multilevel system in which each higher-level structure is composed of several lower-order units. It can be represented as an inverted tree where the trunk is the highest-order network that contains several branches (lower-order networks) that, in turn, branch into several still smaller structural units; ultimately, we reach lowest-level networks composed directly of elementary nodes, e.g., human individuals in social networks. In HS terms, each elementary node (each leaf of the tree) is assigned an HS value of $i = 1$. For each branch that ramifies into two branches with indices i_1 and i_2, the index i is calculated as follows: $i = i_1+1$ if $i_1= i_2$, and, alternatively, $i = \max(i_1, i_2)$ if $i_1 \neq i_2$ (Guimerà et al., 2006).

"The number of branches N_i with index i can be determined once the HS index of each branch is known. The bifurcation ratios $Б_i$ are then defined by $Б_i=N_i/N_i + 1$ (by definition $Б_i \geq 2$). When $Б_i \approx Б$ for all i, the structure is said to be topologically self-similar because the overall tree can be viewed as being comprised of $Б$ sub-trees, which in turn are comprised of $Б^{32}$ smaller sub-trees with similar structures and so forth for all scales" (Ibid., p.664).

Conversely, *heteronomous* systems are characterized by different structures at different levels. Some of them, nonetheless, may possess *fractal/self-similar* properties. This is exemplified by eusocial systems formed by ants (see below). Their *networks consist of hierarchies* that, in their turn, are made up of networks composed of hierarchies. Overall, there is a repetitive interlevel pattern that makes the entire system self-similar.

> In ant societies (revisited in 2.4 below), structures with predominantly network-style and hierarchical organization exist at different structural levels. Worker ants form a *clan*, a hierarchically organized group of interacting workers that have a leader and engage in doing household tasks (Zakharov, 1991, 2005, 2009). Several clans combine to form a *column* that is based on the network principle, so that each clan leader is a partial leader within this second-order structure. The network principle also applies to a higher level: several columns make up a horizontally structured *pleiad (family)* inhabiting a multisection nest, with each column occupying one nest section. However, a still higher structural level is characterized by hierarchical organization. Several ant pleiads form a *genuine colony* in which one of the pleiads plays the role of the central (dominant) nest; there is a radial system of connections between the nests, as is the case with craters in which the nests of the Australian ant *Iridomyrmex purpurea* are located (Zakharov, 1991, p.161). Several genuine colonies establish a stable higher-order decentralized network referred to as a *federation*.

[32] Note. I have substituted the Cyrillic character *Б* (the initial letter of the Russian term *Бифуркация*, which means "bifurcation") for the letter *B* used by the authors cited. The reason is that the letter *B* was used in this work to denote a different variable, *betweenness* centrality (see 1.2.4.1 above).

Heteronomous organization can be quite stable in eusocial systems. However, a large number of other heteronomous systems are characterized by instability and a tendency towards "homonomization", i.e., the dominance of the same structural pattern at all organization levels.

In human society, the "homonomization" trend may struggle with the opposite trend favoring different structural patterns at different levels. Under the regime established in the Soviet Union under Joseph Stalin, there were two different types of Russian families: (1) "reduced copies" of the political system that were characterized by manifest dominance-submission relationships between family members and (2) families with compensatory horizontal relationships that, in psychological terms, provided an escape from the rigorous hierarchical political regime. Many other families were located between the two extreme points of the continuum or alternately approached each of them during their "life cycle" (personal observations and interview data).

In sum, the three types of structures considered in this work—networks, hierarchies, and (quasi-)markets—can interconvert. If network structures are advantageous in a particular situation, it makes sense to prevent their conversion into other types of structures, e.g., their hierarchization. The options discussed above include (1) isolation and the development of a protective matrix (degression in Alexander Bogdanov's words) and (2) interaction with structures of other types and the establishment of polystructural (tandem) systems in which antagonism is overridden by cooperation, and where each component performs specific functions for the benefit of the whole tandem. In network-hierarchy tandems, specific measures should be taken for the purpose of mitigating the effects of the negative properties of both components and enhancing their positive features. Mediation structures functioning between networks and hierarchies hold much potential value in this context. Similar mediators can facilitate interactions between a network and a (quasi-)market structure or between networks within network-network tandem systems.

Chapter II

Network Structures in Biological Systems

While the preceding chapter dealt with general organizational characteristics that apply to structures existing both in biological systems and in human society, this chapter focuses on biological systems alone. The goal is to look for analogs of human networks—as well as, for comparison, hierarchies and quasi-markets—in biological systems. In a similar fashion, Axelrod and Hamilton (1984) suggested using data obtained in studies on *human* cooperation to provide guidelines for research on cooperation in nonhuman life forms in their chapter of the book *Evolution of Cooperation*.

Although this work is concerned with biological and human social systems, many concepts used in the text also apply to nonliving systems. Of relevance are complex *chemical systems* that are characterized by decentralized organizational patterns and, therefore, can be considered in terms of network structures. These systems include catalytic networks, i.e., "chemical reaction systems containing molecular species that catalyze their own replication or the synthesis of other species of the network" (Sardanyes et al., 2012, p.795). Catalytic networks, or catalytic sets, are typified by hypercycles (see below).

Hierarchical structures with central pacemakers, or active centers, also occur in nonliving systems; the point was stressed by Alexander Bogdanov (1980 [1921]) in relation to what he called egression. These structures are exemplified by phase transition processes (e.g., water freezing) that spread from a single active center such as a microscopic ice crystal. In some esoteric doctrines of creation, including those of Kabbalah teachings, the emergence of an active center in originally homogeneous matter is postulated; the whole creation process resembles the development of a chicken embryo initiated by active cell division in a small area on the surface of the otherwise homogeneous egg yolk.

Besides, complex chemical systems can establish *quasi-market* relationships that are based on competition among developing structures or parallel processes that share the same pool of substrates. However, some of these structures do not only compete. They also cooperate, forming analogs of network structures.

Cooperation in conjunction with competition occurs in such complex systems as coacervates, i.e., small spherical droplets of organic molecules that are held together by hydrophobic forces from a surrounding liquid; as well as in lipid microspheres and other molecular structures considered in the literature as potential precursors of primitive living cells.

Of interest are hypercycles originally described by Eigen and Schuster (1977, 1979). Hypercycles are "sets of self-replicating species able to catalyze other replicator species within a cyclic architecture" (Sardanyes et al., 2012, p.795). As far as the origins of life on the planet are concerned, particular attention has recently been given to RNA molecule-containing hypercycles, because these informational molecules are capable of operating as ribozymes, i.e., they can catalyze chemical reactions, including their own replication, if structural units (nucleotides) are available in the environment. Let us consider a system containing several (*n*) different RNA molecules. It is assumed that (1) each RNA catalyzes its own replication using the nucleotide pool, and (2) each RNA also catalyzes the synthesis of another RNA according to the following algorithm: RNA #1 catalyzes the synthesis of RNA #2, RNA #2 the synthesis of RNA #3, ... RNA #*n* that of RNA #1, so that the entire system is cyclic. The RNA species actually form a cyclic one-dimensional network where they cooperate with one another. The fact that each RNA catalyzes the synthesis of the next RNA molecule is interpreted in the literature in terms of "altruistic behavior" of all these RNA molecules. Therefore, like animal biosocial systems composed of altruistic cooperators (as exemplified by ant societies), many hypercycles face the danger that free riders (parasites) can take advantage of others' cooperation without cooperating themselves. Two types of parasites can emerge in a hypercycle: (1) selfish parasites whose synthesis is catalyzed by a hypercycle component, but that do not catalyze any process themselves, and (2) shortcut parasites that bypass some stages of the cycle, thereby threatening to disrupt it (Bresch et al., 1980; Cronhjort & Blomberg, 1997; Sardanyes & Sole, 2007). The danger of parasitism is mitigated if the substances involved in the operation of the hypercycle are incompletely mixed (the system is not homogeneous; Boerlijst & Hogeweg, 1991) or, more generally, if the hypercycle is spatially organized (Cronhjort & Blomberg, 1994; Attolini & Stadler, 2006) with its stages compartmentalized, as is the case with living cells and their numerous organelles and channels.

Importantly, while the components of one hypercycle "altruistically" cooperate, several hypercycles sharing the same pool of structural units and other materials will inevitably compete with one another, and those characterized by the highest catalytic activity will outcompete others, i.e., the system is characterized by primitive evolution based on natural selection (Es'kov, 2004).

It is evident from the above that I have arbitrarily set the boundaries of the realm in which network structure studies are of relevance by addressing biological and human social systems alone. Additional studies can be conducted in which the notions and terms used in this work are to be applied to physical, chemical, technological, and other kinds of systems.

Interestingly, many nonliving systems with manifest network, hierarchical, or quasi-market structures are characterized by *biomorphism*, i.e., they exhibit similarities to biological systems. Stars form centralized or decentralized clusters; importantly, they develop in accordance with quasi-biological evolutionary principles. It is pertinent that one of the stages of a star's evolution was metaphorically denoted as a cocoon star, "a newly-formed massive star... surrounded by dense gas and dust" (Davidson, 1970, p.422). Some biomorphism is also characteristic of elementary particles whose birth, aggregation, and decay are discussed by nuclear physicists in terms of "village stories". Crystals can grow and engulf one another.

Obviously, biomorphism is of relevance to the behavior of the prebiological systems discussed above, including coacervates and hypercycles (see above). It was hypothesized (Es'kov, 2004) that life could emerge in the universe before the first living organism came into being: it could exist in the form of hypercycles.

In this chapter, the main patterns of network organization in biological systems will be discussed. Each of them is used by several different kinds of biological systems. For example, *modular* organization occurs in several groups of invertebrate animals, such as cnidarians, bryozoans, and ascidians. *Eusocial* organization is typical, apart from social insects, of naked mole-rats and, to an extent, of the chough and some penguin species. Each pattern represents a *paradigm* from whose perspective researchers consider the actual organization of a biological structure. These paradigms may overlap in the sense that a biological structure can be described using more than one paradigm.

Biological systems vary in their degree of complexity, ranging from simple conglomerations of anonymous individuals sharing the same resource pool to highly integrated and differentiated systems. While simple aggregations are predominantly characterized by individual competition and represent analogs of *quasi-market* structures, systems with a higher integration level can be structured vertically or horizontally, resulting in the predominance of either *hierarchical* or *network* organizations, or their combination.

Decentralized network structures based on cooperative interaction among their nodes exist at many different levels of biological systems. Such structures are formed by biological macromolecules, e.g., proteins serving as enzymes or regulators, or by intracellular components such as organelles. Network structures are also composed of whole cells, either as free-living organisms (unicellular life forms) or within tissues forming part of a multicellular organism. Such organisms, in their turn, often form colonies, groups, families, and other supraorganismic (biosocial) systems; some of them represent *sensu stricto* networks in organizational terms. Many biosocial systems made up of organisms display self-similar/fractal structural patterns, i.e., they represent multilevel structures whose parts (lower-order structures) look like reduced copies of the whole system. Networks occur at still higher levels; for instance, multispecies communities and ecosystems are often characterized by a network organizational pattern. The whole biosphere represents a decentralized, distributed system, and cooperation among its components seems to be much more important in global terms than their inevitable competition.

This chapter is chiefly concerned with network-type biosocial systems composed either of single cells (sections 2.1 and, for neural networks, section 2.5) or multicellular organisms as their "nodes" (sections 2.2–2.4 and 2.6 of this chapter). These biological systems exhibit some structural features that are directly comparable to those of network structures in human society.[33] Section 2.1 contains subsection 2.1.10 that deals with intracellular and molecular networks. Network structures existing at the multispecies community, ecosystem, and biosphere levels are beyond the scope of this work, despite their obvious importance. They will constitute the subject of another book produced by the author (O. A.).

---

[33] Nonetheless, there are a large number of important differences between biological and human social structures.

2.1. Cellular (Microbial) Paradigm

Humans and animals are not the only forms of life that engage in various forms of collective interaction (social behavior) and form more or less integrated associations of individuals, i.e., *biosocial systems*. Recently, it has been revealed that such behavior is also relatively widely spread in seemingly primitive organisms, including unicellular forms of life. Remarkably, analogs of social behaviors occur, apart from free-living unicellular organisms (fungi, microalgae, protozoans, and bacteria), in the cells from which tissues of multicellular animals and plants are composed. What kinds of organizational structures do living cells form?

Competitive (quasi-market) interactions are widespread in the microbial world; they are supplemented by chains of metabolic processes in which products generated by one organism serve as raw materials for others. Some bacterial colonies growing on Petri dishes are comparable to disorganized human crowds (or "anonymous conglomerations", resembling those of invertebrates such as locusts) where autonomous individuals only compete for nutrients, oxygen, and other resources[34].

Some unicellular organisms form at least temporary, situation-dependent hierarchies. The difference between dominance and leadership considered above (1.2) in general terms, applies to cell structures. Populations of starving bacteria, e.g., *Escherichia coli* or *Bacillus subtilis,* include two subgroups (Akaizin et al., 1990; Ellermeier et al., 2006; González-Pastor, 2011): (1) autolysing cells that die and release their nutrients into the culture fluid; and (2) actively growing cells utilizing these nutrients. This hierarchy is based on the very definition of dominance presented above, which indicated that it was the preferential utilization of available resources by special (dominant) individuals. In *B. subtilis,* subgroup 2 consists of cells in which the sporulation regulator gene *Spo0A* is active; they release two toxins that kill the cells in which *Spo0A* is inactive (they constitute group 1). The cells of subgroup 2 develop immunity to their own toxic products (Ellermeier et al., 2006; González-Pastor, 2011).

Some of the structures that are formed by plant or animal cells, either inside a developing organism during its ontogeny or in cultures grown under laboratory conditions, contain a central (primary) pacemaker. Therefore, the structures are at least temporarily *hierarchical*. A classic example is the heart, where the tissue contains the sinuatrial node (see also the example concerning epidermal cells, subsection 1.7.1). Its cells are the main pacemakers of the myocardium (heart muscle). They spontaneously depolarize, generating impulses that spread and impose their rhythm on the whole heart. This kind of hierarchy has leaders (pacemakers) rather than dominant elements.

Leader cells that produce chemical signals were revealed in groups of free-living unicellular eukaryotes, which form organized multicellular structures. Large groups of ameboid cells of the slime mold *Dictyostelium discoideum* form a migrating "slug", and thereupon, a mushroom-like fruiting body under the influence of cyclic adenosinomonophosphate (cAMP) released by a number of cells (Devreotes, 1989; Mutzel, 1995) that function as partial leaders (see 1.2.6). However, the very fact that there may be a

---

[34] This market-style competition among unconnected individuals is characteristic of some "domesticated", i.e., those cultivated in a lab for a long time, bacterial cultures. In contrast, many "wild" bacteria taken from their natural habitats are likely to form sophisticated cell structures representing true cooperative networks (see Aguilar et al., 2007).

large number of partial leaders suggests, as was pointed out above, that the structure is a three-dimensional network rather than a centralized hierarchy. Such leader cells induce the secondary cAMP synthesis by cAMP signal-receiving cells that activate other cells and, therefore, become secondary leaders, making the entire system still more decentralized.

Generally, in the realm of unicellular organisms, decentralized network structures are more widely spread than even temporary hierarchical systems. Network structures are characteristic of colonies and biofilms of a large number of microorganisms. A large number of bacteria form almost ideally flat network structures.

Despite a lack of any vertical (rank) differentiation among the cells in these structures, these cells may be horizontally differentiated, i.e., functionally specialized. For instance, myxobacteria form dense aggregations, in which a part of the cells converts into myxospores, while some other cells form the walls of the fruiting bodies.

Bacteria and many other unicellular organisms often form highly clustered networks that consist of dense subnetworks connected by sparse links.

2.1.1. Microbial Biofilms as Network Structures

Of special interest are *biofilms*, i.e., "matrix-enclosed microbial accretions that adhere to biological or non-biological surfaces" (Hall-Stoodley et al., 2004, p.95). The diversity of biofilms is fascinating: they range "from patchy monolayers on some surfaces through very thick gelatinous masses associated with cooling systems to filamentous accretions near sewage outlets" (Whimpenny et al., 2000, p.662). Biofilms can be formed by the cells of a single bacterial species (and even single-species films are likely to be heterogeneous because they often include cells with different phenotypes), or they might represent multispecies systems. Some biofilms include representatives of different kingdoms. For instance, the film of a methanogenic association is composed of cells of eubacteria and archaea.

> The typical structure of a biofilm is formed stepwise. Initially, a transient attachment of microbial cells (primary colonizers) occurs, which is due to their interactions with the substratum involving flagella, pili, fimbria (type I fimbria enabling the initial attachment and type IV and *curli* fimbria involved in the subsequent stages), and the proteins of the outer membrane of gram-negative bacteria (reviewed, Monroe, 2007; Lemon et al., 2008; Karatan & Watnick, 2009; Otto, 2010; Smirnova et al., 2010). This stage is followed by the permanent attachment of microbial cells to the surface. Subsequently, microbial cells spread on the substratum that is colonized by them. This is accompanied by the formation of local cell groups (rafts, clusters, or microcolonies) and the intracellular matrix with its characteristic cavities, and the biofilm cover including a part of the matrix as well as lipid bilayer structures (Tetz et al., 1993, 2004; Greiner et al., 2005; Pavlova et al., 2007). Eventually, a single- or multiple-species biofilm with a developed structure is formed; this biofilm can display a lamellar structural pattern or contain mushroom- or pillar-shaped formations. The final stage of a biofilm's life cycle involves its dispersal; its microbial cells return to the planktonic mode of existence. This involves the disruption of the attachment of the cells to the substratum, which is caused by the synthesis of surfactants and enzymes that degrade the matrix, including its components that are directly responsible for cell attachment to the substratum and to one another.

Biofilms can be considered a triumph of decentralized network organization. Processes that are carried out by biofilms during the whole life cycle, from their formation to their dispersal, are coordinated without a central pacemaker, in an almost ideally flat network structure.

In the literature, biofilms are often contrasted with *colonies,* i.e., microbial biosocial systems located on a solid substratum and characterized by a sharp boundary and a distinct shape. While most bioflims are potentially capable of unlimited growth, colonies reach their maximum size; thereupon, they may divide or bud off small daughter colonies. The matrix, which is composed of extracellular biopolymers, forms the colony's strong exoskeleton.

However, these characteristics do not apply to the subtype of colonies that constantly spread over a surface and may lack a clear-cut boundary (reviewed, Kaiser, 2007). They often include rapidly moving swarmer cells. Their movement depends on rotating bundles of flagella (e.g., in *Proteus mirabilis*), type IV fimbria that pull cells, or slime projected from a nozzle, which ultimately pushes them (myxobacteria have both type IV fimbria and slime nozzles).

Colonies can be considered as *isolationist* networks (except swarmer-containing colonies), whereas most biofilms and colonies that spread over the surface are expanding *interactive* networks, in terms of the classification given in 1.6.2 above. Even the final stage of a biofilm's life cycle, i.e., its dispersal, can be regarded as a continuation of the expansion strategy because it enables the biofilm's cells to migrate independently.

The advantages of biofilms include "protection from predators or other external dangers" such as antibiotics and high temperatures, "access to resources" ("staying in favorable environments without being swept away"), and genetic diversity" (Johnson, 2008, p.24). However, like human network structures, microbial biofilms have both advantages and disadvantages for the individuals (nodes) involved. For example, many biofilms tend to slow down the reproduction rate of their cells[35]: the "doubling rates of individuals in a community are generally lower than those living outside"; therefore, living in a network structure "often represents a trade-off between reproduction and survival" (Ibid.).

2.1.2. Multilevel Structures

Multilevel structures are typical of many cell networks including microbial colonies/biofilms that are characterized by a high degree of clustering. They consist of compact subnetworks (clusters, microcolonies, rafts, modules) that contain several dozens or hundreds of cells. These are the building blocks of which the whole colony is made (Ben-Jacob et al., 2004). Cells inside them behave in unison. In colonies of motile bacteria, cells belonging to one subnetwork (a raft; Shapiro, 1995) tend to synchronously move over the agar surface.

In many colonies and biofilms, local dense subnetworks are linked by relatively few cells that establish contacts with other cells using surface cell structures such as fimbria and pili. I observed extraordinarily long cells that span the gaps between two or more subnetworks in an

[35] In a somewhat similar manner, students that form network groups in a classroom (during an interactive class) tend to solve problems in a relatively slow tempo, even though their problem-solving strategy tends to become more creative (see 3.6.1 below).

E. coli colony (Oleskin et al., 1998). In organizational terms, networks characterized by dense clusters and a small number of cells that connect subnetworks (microcolonies) cannot be considered completely flat. The few cells located between dense subnetworks are characterized by high *betweenness centrality* values, and this is one of the several criteria used to estimate network hierarchization (see 1.2.4.1).

Multilevel structures are also characteristic of higher organization levels: whole colonies are the building blocks of multicolonial communities and associations; these associations may include cells of different species (Ben-Jacob et al., 2004, p.369). Such associations are exemplified by the microbial biofilms of the gingival pocket in the mouth cavity or of the large intestine; these biofilms include several hundred or even thousand bacterial species. Each of the species forms its own colonies, which merge into multispecies biofilms.

2.1.3. Coordination Mechanisms in Cell Networks

Some microbial network structures display highly coordinate behavior. A colony of *Myxococcus xanthus* can move as one unit (as a "rank of soldiers") over the agar surface, encircling food particles and collectively digesting them using extracellular enzymes (Shapiro, 1988). Decentralized coordination mechanisms enable networks of filamentous cyanobacteria to form complex patterns including strands, rings, multiradiate structures, branched filaments, etc. (Sumina, 2006). Behavior coordination in cell networks is based upon all the three mechanisms, considered in general terms in 1.6, involving local interactions between nodes, distant regulatory agents, and the matrix, respectively.

2.1.4. Local Cell–Cell Interaction (Short-Range Order)

This coordination mechanism predominantly operates at the subnetwork level, i.e., within the boundaries of microcolonies (cell clusters) that make up a higher-order network such as a biofilm or (macro-)colony. Subnetworks carry out processes that are regulated by nondiffusing chemical signals. The ability of the signals to bind to the receptors of other cells requires direct intercellular contact. In a starving *Myxococcus xanthus* population, cells aggregate and form fruiting bodies that contain spores. The late stages of this process include tight cell packing for the purpose of spore formation. These stages are controlled by the nondiffusing cell surface-attached factor C encoded by the *csgA* gene (Brandner & Kroos, 1998; Søgaard-Andersen, 2004).

"Contact-dependent signaling is probably a common strategy by which bacteria in close contact, such as within biofilms, can modulate the growth and behavior of both siblings and competitors" (Blango & Mulvey, 2009, p.177). For instance, in bacterial biofilms, behavior coordination is facilitated by *contact-dependent inhibition*, which prevents unwanted cell growth, division, and movement because such processes can disrupt the biofilm's complex architecture. The surface-attached protein, CdiA, of an *E. coli* cell interacts with the BamB and AcrB proteins of other cells and inactivates these cells by downregulating their metabolism (Aoki et al., 2009; Otto, 2011).

Similar contact inhibition is characteristic of eukaryotic cells in the tissues of an animal organism.

However, there are mechanisms that counterbalance contact inhibition and cause mutual cell attraction. They enable collective cell migration during an organism's development, as well as the spread of a malignant tumor (cancer metastasis). This process appears to implicate specific signal factors such as the complement fragment, C3a, which binds to the receptor protein C3aR in the neural crest of vertebrates (e.g., of the African clawed frog *Xenopus* and the zebrafish; Carmona-Fontaine et al., 2011).

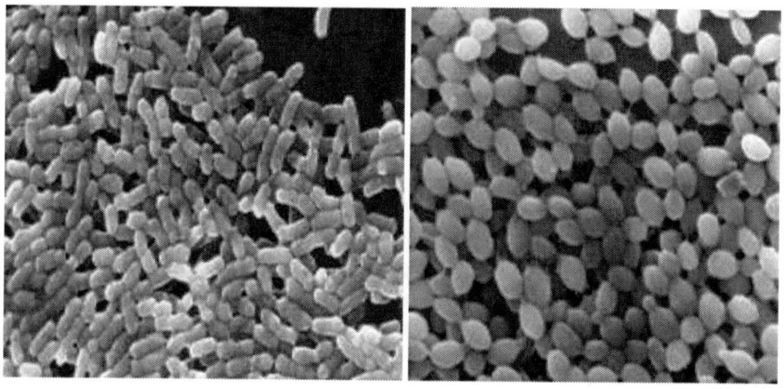

A B

Figure 2.1. Microorganisms establish contacts in order to form complex network structures, as exemplified by microcolonies of (A) *Shigella flexneri* (which causes dysentery) and (B) *Escherichia coli*, a normal inhabitant of the gastrointestinal tract. The pictures are a gift from Dr. Oxana Rybalchenko.

There is much evidence that cell–cell communication and behavior coordination involve cell surface structures, including flagella, fimbria, cell wall evaginates, and glycocalyx (reviewed, Oleskin & Kirovskaya, 2006; Oleskin et al., 2000, 2010). Surface structures are also implicated in forming cell–cell contact sites such as: (1) cytoplasmic bridges (plasmodesmata) hypothetically denoted as intercellular "wave conductors" (Vysotsky et al., 1991); and (2) points at which outer membranes (in gram-negative bacteria) or cell wall peptidoglucan layers (in gram-positive bacteria) of two different cells merge (Tetz et al., 1990). Contacts between bacterial cells are characteristic, for example, of cell networks formed by the bacteria *Shigella flexneri* and *E. coli* (Figure 2.1). It has recently been demonstrated that bacterial cells are connected by nanotubes that transfer molecules between them. Such nanotubes form between the cells of the same species (*Bacillus subtilis)* and those belonging to different species, e.g., *B. subtilis* and *E. coli* (Dubey & Ben-Yehuda, 2011). In a similar fashion, networks of intercellular membrane nanotubes connect mammalian cells. All of these contact types are apparently involved in the synchronization of processes in neighboring cells, and they may facilitate the relay transfer of information to more distant cells (cf. the relay transfer of information in fish shoals, see 2.3).

The behavior of many cellular networks can be approximated in terms of either random or scale-free network structures (see 1.1.3.2). Both kinds of networks have small-world properties; local interactions with neighbors, such as the transfer of signal molecules via intercellular nanotubes, may result in rapid information transmission within the whole network because even distant nodes are separated by a disproportionately small number of links. Even though the small-world phenomenon enhances the global impact of short-range communication, this does not diminish the importance of long-range communication.

2.1.5. Distant Regulatory Chemicals (Long-Range Order)

Distant regulatory chemicals (long-range order) enable global behavior coordination and pattern formation within a cell network (colony, biofilm, etc.). An important example concerning *E. coli* colony patterns was discussed above (see 1.6.3).

Sufficiently well understood systems of chemical regulators allow microbial populations to estimate their own density (quorum). These are *quorum-sensing (QS) systems* (reviewed, e.g., Fuqua et al., 1994; Waters, Bassler, 2005; Long et al., 2009; Shpakov, 2009; Oleskin et al., 2011). QS systems involve chemical signals (*pheromones,* or *autoinducers*) that are released by each cell in a population and act on other cells. If the pheromone concentration and, therefore, the cell density in a bacterial population is sufficiently high, the bacterial cells carry out synchronized processes. For example, the marine bacteria *Vibrio fischeri* and *V. harveyi* start synchronously emitting light quanta. Quorum-dependent processes also include the biosynthesis and secretion of antibiotics and protein factors, e.g., enzyme complexes and virulence factors; cell–cell transfer of genetic information (transformation, plasmid transport, and conjugation); and collective activities, including cell aggregation, biofilm formation, and sporulation. Pheromones bind to receptor proteins (R proteins) and, as a result of a series of biomolecular interactions, the pheromone–receptor complex activates the transcription of genes that are responsible for the processes listed above. The complex also activates the transcription of genes encoding the enzymes that catalyze the pheromones' synthesis, so that many quorum-sensing systems are autocatalytic, i.e., they have a positive feedback loop.

Quorum-sensing systems exist not only in bacteria; they have been revealed in the yeast *Saccharomyces cerevisiae* (Chen & Fink, 2006). If the concentrations of two pheromones (tryptophol and phenylethanol) produced by all yeast cells reach threshold levels, the expression of the *FLO11* gene is activated. Its product, a surface protein, stimulates cell cohesion and inhibits the separation of the parental and the daughter cell upon budding. In this fashion, it causes a morphogenetic event: the yeast starts forming branched filaments (the pseudomycelium). Interestingly, this QS system only operates under nitrogen limitation. High nitrogen concentrations repress the genes whose products are involved in the biosynthesis of the pheromones. Hence, morphogenesis is controlled both by the yeast population density and by the nitrogen level, i.e., the stress caused by the nitrogen deficit.

Animal cells release retinoic acid and respond to it, i.e., retinoic acid is an analog of microbial QS pheromones. It represents an important regulator of cell proliferation, morphogenesis, and cell differentiation during the individual development of animals (Mangelsdorf et al., 1995).

At this point, I reemphasize the difference between *controlling* and *communication* (cf.1.5.1), implying the *stimulus* → *preprogrammed response* and the *stimulus* → *receiving individual* → *variable (chosen) response* pattern, respectively. This distinction applies to unicellular systems, even though many of them only exercise *control*. For instance, purely mechanistic interactions based upon the "the key fits the lock" principle are characteristic of enzymatic processes. In bacterial colonies/biofilms, mechanistic interactions are also typical of processes involving *autoregulatory* substances that are exemplified by D_1 factors (resorcinols). D_l factors convert microbial cells into dormant forms, e.g., into cyst-like cells (Duda et al., 1982; Plakunov & El'-Registan, 2004; Mulyukin et al., 2013).

In contrast, at least some quorum-sensing systems seem to involve communication-like processes because the responses of microbial cells to QS pheromones are stochastic, probabilistic, and not mechanistically preprogrammed. They vary depending on the functional state of each cell and the entire microbial network (colony, biofilm).

Overall, quorum sensing is a "democratic" decision-making mechanism that is widely spread in the microbial realm. It enables the coordination of the activities of the nodes of microbial networks on the basis of a peculiar "voting procedure" (Markov, 2009, http://elementy.ru/news/43): all cells in the population release a pheromone and, once its concentration reaches the threshold level, the cells drastically change their behaviors.

Apart from the microbial world, analogs of quorum-sensing systems apparently function in a wide variety of biological network structures including, e.g., fish shoals or schools (see 2.3). They enable the choice of collective behavioral strategies in the absence of a centralized hierarchy.

It was hypothesized that cancer cells use analogs of quorum-sensing systems while forming metastases (Hickson et al., 2009). In an analogy to microbial colonies or biofilms, cancer cells form complex communities (primary tumors) that are characterized by cell differentiation and distribution of tasks and they "utilize intricate modes of communication, both among themselves and with the surrounding stromal tissue, distant organs, and the immune system" (Ben-Jacob et al., 2012, p.405).

2.1.6. Distant Physical Communication Factors

Information between cells can be transmitted by electromagnetic and sound waves. In the 1920s–1930s, Alexander G. Gurwitsch and his associates studied "mitogenetic radiation" (ultraviolet light), where the quanta were emitted by living cells, and which stimulated the division of other cells including bacteria (Sewertzowa, 1929). Yuri A. Nikolaev (1992) presented evidence that a bacterial culture (*Vibrio costicola*) treated with a lethal dose of the antibiotic chloramphenicol emits a signal that stimulates the growth of another culture that is separated from it by a quartz glass layer. These data were confirmed by Matsuhashi et al., (1996a, b) who suggested the involvement of ultrasonic waves in transmitting messages from culture to culture. Chemical communication factors can potentiate the communicative effects of physical signals and vice versa. This apparently follows from the data on the influence of one bacterial culture (*Pseudomonas fluorescens*) on the adhesive properties of the cells of another culture of the same species (Nikolaev & Prosser, 2000).

Hypothetically, physical communication factors could be involved in the functioning of other network structures, including those in human society. Mood alterations, panic, and anger spread in human crowds so rapidly that this possibility seems quite feasible.

2.1.7. The Matrix

In the microbial and cellular world, the term "matrix" refers to a material object, the biopolymer substances that cement microbial cells in a colony/biofilm, as well as animal or plant cells within a tissue in a multicellular organism. In bacteria, the matrix results from the merging of the external layers of cell envelopes, including the cell capsule. The structure of the matrix can be fibrillar, lamellar, or tubular. It contains extracellular polysaccharides including β-1,6-N-acetyl-D-glucosamine (PNAG), cellulose, colanic acid (e.g., in *E. coli*), alginate (in *Pseudomonas aeroginosa*), glycoproteins, peptides such as polyglutamic acid (e.g., in bacilli), and extracellular DNA molecules (Botvinko, 1985; Sutherland, 2001; Jurcizek & Bakaletz, 2007; Smirnova et al., 2010). The matrix represents a functional organ of a microbial colony/biofilm. It performs a number of important functions in microbial network structures:

- *Structural function.* The matrix coats the entire microbial network structure, represents both its external and internal skeleton, and separates it into a large number of small compartments where dense subnetworks (cell clusters, microcolonies) are located. This promotes local interactions between the cells inside each of the compartments.
- *Integrating function.* Since all compartments are interconnected within the matxix, all subnetworks are integrated into one coherent entity, i.e., the colony or the biofilm.
- *Adhesive function.* The matrix is involved in the attachment of bacterial cells to a surface, which is an important stage of biofilm formation. The polysaccharide-(glucan-)containing matrix of oral streptococci promotes their adherence to teeth and gums (Botvinko, 1985).
- *Adsorbing and concentrating functions:* matrix exopolysaccharides adsorb metals, mineral salts, and dissolved organic substances; the matrix concentrates nutrients, enzymes, and growth factors (Smirnova et al., 2010).
- *Protective function.* The matrix protects colony/biofilm cells from dehydration, heating, hydrolytic enzymes, antibiotics, and other detrimental factors; the matrix formed by pathogenic bacteria masks their antigens targeted by the host's immune system (reviewed, Oleskin, 2009; Oleskin et al., 2011).
- *Communicative function.* Signal chemicals, including pheromones used by quorum-sensing systems, are excreted by cells into the matrix; hydrophilic matrix components promote their spread within a microbial network structure.
- *Synchronizing function.* By helping signals spread inside a colony/biofilm, the matrix also facilitates the synchronization of the processes that are carried out by partly autonomous subnetworks (cell clusters, microcolonies).

Apart from the microbial world, some of these functions are performed by the matrices of diverse network structures in biological systems, as well as the largely immaterial matrix of human social networks. Regardless of the kind of network involved, the matrix typically plays structural, protective, and communicative roles.

2.1.8. The Network Structures of Unicellular Organisms as Coherent Biosocial Systems with Cognitive Capacities

It follows from the above that microorganisms form complex matrix-coated cell network structures that display coordinate collective behaviors exemplified by concerted cell movement, swarming, and fruiting body formation. Biofilms, like animal or plant individuals, are characterized by life cycles (ontogenies). Many biofilms include functionally differentiated cell subtypes, e.g., actively growing aerobic cells, fermenters, and dormant persisters that increase the biofilm's viability during stress (Rani et al., 2007)[36]. Some bacterial colonies on the agar surface seem to include cells of at least two types, which are characterized by quick and slow speeds, respectively. This apparently accounts for the fact that a colony's velocity decreases with increasing nutrient concentrations, such as glucose (which causes the *quick type → slow type* conversion; Sekowska et al., 2009).

Interestingly, the complex behavior of a large number of decentralized network systems can be adequately described in terms of mathematical models and algorithms (cf. subsections on fish and social insects below). These models can be subdivided into three basic subtypes: (1) Eulerian models that focus on the fluxes of individuals in dynamic networks; such models have been applied to slime mold aggregation, bacterial chemotaxis, midge swarming, and group formation in animals (reviewed, Johnson, 2008); (2) cellular automata that consist of boxes; the boxes' functional state can be influenced by their own history and the state of neighboring boxes; and (3) the recently developed (Johnson, 2008) Individual Based Models (IBMs) that demonstrate how individual behaviors produce large-scale patterns in biosocial systems.

Microbial network structures make good use of quorum-sensing systems and other communication facilities that enable them to "collectively glean information from the environment and process it… learn from experience, solve problems, and engage in group wide decision making" (Ben-Jacob et al., 2012, p.405). In bacilli, a collective decision is to be made whether or not to allow cells to exchange genetic information by taking up DNA molecules released by other cells. The chief regulatory role is performed via the "two peptides COMX and CSF (competence stimulating factor), which control the transition to DNA uptake competence in *Bacillus*... The same system is involved in the alternative decision to sporulate. Apparently, cells interpret these factors in conjunction with internal state information so as to decide upon their fate… This decision must then be relayed to other cells so that they can act accordingly" (Ben-Jacob et al., 2004, p.366).

> "Learning from experience" is illustrated by the following findings. The cells of *Paenibacillus vortex* were treated with two antibiotics (septrin and ampicillin); the bacteria responded to the addition of these antibiotics by activating their protective mechanisms: they formed enlarged vortex-like cell aggregations (which presumably were coated by a larger matrix layer and, therefore, were more resistant to the antibiotics) and sped up their movement away from the antibiotic-containing zone[37] (Ben-Jacob et al.,

[36] Such dormant cells are often characterized by a thickened cell wall and an enhanced resistance to harmful external factors (Shleeva et al., 2010, p.3).

[37] In the experiment described by Ben-Jacob et al. (2004), escaping from the antibiotic was impossible because it was contained in the whole agar layer on the Petri plate. However, the authors assume the bacteria "did not

2004, Figure 2). It was revealed that, if bacteria are repeatedly treated with the same antibiotic, they respond "more efficiently; however, this effect is erased if they are exposed to neutral conditions… in between stress encounters. It appears that the bacteria can *generate an erasable collective memory, as if to learn from their experience*" (Ibid., p.369, emphasis added – O.A.).

The formation of the fruiting bodies of the myxobacterium *M. xanthus* is regulated by factor A apart from factor C, which was mentioned above. Factor A is responsible for the quorum-dependent initiation of cell aggregation resulting in fruiting body formation. Factor A is a mixture of amino acids produced by the degradation of cell surface proteins by extracellular proteases (Kaplan & Plamann, 1996). The factor binds to receptor SasS, which activates transcription regulator SasR. If factor A concentration exceeds the threshold level, it drastically stimulates the expression of genes implicated in cell aggregation and fruiting body formation (Kaiser & Dworkin, 2008). The factor A concentration is increased both by (1) a lack of nutrients in the medium and (2) an increase in bacterial cell density. Hence, this collective "decision-making" system combines messages obtained from several different information inputs and generates an integrated response to them.

It seems likely that malignant tumor cells are also capable of "collective decision-making" (Ben-Jacob et al., 2012); these can include the decision of metastatic cancer cells to accelerate their growth. This decision can be encouraged by chemical messages that are sent by the primary "mother" tumor.

2.1.9. The Organismic Approach to Cell Network Structures

One controversial issue is whether biosocial systems of unicellular organisms represent multicellular organisms (Shapiro, 1988, 1995, 1998; Ben-Jacob et al., 2004, 2012) or groups of independent individuals, otherwise known as "cities of microbes" (Watnick, Kolter, 2000; Nikolaev & Plakunov, 2007; Karatan & Watnick, 2009).

Starting from the publication of James Shapiro's article (1988), "Bacteria as Multicellular Organisms," the "concept of bacterial multicellularity has evolved from being considered an adaptive strategy of a few bacterial taxa, for example, the *Myxobacteria* and the *Actinomycetes*, to today's pervading view that nearly all bacteria are capable of multicellular behaviors" (Aguilar et al., 2007, p.638).

Microbial cell networks possess some organismic features. Drawing upon the theoretical work of Herbert Spencer, the Russian zoologist Vladimir Beklemishev (1950) singled out these features (cf. subsection 1.5.2); his ideas are reformulated below in order to apply them to cell networks:

- *The formation of organs that belong to the whole collective system.* The matrix is probably one of the most important organs of cell networks. There are also more specialized colony/biofilm-level organs, e.g., a primitive circulatory system including water-filled cavities and membrane sacs, which contain a hemoglobin-like protein

know" that there was an antibiotic not only behind, but also ahead of them; therefore, the bacteria "thought" that they could avoid it by moving faster over the agar surface.

and are presumably involved in oxygen transport (Duda et al., 1995); membrane vesicles that are filled with enzymes that destroy a part of bacterial cells and enable other cells to use their nutrient substances and growth factors ("altruistic cell death;" Beveridge, 1999); large membrane structures (Greiner et al., 2005) that traverse a whole colony and could perform protective and communicative functions; and aerial projections where sporulating cells concentrate in *Bacillus subtlis* colonies (Aguilar et al., 2007). The formation of these collectively used organs is related to another phenomenon noted by Beklemishev: functions that were originally performed on the individual level are delegated to the colony level. In particular, the protective properties of the matrix and those of the more specialized organs of the whole colony/biofilm may secure the survival of individual cells under stress; these cells do not need to develop individual protective organelles.

- *A general developmental program concerned with the whole system.* As mentioned above, a colony or biofilm has its life cycle, although it is less precisely preprogrammed and more variable, depending on environmental factors, which is in contrast to the ontogeny of a multicellular animal (Yerusalimsky, 1952). Importantly, microorganisms contain gene complexes that enable the cells of one colony to move concertedly and, in many microbial systems, synchronously. These gene complexes are typified by the *S* genes (*S*, as in "social"), including the *bsg* gene in myxobateria (Kroos et al., 1990). These are necessary for swarming and, subsequently, for fruiting body formation. In conformity with the unitary rhythm of the whole colony, some bacteria (*Proteus mirabilis, Serratia marcescens, E. coli,* and others) synchronously convert from vegetative cells to highly motile swarmers with a large number of flagella (reviewed, Kaiser, 2007); swarmers revert to vegetative cells after a colony-specific period of time (1–3 hours with *P. mirabilis*; Harshey, 1994). Each interconversion cycle results in forming an additional concentric circle on the colony surface (Rauprich et al., 1996).
- *The capacity for reproduction and regeneration* manifests itself in a large number of unicellular systems. Some bacteria form colonies that bud off new, smaller colonies. On the periphery of the original colony, the matrix forms tubes, and some of the cells migrate through them to a vacant part of the nutrient medium. The cells form microcolonies that develop into full-scale daughter colonies. This process is characteristic of bacteria that form part of the microbiota of humans and animals (Pavlova et al., 1990). The capacity to regenerate after an injury (Figure 2.2) was demonstrated in studies with cyanobacterial biofilms (Sumina, 2006).
- *The "shrinking" and the structural simplification* (*devolution*; see Corning, 2003b) *of the elements of the system,* i.e., of the cells making up a network structure. For instance, matrix-cemented bacterial structures include abnormal L forms, i.e., cells that are partly or completely devoid of cell walls, whose protective role is delegated to the matrix, as Vysotsky et al. (1991) emphasized. Many biofilms contain morphologically altered bacterial cells that are exemplified by the unusually small cells of *Bacillus cepacia*, which occur in its biofilm (Smirnova et al., 2010).

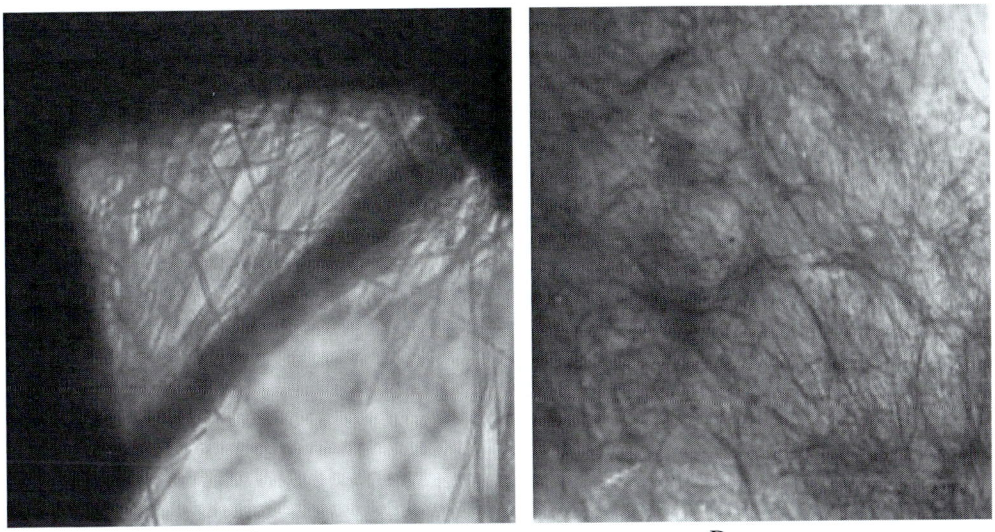

A B

Figure 2.2. The biofilm of the cyanobacterium *Oscillatoria terebriformis* regenerates after rupturing it with a needle (A) and spreading $CaCO_3$ powder over its surface (B). *O. terebriformis* filaments grow across the "wound" on the biofilm surface in (A), and they gradually overgrow the $CaCO_2$ layer in (B). The pictures are a gift from Dr. Eugenia L. Sumina.

- As mentioned in Chapter one (1.5.2), an organism can be unitary or modular. In these terms, multicellular microbial systems are to be considered *modular* organisms composed of a large number of uniform parts, which provides a link to section 2.2 that also deals with modular organisms.

The opponents of the organismic approach to the structures that are formed by unicellular organisms point out that the structures' developmental program (unlike that of a true multicellular organism) can easily change under the influence of environmental factors. Besides, cell differentiation in unicellular systems is reversible and less manifest than in plant or animal organisms. However, some *modular* plant and animal organisms are also characterized by a relatively insignificant functional differentiation of their components (see 2.2 below).

2.1.10. Intracellular Network Structures

Research on intracellular networks is not a very recent development in science. "The notion that genes, proteins, and other biomolecules operate in networks is far from a new idea. It can be argued that the study of network biology began several decades ago with the landmark work of François Jacob and Jacques Monod on the *lac* operon in *Escherichia coli*" (Dwyer et al., 2008, p.1153).

Among various types of intracellular networks, "metabolic networks and gene regulatory networks aim to summarize the basic biochemistry and the set of regulatory interactions of biological organisms, respectively" (Stumpf et al., 2007, p.367).

Living cells contain semiautonomous "biological functional units" (Lengeler, 2004), also termed "morpho-functional units", whose interactions are aimed at performing or regulating a specific function (Chernyshova, 2009, p.344). In bacterial cells, such units include chemotactic systems and the flagellar apparatus. Analogous intracellular units, e.g., the Ras pathway system or cytoskeleton components, are present in eukaryotic cells. Such systems cooperate according to the decentralized network's organizational pattern, as is the case with cytoskeleton microtubules that form various intracellular structures (reviewed, Isaeva, 2012).

Network relationships among gene systems in a living cell are consistent with a relatively new paradigm in molecular genetics. Contrary to the "central dogma of molecular biology" of the 1960s, information need not flow along the traditional *DNA → RNA → protein* pathway. With respect to the genetic machinery of a cell, the network scenario implies the parallel functioning, cooperation, and competition of a large number of DNA sites, RNA molecules, and protein factors. "Genetic regulatory networks include numerous genes coding for transcription factors, ligands, and receptors involved in cell-cell signaling[38], as well as nucleotide sequences controlling the expression of each of these genes" (Isaeva, 2012, p.150).

Of paramount importance in this list are regulatory networks involving transcription factors. "… Transcription factors are themselves produced by trasnscription[39] from genes. Thus the protein encoded in a given gene can act as a transcription factor promoting or inhibiting production of one or more other proteins, which themselves can act as transcription factors for further proteins and so forth. The complete set of such interactions forms a *genetic regulatory network*" (Newman, 2012, p.92).

Intracellular networks, including those consisting of metabolic chains, often exhibit the features of *scale-free networks* (Guimerà & Nunes Amaral, 2006); such intracellular networks seem to obey the power law of node degree distribution and to contain hubs, i.e., nodes with a disproportionately large number of links with their neighbors (see 1.1.2). In eukaryotic cells, such components include key protein kinases and their complexes, including the MAPK cascade, as well as transcription factors (e.g., CREB). Such interunit hubs are analogs of partial network leaders in the animal kingdom and in human society. In some of the networks, one of these influential nodes can claim the role of the central leader, and this results in an at least temporary hierarchization of the network structure. For instance, bacterial cells contain master regulons, and the main stress response–controlling *rpoS* regulon (in *Escherichia coli* and other proteobacteria) can assume dominance under stress. Such intermediates are characterized by a high degree of betweenness centrality (see 1.2.4.1 above), and the networks containing them display a clear-cut *community structure*, characterized by subnetworks with a high degree of clustering connected by relatively few links to one another.

Of interest are protein interaction networks that "consist of all protein-protein interactions in an organism" (Stumpf et al., 2007), which enable us to elucidate the evolutionary relationships among various forms of life. Protein interaction networks in yeast (*Saccharomyces cerevisiae*) contain a few highly connected nodes that hold the networks together; most mutations that result in a lack of these hubs are lethal (Jeong et al., 2001).

---

[38] The factors involved in cell–cell signaling provide a link between networks consisting of intercellular components and those involving whole cells as nodes.

[39] *Transcription* is the process of synthesizing an RNA copy on the basis of a specific DNA gene(s) as the template; the RNA thereupon is used as a template for synthesizing a protein; the latter process is referred to as *translation*.

Protein interaction networks in yeast, a worm (*Caenorhabditis elegans*), and a fly (*Drosophila melanogaster*) display surprisingly similar structures, and their most important hubs are highly evolutionarily conserved, probably because of their vitally important functions (Hahn & Kern, 2005). The work cited demonstrates how numerous the nodes and links of these molecular networks can be; the researchers analyzed 20,252 interactions among 2,434 yeast proteins, 5,977 interactions among 1,977 worm proteins, and 16,002 interactions among 5,082 fly proteins.

The nodes of these intracellular scale-free networks can be classified into a number of subtypes (which is of much interest in terms of network behavior in general) including, among others, (1) hubs of modules that have many connections within their community or "module" and are actually responsible for the module's coherence; and (2) nonhub connectors that are characterized by many links to other models and thus provide for the structural integrity of the whole modular network (Guimerà & Nunes Amaral, 2006).

In conformity with Putnam's (2000) views, these hubs and "nonhub connectors" perform the bonding and the bridging function respectively (see 1.7.2 above). Connectors are highly conserved because of their paramount importance in biological systems. For example, a large number of metabolic pathways share some common intermediates situated at their crossroads (e.g., acetyl-CoA).

Biomolecular networks in living cells exhibit small-world (or even ultrasmall-world)[40] features, i.e., the paths between their nodes are disproportionately short, compared to their numbers of nodes. For instance, within the living cell, "paths of only three to four reactions can link most pairs of metabolites" (Barabási & Oltvai, 2004, p.106).

Intracellular units, e.g., quorum-sensing systems in bacterial cells, do not only cooperate, but they also compete with one another, so that quasi-market-type structures can, at least temporarily, form inside a cell.

It should also be noted that "not all networks within the cell are scale-free. For example, the transcription regulatory networks of *S. cerevisiae* and *Escherichia coli* offer an interesting example of mixed scale-free and exponential characteristics" (Barabási & Oltvai, 2004, p.106); they include parts that resemble random networks. In biological evolution, networks behaving similarly to random network structures can be produced by the duplication of existing nodes, with new copies inheriting part of the links of the original nodes (Stumpf et al., 2007).

Studies on protein interaction networks and other kinds of intracellular network structures currently represent a rapidly developing area of research. However, two critical remarks that were made in 2007 concerning this area still appear to be valid in 2014: "(i) The evolution of proteins is known to be influenced by many factors; studying the effects of network structure on the rate of protein evolution without accounting for these potentially confounding factors could be misleading; (ii) Present protein–protein interaction data are noisy and incomplete; failure to account for this might also introduce bias" because currently available "subnets generally have different properties" from the whole networks whose clusters/modules they represent (Stumpf et al., 2007, p.369).

In sum, most biosocial structures made up of cells represent almost completely flat networks. Hierarchization can be considered a temporary or exceptional phenomenon in cell

---

[40] To reiterate: in typical scale-free networks, the path length between their nodes increases as log log n, where n is the node number (Barabási & Oltvai, 2004).

network structures. In some of these structures, cells and their groups (clusters, microcolonies, rafts) can be functionally differentiated. While behavioral coordination inside such subnetworks implicates local interactions between their nodes, the dominant role in regulating the activities of the entire network (colony or biofilm) is played by chemical communication involving diffusing signal molecules that are exemplified by the pheromones of quorum-sensing systems. Behavioral coordination in cell structures is facilitated by the matrix that performs the structural, protective, and communication-promoting function. Cell networks apparently are capable of collective information processing and "decision-making" that are facilitated by quorum-sensing systems and other communication mechanisms. Networks composed of intercellular biomolecules or their complexes may be similar to either scale-free or random network structures; a large number of them contain dense clusters (cliques).

2.2. Modular Paradigm in the Example of Cnidarian Colonies

This section deals with modular systems that display more prominent organismic features than microbial systems. They are typified by colonial cnidarians, e.g., colonial hydroids, siphonophores, and corals, although similar properties are characteristic of a wide variety of other systems, including invertebrate animals (bryozoans and ascidians), plants, mycelial fungi, and other biological systems consisting of *modules*, i.e., interconnected structural units forming regular spatial patterns (Marfenin, 1999).

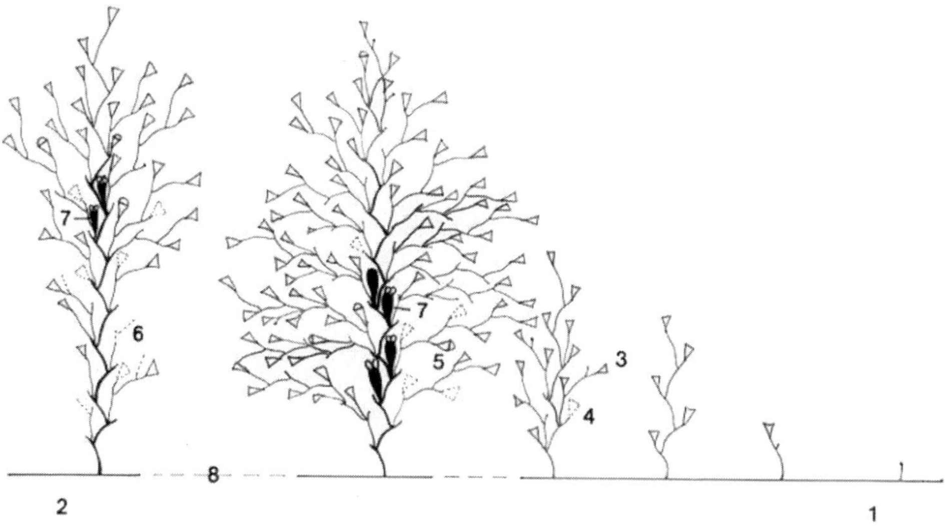

Figure 2.3. The structure of the modular body of the colonial hydroid *Obelia*. Triangles, individual polyps (zooids); they are interconnected via branches that are attached to stems (part of the colony's coenosarc). The numbers on the branches indicate the order in which these polyps grow on the stem. The image is a gift from Prof. Nikolai N. Marfenin.

Cnidarians provide an illustrative example of a multilevel coherent structure composed of structural units (*zooids*; see scheme in Figure 2.3). In colonial cnidarians, each individual zooid that forms part of the whole colony (cormus) resembles an independent solitary polyp (like a hydra) or a medusa. It has tentacles, an oral opening, and a decentralized network of nervous cells spread within its body. However, new zooids are generated via "budding or fission that is not followed by physical separation. Thus, offspring zooids remain attached and physiologically integrated" (Dunn, 2007, p.R233).

The whole group of zooids is attached to a stem or to its branches. Food particles taken in by all zooids are transferred to the gastrovascular cavity of the colony. The cavity is located inside the branched stem (the *coenosarc* coated by a cylindrical sheath called the *perisarc*) of the colony. The entire colony of cnidarians behaves as a coherent system.

2.2.1. Cnidarian Colonies as Modular Organisms

It is currently widely believed that a cnidarian colony is a modular organism like, e.g., a strawberry plant with its runners. Modular systems such as cnidarians display manifest organismic features; importantly, it was cnidarians that were considered by Beklemishev (1950) when he set forth the general notion of the *organism* in his work.

In particular, "*the formation of organs belonging to the whole collective system*" is illustrated by the fact that a polyp involved in swallowing and digesting food is actually an organ of the whole colony—a part of the colony-level digestive system (Marfenin, 1993). Apart from the polyps called gastrozoids that are involved in digesting food and maintaining the flow of the liquid that fills the coenosarc, the colony includes zooids that are responsible for reproduction and other specialized functions. Syphonophores, such as *Physalia*, contain many specialized zooids including gastrozoids, dactylozoids with stinging cells that perform a protective function, gonozoids that form sex cells, gas bubble-containing pneumatophores, and others.

A large number of colonial cnidarians are characterized by elaborate colony-level *developmental programs*, or life cycles "wherein multiple asexually produced zooids, each of which is homologous to a free-living solitary animal, remain attached and physiologically integrated throughout their lives" (Dunn & Wagner, 2006, p.743).

"*The capacity for reproduction*" on the colony level manifests itself in, e.g., corals whose whole colonies bud off "aggregations of small physiologically discreet daughter colonies" (McFadden, 1986, p.1).

As for the "*shrinking and structural simplification of the elements of the system*" also mentioned in 2.1 as a characteristic feature of an organism, it is exemplified by the fate of many gastrozoids in hydroid colonies that can completely eliminate to help the whole colony survive if nutrients are in short supply (Marfenin, 1993).

Modular organisms can be based upon different organizational principles. Pennatularians can form hierarchical colonies that contain a comparatively large leader polyp that guides the movements of the whole colony if it is motile.

However, a large number of other cnidarians are characterized by colonies whose zooids are approximately equal in size; such colonies represent decentralized flat networks. As already mentioned, zooids may be functionally differentiated.

Like many other network structures, decentralized networks formed by colonial cnidarians are multilevel systems. Each colony-level network is made up of smaller structural units (*cormidia*) that can exist independently and contain a minimum number of different specialized zooids; each zooid contains a decentralized nervous system, i.e., a network of nervous cells. In siphonophores, the nervous systems of their zooids merge into one colony-level network, which promotes behavior coordination within the colony.

Apart from decentralization, the network structure of a cnidarian colony is characterized by the cooperative behavior of its nodes. Cooperation among zooids enables the whole colony to optimize its shape, size, and zooid number, adapting them to the conditions under which the colony grows. This is possible because the consumption of nutrients is constantly regulated on the colony level (Leontovich & Marfenin, 1990). Studies in which *Gonothyraea loveni* (Allm.) polyps were given excessive or insufficient amounts of nutrients revealed that the autoregulation of the colony's behavior and development is due to simple mechanisms involving growth arrest and resumption as well as the development or, alternatively, destruction of some zooids, depending on the trade-off between the available nutrient concentration and growth intensity (Marfenin, 2009, p.319).

Of considerable importance in terms of the organismic approach to colonial cnidarians is the fact that individual zooids form groups with preprogrammed, species-specific composition and geometrical patterns; these patterns change during the course of a colony's ontogeny (Kosevich, 2006). In siphonophores, "each colony arises by a highly regulated budding process that arranges polyps and medusae in a precise, species-specific pattern" (Dunn et al., 2005, p.916).

It was emphasized above that cnidarian structures represent modular, not unitary, organisms. However, it is not in all colonies that modules represent uniform, repetitive units. In some hydroids such as *H. falcata,* modules located on the shoot stem and branches differ in their structures and functions, and "these shoot parts could be compared to the organs of a unitary organism" (Kosevich, 2006, p.257).

2.2.2. The Coenosarc as a Functional Analog of the Matrix

Like the matrix of a microbial colony or a biofilm, the coenosarc together with its cover (perisarc) performs structural and protective functions.

The coenosarc enables a colony to survive under unfavorable conditions, e.g., during the seasonal diapause, in an inactive form in which all zooids have been eliminated (Marfenin, 1993). In a similar fashion, microbial cells in a matrix-coated biofilm convert into spores or inactive L forms (microcells, nanocells, etc.), or they may even undergo programmed cell death under stress, including starvation.

The attachment of the coenosarc to a substratum is analogous to the adherence of the biofilm matrix to a surface.

Regulatory molecules can spread within the coenosarcs' liquid-filled cavity, in an analogy to their diffusion via the cavities and channels of the microbial matrix. For instance, the tunica (the perisarc-like outer skeleton) of ascidians enables them to transmit chemical signals that coordinate the zooids' behavior.

Importantly, the outer skeleton of hexacorals stores information about the form and the location of each corallite[41], so that new corallites copy the preexisting ones in terms of their shape and position on the stem.

Although most networks of cnidarians are flat in the sense that zooids do not form hierarchies, branching points on the coenosarc stalks to which zooids are attached can be considered as hubs, making the whole network partly scale-free (if branching points are regarded as nodes, in addition to zooids). Such branching points are characterized by a considerable degree of betweenness centrality: removing a branching point results in the disintegration of the corresponding part of the modular organism.

2.2.3. Decentralized Regulation Scenarios in Cnidarians

In contrast to typical cell networks, most cnidarian systems are predominantly characterized by local interactions between nodes, although distant chemical messages can also play a sufficiently important role. According to Nikolay N. Marfenin (2002, 2009), cooperation among network nodes in colonial cnidarians with decentralized organization is due to the following mechanisms:

1. *Local factors regulating zooid behavior and development.* An example is the local response of a group of zooids to changes in nutrient concentration: they may accelerate or decelerate their growth; new zooids can form, and some of the existing zooids can eliminate. Like cultured animal and microbial cells, zooids display "contact inhibition": their growth slows down if there is not enough room for this process (Marfenin, 1999, p.11).
2. *The coenosarc as a prerequisite for the integration of zooids in the colony.* By connecting individual network nodes (zooids) and forming the backbone of the system, the coenosarc with the perisarc actually exercises control over zooid behavior; therefore, behavior coordination in cnidarian colonies is predominantly based upon controlling actions rather than message communication. This coenosarc-dependent integration of zooids that have fixed positions within the whole system prevents them from competing for a specific territory. It should be reemphasized that it is the dominance of cooperation over competition that is typical of a network, which is in contrast to what is evident in a quasi-market structure.

The behavior of each zooid that, e.g., contracts and causes the hydroplasm (coenosarc-filling liquid) to flow, only weakly influences the whole system. However, its effect is potentiated if its rhythm coincides with that of the majority of other zooids (Marfenin, 2002, 2009), and this was discussed above in the context of oscillator synchronization caused by local interactions within a network (Kuramoto, 1984; Arenas et al., 2008, and see 1.6.2). Synchronized zooid behavior enables distant food particle transfer within the colony and helps zooids coordinate their activities.

In an analogy to cnidarian network structures, it seems likely that the preprogrammed individual development of an animal embryo is actually subject to regulation by local

[41] Corallites are the skeletons of the zooids that form a coral.

mechanisms controlling the division and movement of cells that only respond to factors associated with their microenvironment. To draw a biopolitical comparison, the slogan of a large number of political, environmental, business, or cultural networks in human society is the *act locally, think globally* maxim (the *glocalism* principle, see 3.8.8 and footnote 112).

The decentralized organization involving local processes and the controlling behavior of the coenosarc-like structure (the *matrix* in the sense suggested in this work; see 1.6.4) apparently underlie the activities of other modular systems of colonial invertebrate animals. A bryozoan colony represents a coherent system that is capable of responding to local changes in the environment. The polypoids that form part of a bryozoan colony can coordinately behave without a central leader. They bend towards one another once they locate a food particle of a suitable size; irritating one polypoid with, e.g., a needle, changes the neighbors' behavior. They stretch as if closely watching what is going on (Shupatova, 1999).

2.2.4. Principles of Decentralized Organization of Modular Systems

According to Marfenin (1993, 2002, 2009, 2011), decentralized network structures of modular organisms demonstrate a number of general organizational principles:

1. The presence of *a large number of equivalent components* in a system. Each functional part of this type of system includes numerous uniform components, e.g., cells in a multicellular organism or multicellular modules in modular networks; no single component has the power to dominate others.
2. A *weak influence of each component on the whole system*. In colonial cnidarians, e.g., hydroids, the contraction of an individual zooid per se does not regulate the flow of the hydroplasm inside all parts of the coenosarc, even though this flow results from the activities of all zooids.
3. An *increase in a component's influence caused by the concordance between the rhythm of its activity and the system's dominant rhythm* that results from the superposition of the activities of the components whose phase differences are sufficiently small for them to enable synchronization.
4. The *presence of nearly identical behavioral programs in each component*: the uniform components of the whole system function on the basis of similar built-in instructions.
5. *A feedback loop allowing the system to periodically readjust the rhythms of its components in conformity with the dominant rhythm of the system's activity.* This can be achieved by synchronizing the time of onset of identical processes carried out by different components of the system, e.g., the contractions of several growth apexes within the coenosarc of one colony.

Of general biological and social interest are the advantages of modular organisms related to their decentralized network organization:

- *Plasticity of the body shape* that is often due to the fact that the parts of a modular organism existing under favorable conditions grow faster than those where the growth conditions are less advantageous. In many soft corals, "the particular growth form assumed by a colony is often a function of the flow regime it inhabits" (McFadden, 1986, p.2). In a similar fashion, network structures in human society are more organizationally plastic than are rigid bureaucracies. The flexibility and plasticity of a network structure manifests itself in its capacity to perform several different social roles. A network structure can concomitantly or sequentially represent a scientific research team, a charity foundation, a business firm, an educational institution, a group of artists, and a political pressure group, depending on the aspect of the network's interdisciplinary mission that comes to the forefront in a given situation (Oleskin, 2007b).

- *Variability of body size*. Under favorable conditions, e.g., while abundant nutrient resources are available, a modular organism grows actively and, as its size increases, the growth tends to accelerate. Size changes do not result in significant organizational alterations in the network structure. An analogous phenomenon in network structures in human society is referred to as their *scalability* (Kahler, 2009a, b; see 1.7.2.2 above). Under unfavorable conditions, the size of the network structure can decrease; despite this size reduction, the structure retains its viability and is ready to expand once the conditions improve. For example, a nutrient-depleted colony sustains the growth of only one single stolon, i.e., one shoot with zooids, while the energy spent on other body parts is minimized (Marfenin, 1999, p.13). In human society, an analogous decrease in the size of network structures that retain their integrity and growth potential is referred to as *devolution* (Corning, 2003b).

These properties of modular biological systems exemplified by cnidarian colonies are analogous to the important features of many networks in human society that are referred to as *adaptability*.

In sum, the presence of a large number of uniform components (units) and the parallel processes carried out by them are prerequisites for the coordinate behavior at the level of the whole system (the modular organism), which is based upon (1) the weak influence of each unit on its operation, and (2) an increase in a unit's power in proportion to the concordance between its rhythm and the system-dominating rhythm. Most modular networks have no central leader, but they do contain a matrix-like integrating agent (in the form of the coenosarc with the perisarc in cnidarians), which connects the units and helps them synchronize their activities.

2.3. Equipotential Paradigm in the Example of Fish Shoals/Schools

This section mainly deals with completely flat biological networks typified by shoals or schools formed by fish.

Generally, fish are characterized by a wide variety of organizational structures ranging from loose to highly integrated systems. They include (1) "asocial aggregations that have boundaries defined by physical or biotic factors such as heat, light, or food" (Norris & Schilt, 1988, p.154); (2) "social aggregations in which members are attracted to one another" (Ibid.);

they are also called fish *shoals*; and (3) *schools* where all fish are uniformly oriented and their behavior is effectively coordinated (Pavlov & Kasumyan, 2003, p.6).

Asocial aggregations are characterized by the dominance of *quasi-market* relationships. Fish compete for food, shelter, and other resources; they also enter into trophic relationships with other species, i.e., feed on them or, alternatively, are used as food by them. Such aggregations form under the influence of environmental factors, e.g., fish concentrate in areas where food is plentiful. Typically, asocial aggregations have no internal structure, and there is no distribution of functions among individuals inside them (Pavlov & Kasumyan, 2002, 2003).

Social aggregations represent a more advanced organizational pattern. Its distinctive feature is that "individuals actively seek the proximity of each other instead of co-occurring in the same spot because of an attraction to the same environmental conditions or factors" (Krause & Ruxton, 2002, p.2).

The further evolution of organizational forms results in the formation of schools, which are characterized by uniform fish orientation and, typically, cooperation among their members. Their cooperation enables collective hunting and protection against predators, alloparental behavior (helping parents nurture their hatchlings), and food sharing (see below, 2.3.4), as well as effective intraschool communication.

The increase in organizational complexity can result in the hierarchization of fish shoals or schools. However, there are a large number of fish species that are characterized by advanced social organization but whose shoals/schools retain the horizontal network organizational pattern. This brings us back to the issue whether the interconversion of organizational structures is an inevitable process (1.7).

2.3.1. Hierarchical Aggregations

Shoaling fish include species that form rigid hierarchies. Hierarchical shoals that contain a dominant individual are characteristic of demersal fish species and fish in small water bodies (Radakov, 1972) including the spotted, lace, dwarf, and thick-lipped gurami. Stable hierarchies are established in small-sized shoals where fish stay together for a long time, e.g., in jack fish aggregations that maximally consist of 4 individuals (Pavlov & Kasumyan, 2003). Stable hierarchies also emerge in shoals composed of representatives of different species, which is a relatively rare phenomenon. In a shoal containing a barracuda and several comparatively small jacks, the barracuda dominates the jacks (Ibid.).[42]

2.3.2. Equipotential Shoals/Schools

Special emphasis in this section is placed upon the non-hierarchical organization that is preferred by most fish species. Such biosocial systems are completely flat (leaderless) networks called *equipotential shoals* or, if polarized, *equipotential schools*. Such *"fish*

[42]In primitive human society, one of the mechanisms of political hierarchy formation involved the establishment of a hierarchical structure made up of two originally non-hierarchical groups. For instance, cattle breeders were enslaved by armed hunters. This happened in South Africa in the 19th century.

shoals/schools can be defined as temporary groups of individuals that mostly belong to the same species and have a similar size and physiological status; they lack permanent leadership and intragroup dominance, and they are characterized by intense interrelationships between individuals that tend to aggregate and display coordinate uniform responses to external stimuli" (Pavlov & Kasumyan, 2003, p.143). Apart from a lack of permanent hierarchies, this definition emphasizes other important shoal/school features that will be discussed below. Shoaling/schooling behavior is typical of pelagic fish living in the open ocean, and also of fish characterized by spawning migrations and other long-distance movements.

In the absence of a permanent leader, a chance individual temporarily occupies the foremost position in a moving shoal/school. In schools of young pollacks, the time during which an individual "leads the way" varies from a fraction of a second (0.25-0.5 s) to several seconds… thereupon, "the fish is located in the middle or even in the rearmost part of the school" (Radakov, 1972, p.86).

Fish, in contrast to bacteria within a matrix-cemented colony or biofilm and polyps attached to the coenosarc of a cnidarian cormus, are not connected physically. Of relevance to networks in general—both in biological systems and in human society—is the issue concerning the reasons why independent individuals form more or less integrated networks that are based on cooperation among them. One approach invokes the concepts of kin altruism and reciprocal altruism (see 1.2.1).

The term *kin altruism* (Hamilton, 1964) refers to sacrificial behavior for the benefit of close relatives. An obvious problem with fish shoals/schools is that most of them include genetically unrelated fish; moreover, two originally autonomous fish schools can merge and behave as one coherent school. Admittedly (as pointed out below), fish shoals/schools, and the small clusters they often include, consist of similar individuals in terms of size, food preferences, and even individual features ("personalities"), but it is often an open question as to what extent such similarities are genetically fixed.

Reciprocal altruism (Trivers, 1972), where an unrelated individual may benefit if the beneficiary is ready to perform an analogous act of altruism that benefits the altruist, implies that there exists an advanced nervous system that is able to recognize individuals and memorize one's past interactions with them. Although the concept of reciprocal altruism may be applicable to groups of marine mammals such as cetaceans (and particularly dolphins) that also form shoals/schools, fish behavior appears to be too primitive to involve reciprocal altruism.

2.3.3. Synergies in Fish Network Structures

The formation of shoals or schools in fish and some other animals (echinoderms, arthropods, amphibians) can be considered in terms of Peter Corning's *synergy* model, assuming that all partners benefit from the interaction (Corning, 1983, 2003a, b, 2005). The following kinds of synergies apply to shoaling/schooling fish:

- *Synergies of Scale*. A larger fish shoal[43] is typically more successful in protecting its members from predators by confusing them and quickly signaling danger to the members.
- *Cost and/or Risk Sharing*. This concerns a fish shoal attacked by a predator, in contrast to a solitary fish.
- *Information Sharing and Collective Intelligence*. A fish shoal uses a number of efficient communication channels; the information acquired by individuals in it forms part of the shoal-level information pool.
- *Labor (Function) Division, or their Combination*. This occurs less frequently in generally homogeneous fish network structures. However, predator detection can be facilitated by the functions performed by scout fish individuals (Pavlov & Kasumyan, 2003); besides, labor division may take place on the supraindividual level. For instance, a part of a shoal can perform a specialized function, such as forming a large gap in the middle of the shoal during a fountain maneuver (see below) aimed at confusing a predator.

Different variants of network structures are predominantly characterized by different kinds of synergies; many species alter their structural organization in response to changes in the environment. In particular, "because in daytime the polarized school is a safer place to be and because the aggregation (where the orientation of fish is not polarized – *O.A.*) allows more freedom of movement for such activities as food finding, groups in open space oscillate between these extremes during varying levels of predation" (Norris & Shilt, 1988, pp.149–150). Young *Girella punktata* individuals form equipotential shoals at high tide; at low sea levels, these individuals stay in separate pools of water where they form dominance hierarchies (Mory, 1956, quoted according to: Radakov, 1972). Generally, a homogeneous or unstable environment, a lack of food, and a high predation pressure promote shoal formation. Individual territories are characteristic of a heterogeneous but stable environment (Mikheev, 2006, p.131). In a similar fashion, cetaceans, e.g., dolphins, can easily change their biosocial structures. The changeability of organizational structures is also typical of many other animal species including primates.

2.3.4. Affiliation in Fish Shoals: Minimization of Individual Differences

Fish in a leaderless (equipotential) shoal tend to stay close to one another; a fish that happens to swim away from the shoal is likely to return. Fish monitor their neighbors' behavior using (1) their sight and (2) the lateral line that enables them to detect movement and vibration in the surrounding water (mechanoreception). To an extent, they also rely on their chemical sense (an analog of olfaction) and electrical fields (Pavlov & Kasumyan, 2002; Mikheev, 2006). According to the definition of the shoal/school given in 2.3.2, it consists of "individuals that tend to aggregate", i.e., those that exhibit affiliative behavior. Studies with verkhovkas (*Leucaspius delineatus*) revealed that even a solitary individual attracts other

[43] Here and below, the term "shoal," unless redefined otherwise, is used as a broad term corresponding to both shoals *sensu stricto* and polarized schools.

individuals; increasing the number of "actors" (conspecific fish demonstrated to an individual during an experiment) considerably increased this effect, which reached a maximum when three conspecifics were demonstrated to the tested individual (Pavlov & Kasumyan, 2003). Shoaling fish are attracted by their own mirror images.

Like attracts like: a shoal is composed of individuals that have a similar size and physiological status. In bacterial biofilms and colonies, the individual features of their cells only manifest themselves to an insignificant extent; in fish shoals, individual differences among fish are *minimized.* There is evidence that a fish shoal predominantly consists of fish characterized by the same sex (Mourrier et al., 2012), age, intensity of shoaling behavior (Croft et al., 2005), and other behavioral traits ("personality;" Matessi et al., 2010), etc. The tendency towards eliminating individual differences is partly due to the fact that affiliative behavior is preferentially displayed with respect to similar individuals in terms of size and other characteristics. White and black *Poecilia latipinna* individuals tend to group with other individuals of the same color (Pavlov & Kasumyan, 2003, p.57). Interestingly, some animals with a more advanced nervous system, e.g., dolphins, also form groups composed of similar individuals (Croft et al., 2005).

Fish network structures are, therefore, *assortative* (see 1.1.3.3): they tend to include similar individuals. As Figure 2.4 demonstrates, large and small female Trinidadian guppies tend to form partially autonomous subnetworks within the network of a whole population.

However, individuality is not completely eliminated. For example, members of one shoal of Siamese fighting fish display "personality differences": some of them are characterized by sporadic and others by persistent signaling (Matessi et al., 2010). Large equipotential shoals often include smaller subgroups in which the tendency to eliminate individual differences is more manifest.

Downplaying the importance of individual differences promotes a fish network's adaptation to predation pressure. Predators become confused, i.e., overloaded with visual information, if the members of the network are almost indistinguishable. In network structures where the members are quite similar in terms of size and appearance, predator avoidance mechanisms can work effectively. They include the dilution effect (the probability of being caught by the predator decreases as the number of fish in the network increases) and the shoal compaction effect, sometimes making the predator consider the whole shoal as one large fish (Radakov, 1972; Norris & Schilt, 1988; Pavlov & Kasumyan, 2002, 2003; Krause & Ruxton, 2002; Mikheev, 2006).

A solitary fish has a blind zone that can be used by an attacking predator. In a fish shoal (particularly in a polarized school), this zone is either lacking or it is reduced and located far behind the fish's body because individual visual fields overlap.

A neighbor blocks a part of an individual's visual field. The larger the neighbor, the bigger the part blocked by it. At the level of the whole school, the optimum situation is that all fish have approximately equal sizes, and no individual has to stay near a much larger neighbor (Rountree & Sedberry, 2009).

A predator easily locates a fish whose appearance or behavior differs from those of the majority of other fish in the shoal. Extraordinarily large individuals face a serious risk, although they have important advantages, if the species to which they belong does not form shoals and holds individual territories. Hence, predators prevent the hierarchization of fish shoals that, therefore, retain their flat network organization.

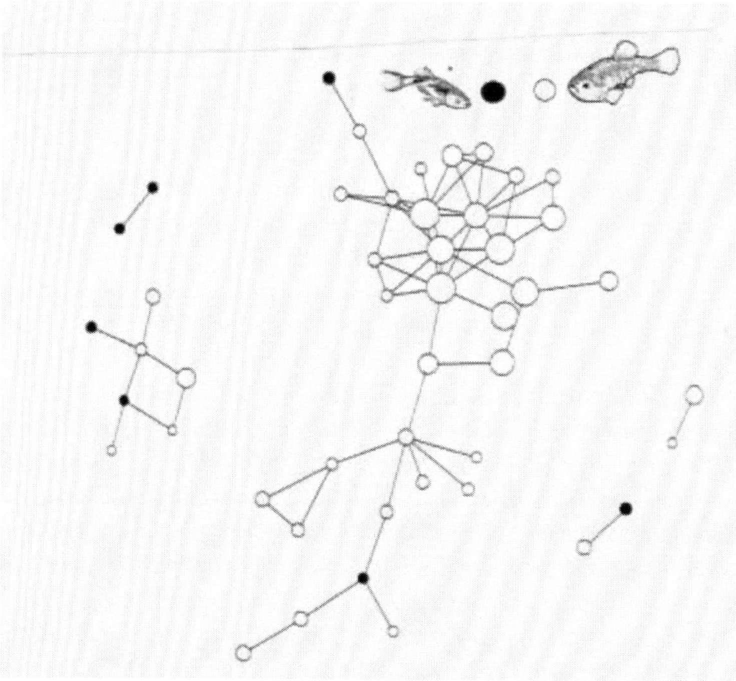

Figure 2.4. The social network of Trinidadian guppies (*Poecilia reticulata*).The nodes that represent females are indicated by open circles, and males are indicated by filled circles.The size of a node is a reflection of the fish's body length; the scale of the node sizes is arbitrary to exaggerate the difference between large and small fish. Reproduced from Figure 3.3 of the work (see Ref. Croft et al., 2008): CROFT, DARREN P.; *EXPLORING ANIMAL SOCIAL NETWORKS.* Princeton University Press. Reprinted by permission from Princeton University Press.

The tendency towards the elimination of individual differences in equipotential networks implies that a newcomer in this type of network should copy the behavior, the activity rhythm, and the diet of the members of the network (*when in Rome, do as the Romans do*). However, the individuality of a fish manifests itself to an extent, while the fish is deciding which shoal it should join. In an experimental setting, stickleback individuals could choose between a courageous and a cowardly group. Both courageous and cowardly individuals tended to prefer the courageous group (Harcourt et al., 2009). Guppies preferentially joined shoals whose members fed on bloodworms and not those feeding on flaky food particles (Morrell et al., 2007). Apparently, natural selection has selected for fish that tend to join the shoals whose members are better adapted to their environment or where they are more successful. In the futuristic network society described in the book *Netocracy...* by Bard & Söderqvist (2002), a human individual is embedded in a world of networks where rigid behavioral norms are imposed on their members. Nevertheless, an individual can choose the network(s) in which it will enroll.

Shoaling is a process that takes place upon reaching the threshold density of a fish population, in an analogy to quorum sensing-dependent processes in microorganisms (see 2.1.5). Spawning herrings form dense aggregations once their density increases to 0.2 individual per 1 m^2. These aggregations generate spreading "waves," resulting in the formation of a 20–30 × 3-4 km shoal (Makris et al., 2009).

2.3.5. Cooperation in Fish Networks

Equipotential networks are characterized by intense cooperation that benefits all their members. Despite the lack of a hierarchy, fish shoals can concertedly move and perform complex maneuvers aimed at protecting the fish against predators or locating food.

For instance, if a predator or an object imitating it appears, fish can form a dense ball-shaped aggregation that confuses the predator who, as mentioned above, may take the aggregation for a single big fish. The predator is also confused by the skittering of visually indistinguishable individuals. In particular, contrasting patterns, e.g., dark and light stripes, on the bodies of the fish can "induce a false sense of self-motion in a predator, thus confounding its attack" (the *advanced confusion effect*; Norris & Schilt, 1988). The predator is confused, and a "reduced attack-to-kill ratio" is observed, "resulting from an inability to single out and attack individual prey in a group" (Krause & Ruxton, 2002, p.19). A fish shoal can form a temporary doughnut-shaped structure so that the predator finds itself in the empty space in the middle (the *fountain* maneuver).

Apart from efficient protection from predators, fish shoals give their members important advantages in terms of food location and preemption, in contrast to solitary fish. Obviously, "the aggressive behavior of solitary fish occupying individual territories cannot stop the avalanche-like attack of a shoal" (Pavlov & Kasumyan, 2003, p.25), and these words emphasize the special power of network structures—including those in human society.

Drawing on Konrad Lorenz' works (Lorenz, 1966), equipotential fish shoals can be regarded as "open anonymous aggregations" because, in addition to downplaying the importance of individual features, they are open to newcomers that cause no aggressive behavior among the shoal members. In our terms, we can classify many equipotential fish shoals as *interactive* networks that are not characterized by isolation and whose boundaries are not closed.

If predators also form shoals, their attack on prey results in a struggle between two networks, similar to a fight between a terrorist network and a counter-terrorist organization that is also built upon network principles. Predator shoals carry out complex maneuvers. For example, a shoal assumes the form of a crescent, trapping the prey between the cusps. Likewise, Hannibal encircled the Roman Army at Cannes during the Second Punic War. An arcuate predator shoal "sends 'waves of influence' in multiple directions through a school (of prey – O.A.), confounding the reactions of school members that receive more than one signal at a time" (Norris & Schilt, 1988, p.160). Predators can also trap the prey near the water–air boundary (this maneuver is typical of predatory jackfish) or in an area of shallow water. Tuna fish perform specialized roles during collective hunting (Partridge, 1982) so that their shoal behaves like a project team aimed at attaining its goal (catching the prey), and similar behavior is characteristic of clans of eusocial insects (see 2.4.4 below).

2.3.6. Coordination Mechanisms

Behavior coordination in equipotential structures, such as fish shoals, partly depends on "long-range short-term order factors", i.e., short-lived diffuse regulatory agents including pheromones and, presumably, electric fields sensed by fish. However, of paramount importance is a coordination mechanism that plays an intermediate role between long-range and short-range regulators. This mechanism is based on monitoring neighbor behavior using several sense organs, primarily the eyes and the lateral line; its effect rapidly spreads within the shoal, resulting in drastic changes in its members' behavior and structure. This is the *relay information transmission* mechanism. It is facilitated by the capacity of individuals to copy others' behavior that is typical of most shoaling fish species. For example, fish that locate food and move toward it encourage other members of the shoal to follow them (Pavlov & Kasumyan, 2002). The excitation wave in the shoal is not attenuated if the stimulus (food, danger, etc.) has been detected by approximately one-third of the shoal (Radakov, 1972; Gerasimov & Darkov, 1984).

2.3.7. Fish Shoals as Analogs of Neural Networks

Excitation waves in fish shoals are functionally analogous to the impulses that spread in neural networks (see 2.5 for details). Both kinds of decentralized structures perform two functions: (1) transmitting signals (impulses) to those parts of the structure that have not directly perceived the primary stimulus; and (2) inhibiting the spread of signals (impulses) caused by stimuli that have lost their importance. A fish that has responded to a signal, e.g., has changed the direction of its movement in response to a dangerous stimulus, temporarily becomes insensitive to analogous stimuli (Pavlov & Kasumyan, 2003). In a similar fashion, a neuron (nervous cell) becomes *refractory* upon transmitting an impulse. In an analogy to neural networks, "individual decisions" in fish shoals "are also affected by the actions of neighbors..., and the collective reaction of all members can create a group-level, or 'emergent,' response" (Viscido et al., 2007, p.154).

Like nervous cells in neural networks, fish in shoals collectively store information. Fish shoals have the conditional reflex pool (CRP): they collectively store learned responses to regularly repeated stimuli (the "collective memory" system). The CRP in a shoal of fish exists thanks to their capacity to imitate their neighbor's behaviors, as long as the distance between them does not exceed 4–5 fish body lengths. This enables effective information exchange between individuals (Krause & Ruxton, 2002).

The sensory organs of all individuals in a shoal form part of the sensory integration system (SIS) that involves the receipt of environmental information by any member of a group and the passage of a reaction to it through the group in all directions within the shoal (Norris & Schilt, 1988). As far as protection against predators is concerned, I should emphasize the *many eyes* effect: "an individual in a group need not detect a predator itself to be warned of an attack as at least one of the other group members" sights the predator and spreads the message in the group (Krause & Ruxton, 2002, p.19).

Hence, apart from the *equipotential* paradigm, fish networks can also behave in conformity with a different organizational pattern, referred to as the *neural* paradigm that is discussed below (2.5).

2.3.8. Algorithmization of Individual Behavior

An additional and similar feature of neural networks and fish shoals is that the behavior of both kinds of structures can be modeled using algorithms, i.e., sets of relatively simple mathematically formulated behavior rules. An example is a mathematical model dealing with the structure and shape of fish shoals. It is assumed that fish follow three rules: (1) they move away from other individuals that are too close to them; (2) they move in the same direction as their neighbors; and (3) they get closer to others if the distance from them is too large. The area around a fish is subdivided into three zones: (1) the *repulsion* zone; (2) the *imitation* zone; and (3) the *attraction* zone. The sizes of the zones differ according to the inequality:*1<2<3*. By varying these sizes, we can obtain a number of different geometric images at the shoal level, including a sparse stationary shoal; a "doughnut" formed during the fountain maneuver, with fish rotating around the mass center; and a polarized school with all fish moving in the same direction (Couzin et al., 2002).

2.3.9. The Fractal Properties of Equipotential Network Structures

Large fish shoals tend to separate into relatively autonomous small groups that are loosely connected to one another (Croft et al., 2005). In other words, fish aggregations, like most other biological networks, are characterized by cliquishness or community structures, i.e., they include subnetworks with a high degree of clustering that are separated by almost empty zones. For instance, the shark aggregation living near the northern coast of Moorea Island includes four relatively tight subnetworks, and one of them tends to separate into two second-order "sub-subnetworks" (Mourrier et al., 2012). The opposite process, i.e., merging several shoals into a larger "supershoal," is also carried out by shoaling fish. Therefore, they form multilevel (fractal) networks consisting of nodes that also represent network structures. Many equipotential networks are characterized by a fine-grained internal structure: subgroups of 3–5 fish are their minimal units. Such "core units," composed of similarly sized fish, are manifest in the Trinidadian guppy network shown in Figure 2.4. The formation of subgroups decreases the probability of a predator attacking each of the individuals involved (Pavlov & Kasumyan, 2003), i.e., it contributes to the protective function performed by the shoal.

The internal structure of a shoal is influenced by the local landscape; each of the parts of a stationary shoal often tends to stay at a specific site. For instance, each subgroup in the shark aggregation mentioned above preferred one of the 10 stations located near the coast of Moorea Island (Mourrier et al., 2012). Clusters of tuna fish are attracted by large floating objects, including artificial ones (fish aggregation devices; see Stehfest et al., 2013).

2.3.10. Fish Shoal Structure: An Analog of the Matrix?

Most shoaling fish do not build any constructions[44], but the supraindividual structure of their shoals performs protective, communicative, and structural functions, in an analogy to the matrix of microbial colonies. Effective shoal maneuvers enable fish to escape from predators and, therefore, to survive, even though their individual means of protection (e.g., spines or venom glands) are relatively weak. In a network, individuals hide behind one another, and "the searching predator must at first deal largely with the envelope of the group" (Norris & Schilt, 1988, p.155). The orderly arrangement of individuals within subgroups forming part of

[44] However, stickleback fish build nests, "houses" used collectively by developing eggs and young individuals.

a single structure promotes communication via relay information transfer. The whole shoal has a distinct geometric image that, to reiterate, may be considered a single fish by predators; this image is a material manifestation of the stable social behavioral rules that constitute the immaterial component of the matrix of a network structure (see 1.5.3). These social norms form part of what Norris and Schilt (1988) called *cooperative disposition,* i.e., "ingrained adherence to essential group rules".

The communicative function of the network's matrix involves the sensory integration system and the shoal-level conditional reflex pool, which are prerequisites for the communicative continuity of the whole network structure. The shoal's spatial structure helps the network support cooperative behaviors and inhibits those disrupting group processes. For instance, a fish's attempt to cross the shoal boundaries increases the risk of a predator's attack on it and, therefore, such individual behavior is discouraged by the signals emitted by most fish shoals.

2.3.11. Quasi-Organismic Properties of Equipotential Networks

Fish are autonomous organisms exhibiting individual behavior. They do not belong to colonial animals, as exemplified by cnidarians, whose individuals are connected and form part of the body of the modular organism (Marfenin, 1993, p.7). Nonetheless, many fish shoals and schools represent coherent networks. An equipotential network structure lacks permanent "colony-level organs," considered by Beklemishev (1950) as an important organismic feature. However, internal empty spaces ("vacuoles") are temporary protective organs[45] formed by a shoal during the fountain maneuver. The behavior of the network is coordinated, which enables it to engage in such activities as collective hunting and spawning. As mentioned above, the whole structure has a distinct geometrical shape. Individual fish move at regular intervals in a migrating school (scheme in Figure 2.5) and form regular patterns (subnetworks) in it. Some important functions, such as protection against predation, are performed at the collective rather than at the individual level. Shoals can divide or bud off small fragments. In sum, individual behaviors in a typical equipotential network are coordinated to such an extent that the network possesses certain organismic properties, although they are less manifest than in cell networks and modular systems of cnidarians.

Figure 2.5. The arrangement of fish in a migrating school. Reproduced by permission from the following work: Pavlov & Kasumyan, 2003 (Figure 4). Copyright © Pavlov, D. S. & Kasumyan, A. O., 2003.

[45] In a similar fashion, empty spaces appear in a microbial biofilm during the late stages of its development. Their function is to promote the dispersal of bacterial cells.

Importantly, the equipotential paradigm works not only in fish shoals. Similar flat networks are formed by marine invertebrates (echinoderms and cephalopods), some birds, and such mammals as seals and cetaceans, including dolphins. Analogous structures are spontaneously formed by agitated human crowds.

The equipotential organizational pattern (termed the equipotential paradigm in this work) is considered in this section in the example of fish networks (shoals/schools) where individual differences are eliminated to a considerable extent. Effective shoal/school-level behavior coordination within an equipotential network is mainly due to relay signal transfer using vision and mechanoreception (detecting movement and vibration in the surrounding water) involving the lateral line. Large networks tend to separate into smaller subnetworks that are relatively loosely connected to one another. The networks often possess quasi-organismic properties because the individuals within them obey network-level behavior rules that integrate the whole structure, in an analogy to the microbial matrix.

2.4. Eusocial Paradigm in the Example of Ant Societies

Eusocial systems (the prefix ευ- means "true" or "good" in Greek) combine hierarchical and network structures. Nonetheless, the network prevails over the hierarchy: these structures lack a central leader.

Eusocial systems combine three features: *(1) overlap in generations; (2) care of the young by adults; and (3) division of labor into reproductive and nonreproductive casts* (Wilson, 1971, 1975; Hölldobler & Wilson, 1990, p.298). Eusocial systems are formed by social insects such as termites, ants, bees, and wasps, as well as some gall-inducing aphids, thrips, and the Australian eucalyptus-boring beetle *Platipus incompertus* (Hölldobler & Wilson, 2009); their analogs are characteristic of shrimps (genus *Synalpheus),* gudgeons, naked mole-rats (*Heterocephalus glaber* and *Cryptomys domarensis*), birds (ox birds, choughs, and some penguins), and spiders that build a collective web.

Interestingly, at least some of these structures were considered in the literature not only in terms of eusociality, but also from the viewpoint of alternative paradigms of network organizations, including the modular paradigm. This paradigm emphasizes the organism-like properties of such systems, and it is actually adopted by the proponents of the superorganism theory, considered in general terms in 1.5.2. The following subsections deal with the three features of eusocial systems singled out by Wilson (1971) and included in the definition given in the work (Hölldobler & Wilson, 1990), as quoted above.

2.4.1. Coexistence of at least Two Generations (The Parents and the Offspring) in the Same Biosocial System

This promotes the coordination of individual behaviors on the basis of an effective communication system. Old worker ants store in their memory the map of routes leading to foraging sites and the location of trees with aphid colonies; they can guide young foragers to these foraging sites (Dlussky, 1984; Zakharov, 1991). Worker ants are activated by pheromones produced by the brood. Interactions between different generations are promoted by structures that represent the network-consolidating *matrix*, in the same sense as discussed in 1.6.4: an anthill, a termite mound, or the system of burrows made by naked mole-rats. In an analogy to the extracellular matrix of microbial colonies, they perform structural, protective, and communication-promoting functions.

The generation overlap is implicit in the *subsocial* scenario of eusocial system formation that was originally suggested by Evans (1958): "a single foundress gains enough longevity to coexist in the same nest as her female offspring. In this case the most primitive colony is an extended family" (quoted according to: Hölldobler & Wilson, 1990, p.27). The offspring subsequently evolve "to workers that feed and protect the mother (queen)" (Hölldobler & Wilson, 2009, p.34).

The alternative *parasocial* scenario based on cooperation among a number of reproductive females (queens) belonging to the same generation (briefly described in 1.7.1) also implies that several generations overlap. The reason behind nurturing the offspring at the initial stage of colony development is that the offspring are expected to join the older generation in doing various kinds of work. This scenario includes a stage at which the several competing reproductive females form a dominance hierarchy that can be based on aggressive interaction. This pattern was observed in the South American ant *Pachycondyla inversa* (Kolmer & Heinze, 2000, cited according to: Hölldobler & Wilson, 2009, p.94). A contest among several pretenders (potential queens) resulted in one of them occupying the lower rung of the hierarchy, i.e., she refrained from reproduction and specialized in foraging for food. All the other pretenders stayed in the nest as reproductrices (queens), so that, curiously enough, the hierarchy had a very broad top and a narrow bottom occupied by only one subordinate.

During the course of the further development of eusociality, this hierarchy develops into a three-dimensional network in which the queen does not play the central leader's role. She only specializes in reproduction that represents one of the many "specialties" existing in the insect colony. Nonetheless, some hierarchical relationships between individuals and their groups often occur even in a mature eusocial system (see below).

2.4.2. Cooperation in Nurturing the Young and in other Collective Activities

It is well known that ants, termites, and other social insects cooperate in various activities (Figure 2.6), including building anthills or termite mounds, rearing the brood, foraging, and defending their nests. Leaf-cutter ants collectively cultivate fungi. Ecitons form large material network structures composed of their connected bodies.

2.4.2.1 Role of Short-Range Coordination

Collective activities often involve short-range coordination factors. Important communicative signals (e.g., concerning food location) are transmitted when ants contact one another with their antennae (Dlussky, 1984; Zakharov, 1991). Short-range coordination is involved in synchronizing individual behaviors, mobilizing inactive individuals, and initiating avalanche-like processes in eusocial systems.

Similar to fish shoals, insect societies are capable of relay information transfer, so that originally "close-range" coordination factors eventually exert a long-range influence on the whole insect social network, demonstrating small-world behavior. Like fish individuals, worker ants that have recently been mobilized for helping their nest mates "become quiescent for a refractory period during which they cannot be reactivated" (Detraine & Deneubourg, 2006, p.167); both fish and ants, therefore, display neuron-like behavior (see section 2.5 on neural networks, below).

2.4.2.2. Role of Long-Range Coordination

Despite the existence of hierarchical structures (clans) at its lowest structural level, a eusocial system as a whole lacks a central leader[46]. Nevertheless, long-range coordination factors such as pheromones enable effective behavior coordination in eusocial systems. In the ants *Lasius niger* and *Messor sancta*, long-range communication involving specific digging pheromones promotes the construction of branched nest galleries (Rasse & Deneubourg, 2001; Buhl et al., 2004, 2005). Such communication is also implicated in collective foraging behavior. "As soon as one ant has succeeded in discovering a food source, it goes back to the nest and lays a chemical trail. The trail pheromone then triggers the exit of additional foragers and guides them as an Ariadne thread to the food source. After feeding, each recruited ant can, in turn, reinforce the foraging trail and stimulate other nestmates to forage" (Detrain & Deneubourg, 2006, p.167).

Figure 2.6. Cooperation in ants: work and risk sharing. Foragers provide the ant society with food, and their raids involve exploring areas and fighting dangerous insects. Reproduced with permission from the following work: Zakharov, 1987.

Similar to bacteria that use quorum-sensing systems, ants can estimate the density of their own population from the number of direct contacts with other ants or, more indirectly, e.g., from the number of their footprint marks (Devigne et al., 2004). Upon reaching a critical

[46] To reiterate, this role is *not* played by the reproductive female (queen).

density, ants more efficiently perform various collective activities, e.g., foraging (reviewed, Detrain & Deneubourg, 2006).

2.4.3. Roles and Castes in a Eusocial System

Only one of the castes is typically involved in reproduction (reproductive females and males)[47], whereas the other castes belong to the "service sector" and include workers and soldiers. In developed eusocial systems, workers and, in some systems, soldiers are subdivided into subgroups that perform different functions. In colonies of the termite *Anacanthotermes ahngerianus*, the worker caste includes the queen's retinue, nurses, grooms, water carriers, and harvesters (Zhuzhikov & Shatov, 1984).

Reproduction is actually one of the specialized functions performed in a system where all kinds of specialists are important. Therefore, the important point is not necessarily that "adult members are divided into reproductive castes and partially or wholly nonreproductive workers" (Hölldobler & Wilson, 2009, p.7). Eusociality, in more general terms, implies the existence of several permanent specialized subgroups among adult individuals (Zakharov, 1991, 2005). Hence, individuals need not be subdivided into reproductive and nonreproductive castes: of paramount importance is the fact that they are functionally differentiated, i.e., an insect society includes subgroups of *specialists* that can have their functional leaders, making the whole structure a three-dimensional network.

Of relevance in this context are studies recently conducted with honeybees (Johnson, 2010). Their workers change their "specialties" as they get older, and a similar pattern is characteristic of ants. In terms of their specialization, honeybee workers typically undergo four main age-dependent stages[48], consecutively specializing in (1) cell cleaning; (2) taking care of the brood and the queen; (3) "middle age functions" including comb building, nectar receiving, and guarding the nest entrance; and, finally, (4) foraging for food. It is only the transition from cell cleaning to nursing that is a preprogrammed stage of individual development. The other role changes depend upon social signals. The transition from nursing to comb building and other "middle age" functions is linked with the emergence of new nurses in the beehive, which reduces the effects of queen- and brood-released pheromones that otherwise block individual development at the nurse stage. Switching from "middle age" functions to foraging is aimed at optimizing the *forager:nectar receiver* ratio. An individual that is physiologically ready for becoming a forager may remain a nectar receiver for a long time if there are sufficient numbers of foragers and if no "unemployed" nectar receivers exist (Johnson, 2010).

These specialization changes during an individual's life are characteristic of various insect societies; they are referred to as temporal *polyethism*, or the formation of temporal subcastes among workers; in many ant species, "young workers attend the brood and queen, while other workers are more active in foraging" (Hölldobler & Wilson, 2009, p.93). The relationship between aging and role/function changing is debatable.

---

[47] There are ponerine ant species that lack the reproductive caste; some of the workers are gamergates: they lay viable eggs along with performing typical worker functions, such as rearing the brood (see 2.4.4).

[48] Winter bee generations do not undergo these stages and represent generalists rather than specialists.

However, there are ant and termite species in which the differentiation of workers into functional subcastes is permanent; the development of a worker individual can follow several alternative routes. Different subcastes display behavioral, physiological and, in some species, clear anatomical differences. Highly complex colonies formed by the ants of the genus *Atta* have four worker subcastes that differ in size and are called minors, mediums, majors, and supermajors; the latter two groups are denoted as soldiers (Hölldobler & Wilson, 2009).

Generally, a eusocial system finds itself in a stress field formed by a pair of opposite tendencies: (1) the tendency towards functional efficiency maximized by permanent role differentiation; and (2) the tendency towards flexibility associated with the capacity to change from one role to another; a similar flexibility and functional interchangeability that enable parallel information processing and self-adjustment are characteristic of neural networks (see 2.6 below).

The ratio between the numbers of different specialists varies depending on the demand–supply ratio in terms of the respective "jobs" and competition among potential incumbents. This implies that there are quasi-market–style interactions among individuals. Therefore, the whole eusocial system is polystructural (see 1.7) because it combines network, hierarchical and, to an extent, quasi-market structures. There are mechanisms that regulate competition among worker individuals for the benefit of the whole system, which is predominantly characterized by cooperation and network organization.

Nodes (cells) in microbial network structures are characterized by incipicnt (or lacking) functional specialization, zooids in modular structures may or may not be functionally specialized, and equipotential shoals/schools predominantly consist of similar individuals in terms of both morphology and social functions. In contrast to the systems considered above, *functional differentiation of individuals to the point of the formation of several castes is a typical feature of eusocial systems.*

2.4.4. Hierarchies and Competitive Interactions in Eusocial Systems (in the Example of Ant Families)

In contrast to the flat networks of colonial cnidarians or shoaling fish, most eusocial systems, exemplified by ant societies (Schmidt-Hempel, 1990; Zakharov, 1991, 2005), are characterized by complex multilevel structures that are partly hierarchical. The contribution of hierarchical organization is sufficiently important at several levels, ranging from small teams (clans) with "team leaders" to whole families (pleiads) that interact in order to form higher-order hierarchical structures (the multilevel "layer cake" structure was briefly described in 1.8.6 above).

Figure 2.7. Team formation in the colony of the ant *Camponotus socius*. The leader recruits the whole "team" by laying a long-lasting orientation trail and a short-lasting recruitment signal. The scheme is based on a modified picture (Hölldobler & Wilson, 2009, Figure 6-24). Reproduced by permission from the work *The Superorganism: The Beauty, Elegance, and Strangeness of Insect Societies*. Hölldobler, B. & Wilson, E. O. New York: W.W. Norton. Copyright © 2009 Bert Hölldobler & E.O. Wilson.

As noted in 1.8.6, functional leaders in small hierarchically organized teams of workers, or *clans,* are responsible for the coordination of their work (Figure 2.7). For example, in an ant nest, there is a clan (a "project team") specializing in collecting honeydew. The leader (the supervisor) monitors the migrations and the functional state of aphids (whose excretions are collected by ants); the supervisor also coordinates the behavior of workers who feed and pasture aphids (Zakharov, 2005, p.47). Clan leaders can either entice or coerce other clan members to do certain jobs (such as collecting aphid secretions) by touching them with antennae or assuming threatening postures. Therefore, clan leaders use two hierarchy-specific coordination mechanisms that are apparently based on communication and controlling action, respectively. Retrieving large food items in *Eciton burchelli,* the army ants of Central and South America requires the collective effort of a "transport gang." "Because exceptionally large objects can be moved only by strong workers, the first member of the gang is usually a supermajor /the biggest worker subcaste – O.A./. After this heavy lifter has moved into position, smaller media workers join in until the prey is moving expeditiously homeward" (Franks, 1986; quoted from: Hölldobler & Wilson, 2009, p.163). Hierarchical relationships can also be established between permanent specialized groups. In a number of ant species, workers staying outside the nest, such as foragers and scouts, have a higher social rank than those working inside the nest (nurses and the retinue of the queen; Dlussky, 1981, 1984).

In the Australian ant species, *Pachycondyla sublaevis,* there is no special reproductive caste. High-ranking ants are *gamergates.* This term means that they combine the reproductive and the worker roles: they lay eggs and care for the brood. The foraging is left to the older, lower-ranking, individuals that are not involved in reproduction (Ito & Higashi, 1991; Higashi et al., 1994). In the subtropical American ant *Odontomachus brunneus*, "high-ranking workers remain in the brood zone /of the nest – O.A./, medium-level workers stay in the nest but away from the brood, and the lowest-ranking individuals, which are also the oldest and hence marked by withered ovaries, evidently conduct the foraging" (Powell & Tschinkel, 1999; quoted from: Hölldobler & Wilson, 2009, p.95).

The role of *quasi-market* relations in the organization of an ant society at the clan level is emphasized by the fact that workers *compete* for the clan leader position. The competition takes the form of a tournament in which the stronger and more aggressive candidate pushes the other candidate down and carries it away from the tournament site. The candidate belonging to the dominant interclan alliance is more likely to assume the status of the clan leader (Zakharov, 1995). Competitive interactions that may involve aggressive behavior also occur in a variety of other situations. Individuals engage in competition for reproductive success (this concerns candidates for the queen position) or for "job positions" including the right to start a new building project, such as the construction of a new anthill.

However, an important feature of this type of system that enables us to consider it a coherent cooperative *network (sensu stricto)* is the existence of specific mechanisms that restrict and redirect competition, promoting cooperation within the whole system. Such mechanisms are implicit in the *stigmergy* principle discussed below (2.4.8). *Intracolony* competition (e.g., for reproduction privilege) is diminished because a colony (a "superorganism") should maximize its chances to successfully engage in *intercolony* ("inter-superorganism") competition (Reeve & Hölldobler, 2007).

2.4.5. Eusocial Structures as Three-Dimensional Networks

Hierarchically organized worker teams consist of a small number of workers, and their makeup varies depending on the situation, like that of temporary creative subunits in a hirama in human society (see 1.2.2 above). The dominant members within these structures are perceived as partial leaders from the viewpoint of the whole ant network structure (Schneirla, 1971). Presumably, special mechanisms are at work in ant families that prevent an increase in the number of members in such "project teams" and, therefore, counteract the establishment of a more rigid hierarchy within them. The leader role represents just one of the many possible specializations.

The leader role is the primary social function performed if an ant family develops according to the subsocial scenario briefly described in this section. The reproductive female that starts building a new nest lives long enough to coexist with representatives of the next generation, who supplement the reproductive function with a variety of other socially important jobs. Accordingly, the foundress that originally dominated the family switches to a more specialized function that is directly related to reproduction.[49]

Predominantly, eusocial systems, including those formed by a large number of ant species, represent *heterarchies*, i.e., split hierarchies lacking a central leader (see 1.2.2. above) "with activity in the lower units feeding back to influence the higher levels" (Hölldobler & Wilson, 1990, p.355). Importantly, "the highest level of the ant colony is the totality of its membership rather than a particular set of superordinate individuals who direct the activity of members at lower levels" (Ibid.). This reminds us of the organizational principles of Israeli *kibbutzim* and other non-hierarchical structures in human society, in which the general assembly of the members is the supreme decision-making body.

A more serious challenge to the network organization principle is posed by the establishment of more long-lived dominance relationships between specialized groups of workers, as was described in 2.4.4. In ants, this hierarchy coexists with the network's organizational pattern.

2.4.6. Eusocial Structures as Multilevel Systems

To sum up, ant families are multilevel systems. Their network structures consist of hierarchically organized worker teams (clans), and not of individuals. The principle where *networks consist of hierarchies* is in contrast to the principle where *networks consist of networks*, which is characteristic of cellular, modular, and equipotential structures considered above. As noted in 1.8.6, the *networks consist of hierarchies* principle also applies to the superfamilial level at which hierarchical *genuine colonies* (made up of families/pleiads) form horizontal networks referred to as *federations*.

Networks and hierarchies have advantages at different levels of ant societies (or other eusocial systems), and this seems to account for the "layer cake" pattern. Hierarchies prove to be more efficient as far as (1) the establishment of primary social structures, i.e., clans with leaders on their top, and (2) the transition from families to suprafamilial structures, i.e., the

[49] As mentioned above, the reproductive female (queen), apart from laying eggs, releases an activating pheromone improving the work performance of worker ants.

formation of small offshoots dominated by the maternal family, are concerned. Nevertheless, the growth of hierarchical structures results in an increase in their instability unless the hierarchies combine to form a higher-level horizontal network (Zakharov, 2005, 2009).

The instability of hierarchies is apparently due to the limited regulatory influence of the central leader if it deals with too many subordinate individuals or structures. This may result in conflicts between the parts of the hierarchical structure. In addition, mobilizing passive individuals to perform socially important functions becomes increasingly difficult. An analogous problem is one that is familiar to specialists in organization management in human society; this problem can be solved using two different strategies:

- *Bureaucratic strategy:* establishing a multilevel pyramidal structure in which the central leader (boss) directly deals with relatively few second-order bosses that interact with third-order bosses;
- *Network strategy* (more specifically, the *hirama strategy*, see 1.2.2): setting up decentralized structures with a number of partial leaders who reach a consensus to coordinate their activities and those of the whole pool of network members.

From the data available in the literature (Dlussky, 1984; Zakharov, 1991; Hölldobler & Wilson 1990, 2009), it follows that ants use the second strategy, i.e., hierarchical structures become the nodes of higher-level networks *sensu stricto*.

2.4.7. Active Specialists and the Pool of Mobilizable Generalists

Ant societies include a pool of worker individuals that normally do not belong to any specialist team (Schmidt-Hempel, 1990) and that can be mobilized for coping with urgent problems. Studies with ants were conducted in which the number of available foraging sites in the form of honeydew-excreting aphid colonies was limited. The ant societies mobilized part of the formerly inactive individuals that became "specialists belonging to small teams taking care of aphid colonies" (Novgorodova, 2003, p.229). In a honeybee colony, "the presence of just a few highly defensive individuals in a hive can incite less defensive colony members to join in an attack" (Wray et al., 2011, p.566).

2.4.8. Organism-Like Properties of Eusocial Systems

At the family/pleiad level, eusocial structures behave as coherent organism-like systems and are termed as superorganisms in a number of works (Wheeler, 1928; Chauvin, 1963; Hölldobler & Wilson, 1990, 2009, 2010; Kipyatkov, 1991); the basic elements of superorganisms "are not cells and tissues but closely cooperating animals"(Hölldobler & Wilson, 2009, p.4). Advanced eusocial systems formed by a large number of ant and termite species possess a number of properties that conform to the general characteristics of the superorganisms listed in 1.5.2:

- The eusocial systems of ants or termites form "colony-level organs." According to Hölldobler & Wilson (2009, p.85), reproductive, worker, and defensive (soldier) castes are comparable to gonads (reproductive organs), somatic organs, and the immunity system, respectively. Table 5.1 in the work cited also includes analogs of the circulatory and the nervous system, but they do not represent any material structures; instead, they correspond to the "food distribution" system in a colony and "communication and interactions among colony members," respectively[50]. In terms of the *matrix* concept suggested in this work (see 1.6.4 above), these "superorganismic" features form part of the *immaterial component* of the matrix of a eusocial system that represents a special type of network structure. Eusocial systems also contain the *material* component of the matrix that is exemplified by anthills or termite mounds that in many respects perform the same functions as the extracellular biopolymer matrix of microbial colonies/biofilms.
- Indisputably, there is a general colony-level plan concerning the development pattern and structure of the whole eusocial system. This is termed "sociogenesis," i.e., the programmed "growth and development of the colony" that is comparable to the "growth and development of the embyo" (Hölldobler & Wilson, 2009, p.85).
- The system can reproduce like an organism. For instance, upon reaching a certain maximum size, an ant family tends to bud off daughter families, forming a mother colony-dominated hierarchy (a *genuine colony*). To reiterate, this results in the formation of a superfamilial network composed of hierarchical genuine colonies.
- Some individuals are characterized by a reduced size and a simplified structure, behaving like functional organs of the whole system. Some ant families include individuals functioning as honey pots, i.e., they store honey in their abdomen.
- Whole eusocial systems are characterized by collective personalities. For example, honeybee colonies differ in terms of aggressiveness, defensive responses, foraging behaviors, activity levels ("runniness"), and other collective features (Wray et al., 2011).

"To speak of the colonies of driver ants—or other social insects, such as the gigantic colonies of leafcutter ants, the honeybee societies, or the termite colonies—as more than just tight aggregations of individuals is to conceive of superorganisms and invite a detailed comparison between the society and a conventional organism" (Hölldobler & Wilson, 2009, p.xxi).

The work cited stresses that cooperation prevails over competition within a "superorganism." It should be reemphasized that the dominance of cooperation over competition is considered one of the distinctive features of networks (*sensu stricto*) in this work. In these terms, the "superorganism" represents a highly integrated subtype of the network structure that is the "protagonist" of this work. According to Hölldobler & Wilson (2009, p.84), the superorganism is "united by a closed system of communication;" therefore, is can be classified as an *isolationist* network that is characterized by closed borders and unlimited communication inside these boundaries (see 1.7.2.1).

---

[50] The colony-level analog of the organism's immune system also includes, apart from defensive castes, a component based on alarm–defense communication (Hölldobler & Wilson, 2009).

The superorganism concept is not supported by some other researchers (see e.g., Zakharov, 1991; Ratnieks & Reeve, 1992) who stress the individual features of each ant or termite, as well as the level of competition between them. Workers differ in terms of their activity/inactivity (Schmidt-Hempel, 1990). Generally, “no ant looks like another and each individual is characterized by a specific behavioural profile” (Detrain & Deneubourg, 2006, p.177). Interactions between individual worker ants include not only control but also communication processes that allow for stochastic, probabilistic responses, which may vary depending on the signal receiver’s individual traits and functional state.

Eusocial systems make good use of social mechanisms that mitigate interindividual competition, making it beneficial for the whole system and implementing the *competition for cooperation* principle. Ants simultaneously start building a new anthill at several points, so that a number of clans are involved. There often is a pool consisting of a “redundant workforce” that tends to join the most efficient teams. They behave in conformity with the *stigmergy* principle suggested by Pierre-Paul Grassé (1959): the results achieved by working ants stimulate further work at the same site. Therefore, competition among teams speeds up the construction of the anthill. The stigmergy principle is of relevance to network structures in general because it works not only with ant families and other eusocial systems. “Stigmergy has been observed in ants, bees, fish, birds, artificial life boids,… and robots equipped with flocking and foraging software” (Lewis, 2013, p.8). This principle is also used by network structures in human society including, unfortunately, criminal and terrorist networks that improve the performance of subnetworks (clusters, cells, modules) inside them by creating a competitive environment, with special rewards given to the most efficient subnetworks.

2.4.9. Isolationist or Interactive Networks?

Insect families typically tend to form closed (*isolationist*) networks: they delimitate their territories and defend them. Nonetheless, they can use a strategy enabling them to overcome the borders between two families. The families build small buffer nests along the border line. Both families transfer part of their brood to these buffer nests. These nests are, therefore, inhabited by mixed brood with mixed odors. The mixed brood migrates to the two original families, and the ants become unable to use olfactory cues to distinguish “insiders” from “outsiders.” The two originally independent ant families eventually merge into a single network structure (Zakharov, 1991).

2.4.10. Algorithmization of the Behavior of Eusocial Systems: Their Quasi-neural Properties

The behavior of eusocial systems has been recently investigated using mathematical models based upon relatively simple algorithms (Bonabeau et al., 2000; Detrain & Deneubourg, 2006). Models using quasi-eusocial principles have been applied in the field of technology (Deneubourg & Goss, 1989; Bonabeau et al., 2000; Sumpter, 2006). Typical features of network structures of social insects are currently used for developing artificial Swarm Intelligence systems with distributed controlling units. Generally, systems simulating social insect networks are characterized by considerable flexibility and reliable operation in a dynamic environment. Similar to fish shoals (and to people in a state of panic), each individual insect does not possess all of the information concerning the activities of the whole system, and it follows relatively simple behavior rules. In eusocial systems, "the requisite information is distributed among the colony members. Thus, a distributed colony intelligence is created greater than the intelligence of any one of the members, sustained by the incessant pooling of information through communication" (Hölldobler & Wilson, 2009, pp.58-59).

Models that simulate the behavior of eusocial systems are exemplified by the *ACO* (*Ant Colony Optimization*) system. It is aimed at simulating the behavior of ant foragers searching for an optimum route leading to a foraging site. It is assumed, in conformity with the *stigmergy* principle outlined above, that (1) the ants initially test accidental foraging routes; (2) each ant marks the route used by it with pheromone that attracts other ants; importantly, the better the route (e.g., the shorter the distance), the more pheromone is released; (3) other ants choose the route marked with the maximum amount of the pheromone; and (4) the pheromone rapidly evaporates and its concentration decreases, unless new pheromone amounts are released. In many situations, this simplistic model provides an adequate solution to the problem of enabling a majority of ants to find the optimum path between the nest and the foraging site (Sumpter, 2006). Similar models are used for simulating the activation of bees in a hive by the dancing of workers. Likewise, termites place pheromone-scented mudballs on those routes that lead to foraging sites. Other termite workers choose the routes covered with the maximum number of mudballs.

Such models provide much food for thought with respect to a number of problems concerning human society. For instance, ACO-like models can help us solve the Traveling Salesman Problem (Bonabeau et al., 2000)[51] in which a salesman visiting a number of cities aims to find the shortest itinerary.

Synergetics provides powerful conceptual tools for the investigation of self-organizing systems. Amazingly, the behavior of living organisms and their groups (superorganisms) often reveals similarities to that of nonliving systems. Suffice it to mention "striking patterns in the non-living world, such as the ripple marks on the surface of sand dunes, the hexagonal Bénard convection cells formed by heated oil or the swirling spirals of chemical compounds produced by Belousov–Zhabotinsky reactions. All these examples share the same basic means by which they acquire their structure. Under a particular set of initial conditions and parameter values, patterns which extend well beyond the scale of their individual subunits, can arise spontaneously through physical and/or chemical interactions internal to the system, independently of any external ordering influence" (Detrain & Deneubourg, 2006, p.163). Cybernetic and synergetic concepts of feedback loops and fluctuations apply whenever ordered systems spontaneously emerge in an initially chaotic environment or a biological system has a choice between several alternative states (e.g., ordered versus disordered foraging patterns in insects).

The development of the caste system in an insect's biosocial system can be studied using algorithmization as well. Suffice it to mention that the development of a young individual insect includes "decision points": depending on the conditions, a larva may develop into a

[51] This computationally difficult problem can be formulated as follows: if we know the list of cities to be visited by the salesman and the distance between each pair of them, how can we find the shortest possible route that visits every city exactly once and returns to the origin city.

reproductive or a worker individual. In some ant and termite species, there is a second "decision point": an individual worker may become either a minor worker (a generalist that can potentially perform several different functions) or a major worker (a soldier or a storage caste member). A complete sequence of decision points that produces a caste can relatively easily be described using algorithms (Hölldobler & Wilson, 2009, p.54).

Generally, relatively simple algorithms concerning individual-level behavior combine to produce complex patterns at the level of the whole network structure. In terms of algorithmization, individuals (e.g., worker termites) often make binary choices. For instance, a termite can either walk along a pheromone-scented route or not, and the decision largely depends on the level of pheromone concentration on it.

However, it should be emphasized that the behavior of a biological system, e.g., an ant's eusocial structure, should not be straightforwardly compared to that of a nonliving system, even if they appear to follow similar sets of algorithms. Unlike atoms or molecules, all living organisms, e.g., ants, have individual features that renders the collective biological systems heterogeneous. Importantly, the behavior of a biological system is *adaptive*: it represents the result of natural selection.

Living organisms (including insects) are similar to neural networks because they can make decisions using *intelligent criteria* (this capacity has been attributed even to microorganisms that form colonies or biofilms by Ben-Jacob). These criteria include, e.g., the following: *If I cannot retrieve this food item alone, then I will lay a recruitment trail* (Detrain & Deneubourg, 2006).

The "intelligent criteria" can be incorporated in the ACO model by making additional assumptions. For instance, ants are endowed with the capacity to estimate the positive or negative effects of small changes in their routes, which enables them to further optimize the route by the trial-and-error method.

The above scenario of network organization involving active specialized "project teams," in addition to a relatively inactive pool of generalists, conforms to a theoretical model in which worker ants differ in their sensitivity to pheromones[52] that stimulate their work. Normally, the available pheromone concentration is only sufficient for stimulating a minority of individuals whose sensitivity is particularly high; the rest of the individuals function as an inactive pool of generalists. In a critical situation, additional amounts of the pheromone are released. They activate the pool of generalists (Bonabeau et al., 2000).

Many models simulating social insect behavior implicitly or explicitly assume that groups of insects are capable of parallel information processing and associative learning. These features are also characteristic of neural networks. It seems likely, therefore, that neuron-like behavioral traits are displayed by ants or other species that form eusocial systems.

Importantly, eusocial systems and some of their artificial analogs combine the serial and the parallel strategy of performing tasks/solving problems. Not only is the same subproblem or problem-solving stage simultaneously tackled in a parallel fashion by several workers or teams (which enhances the system's reliability and resistance to disturbing factors), but a single individual, in many eusocial systems, can change from one role to another and, therefore, "perform whatever task is closest to hand or fill in for another worker that has halted" (Hölldobler & Wilson, 2009, p.120). This *series-parallel* strategy increases the

[52] Activating pheromones are released by the queen and the brood.

efficiency of the entire eusocial system. The same strategy is also widely used by the elements (neurons or their analogs) of neural systems.

Eusocial network systems are characterized by a high degree of functional differentiation of their nodes (individuals) and quasi-organismic(supraorganismic) properties as well as by the formation of specialized "project teams" and permanent functional groups. Due to the existence of hierarchical dominance–submission relationships within eusocial systems, they are to be considered as three-dimensional networks. However, hierarchies are embedded in a predominantly network-type organizational environment. Hierarchical worker teams form part of a larger decentralized network structure. Workers that are not involved in the activities of specialized teams belong to the relatively inactive pool of generalists. The pool can be mobilized for doing urgent tasks. Eusocial systems (including the suprafamilial levels) are multilevel structures: they represent networks consisting of hierarchies made up of networks that, in turn, consist of hierarchies. Many eusocial systems and their artificial analogs display neural network-like behavior.

2.5. Neural Paradigm

While flat network structures prevail in microbial colonies/biofilms and in a large number of other cell systems (see 2.1), structures composed of nervous cells (neurons) display more complex organizational patterns. Not only do most of them represent three-dimensional networks with partial/temporary leader cells (or leader subnetworks), but a large number of *neural structures* are actually polystructural systems (as defined in 1.7 above): they are characterized by a complex combination of *sensu stricto* networks, hierarchies, and—in some systems—quasi-market structures. Another important feature of neural structures is that the nervous cells within them are subdivided into many subtypes and are functionally differentiated. In this respect, neural structures are, to an extent, similar to eusocial structures that include several castes. Importantly, connections between nodes (neurons) are directed (feed-forward) links; in many neural networks, they are supplemented by other links that transfer information in the opposite direction (feedback loops that are peculiar to recurrent neural networks).

Neural networks and their analogs have recently received much attention in the literature, and one of the reasons behind the interest is that the *brain* is made up of multilevel, three-dimensional neural networks. The human brain contains an estimated 15–33 billion nervous cells (Pelvig et al., 2008) and 10–50 times more cells of a different type called glial cells, or glia (Kandel et al., 2000). Glial cells perform some of the functions that are characteristic of the microbial matrix (see below). Importantly, "all behavior is a result of brain function… The actions of the brain underlie not only relatively simple motor behaviors such as walking and eating but all the complex cognitive actions that we believe are quintessentially human, such as thinking, speaking, and creating works of art" (Kandel et al., 2000, p.5); all these processes are ultimately based on the operation of neural network structures.

Neurons, the functional nodes of neural network structures, are connected by links including (1) *axons*, long thin processes transmitting electrical impulses from a neuron's body (soma) to a *synapse*, a narrow cleft between the neuron and its neighbor; and (2) branched *dendrites* transferring the electrical signal from a synapse to the soma.[53] Transmitting impulses across the synaptic cleft involves *neuromediators (neurotransmitters)* that cross the cleft and bind to the receptors of the postsynaptic neuron's dendrite or soma. This either stimulates (a stimulatory neurotransmitter) or inhibits (an inhibitory neurotransmitter) the postsynaptic neuron. The postsynaptic neuron summates all stimulatory and inhibitory influences and generates an impulse (is depolarized) if the summated excitation exceeds the threshold value. After transmitting an impulse, a neuron temporarily becomes *refractory*, i.e., unresponsive, to stimulatory agents. An analogous effect is characteristic of fish within equipotential network structures (shoals or schools, see 2.3).

2.5.1. Artificial Neural Networks

The organization of neural networks is simulated using artificial analogs of neurons. Similar to fish shoals or schools (Couzin et al., 2002) and ant families in ACO-like models, the operation of neural networks is modeled using behavioral algorithms that are followed by each neuron analog.

Generally, an artificial neural network (an ANN) can be defined as "a parallel, distributed information processing structure consisting of processing elements (which can possess a local memory and carry out localized information processing operations) interconnected together with unidirectional signal channels called connections. Each processing element has a single output connection which branches ('fans out') into as many collateral connections as desired (each carrying the same signal—the processing element output signal)" (Simpson, 1990, p.3). The ANN developed in the seminal work by McCulloch and Pitts (1943) contained three types of processing elements (artificial neurons): (1) several *input units* that perceive external information; (2) several *hidden units* that process the information obtained from the input units; and (3) a single *output unit* that is connected to all hidden units; it generates the output signal of the whole neural network.

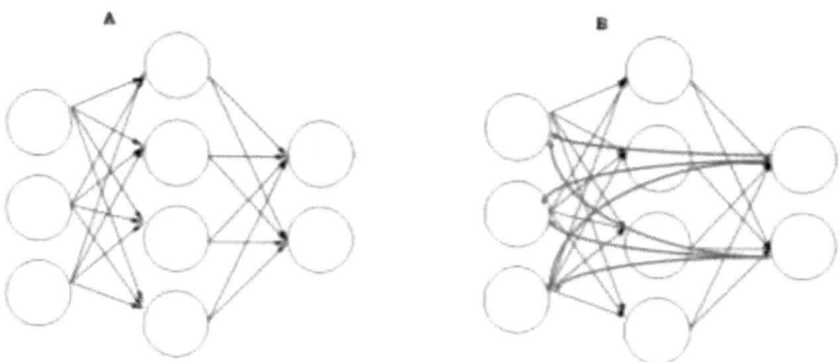

Figure 2.8. Simple neural networks that combine consecutive (within the *input → hidden layer(s) → output* sequence) and parallel (due to the operation of several nodes in at least some of the layers) information processing. (A) A classical perceptron; (B) a recurrent neural network with additional feedback links running from the output to the input layer.

[53] In some synapses, the axon of one neuron directly interacts with another neuron's soma, in the absence of a connecting dendrite.

McCulloch and Pitts' network was substantially modified, as neural network theory was developed in parallel to research on real-life neural networks including those of the human brain. The classical perceptron designed by Rosenblatt (1957) contained several output units where the data processed by the hidden layer were summed up. Interestingly, the structure of a perceptron (Figure 2.8, A) is similar to that of a human social network such as a hirama (Figure 1.4 in section 1.2) that includes several partial creative leaders that collect and sum up ideas suggested by network members. There are ANNs called multilayer perceptrons that contain several hidden unit layers[54].

Artificial neurons in ANNs summate stimulatory and inhibitory inputs. Like brain neurons, artificial units should be trained in order to solve a problem correctly, e.g., to recognize an image demonstrated to an ANN. Training implies adjusting the coefficients that regulate the strength of the links between interacting neurons, e.g., between the hidden layer and output neurons in the perceptron, schematically represented in Figure 2.8, A. These coefficients correspond to what is referred to as *link weights* in network science in general[55]. For example, let us assume that the link between a hidden layer unit and an output unit has a weight of 0.2 and the output unit is excited (fires) if the total input reaches unity. This means that this hidden unit should send five consecutive signals (or there should be four other units with the same weight working in parallel) to cause the discharge of the output unit. The overarching goal of training a neural network is to adjust the link weight values in such a way that their sum total reaches the threshold level, enabling neuron firing (impulse generation) whenever the network finds an adequate solution to a problem, e.g., recognizes an image.

A step was taken towards approximating the actual properties of biological neural networks by developing *recurrent* ANNs (Figure 2.8, B). This area was pioneered by J. J. Hopfield (1982) who developed networks where the output was linked by feedback connections with the input-processing nodes of the network, such as the "hidden layer(s)" of a perceptron or its modified version. In a similar fashion, neurons in the secondary visual cortex of the brain collect input data from the primary visual cortex (V_1) and recognize the patterns in them; in addition, the secondary visual cortex sends messages back to V_1 via feedback links (Angelucci & Sainsbury, 2006).

In many biological neural networks, nodes can compete with one another, and similar types of competition are characteristic of the nodes of the network suggested by Kohonen (2001 [1989]). The network contains a number of units that function in parallel. They obtain the same input signal and generate response signals. Among the response signals, the maximum signal is selected; this signal is considered equal to 1 ("The winner takes it all"), and all other response signals are ignored (=0). Kohonen neurons can be compared to a set of lamps; only one of them turns on in response to any input data (Wassermann, 1992). Importantly, competition between network nodes (i.e., quasi-market interactions between them) is a widely spread phenomenon in biological neural network structures. The activity of one group of neurons can suppress that of another group, and vice versa (Dubynin et al., 2003).

---

[54] The *Adalin* network developed by Professor Bernard Widrow and his graduate student, Ted Hoffat (Stanford University), in 1960 contained only one layer of artificial neurons that created the output.

[55] As emphasized in 1.1, links between nodes in many networks differ with respect to their strength or efficiency; this difference is referred to as *link weight* (w).

Currently, ANNs ("artificial intelligence" systems) are widely used for analyzing speech, recognizing images, solving statistical problems, and optimizing the operation of a variety of technical systems. They also serve as models of natural (biological) neural networks.

2.5.2. The General Properties of Neural Networks

Research on ANNs has recently provided important insights into the general features of neural networks that are approximated by their artificial analogs. Moreover, it has become obvious that many of these features are actually characteristic of biological network structures in general, irrespective of their specific type or paradigm. To an extent, most network structures in biological systems can be regarded as *quasi-neural* structures, even though their behavior may be more adequately described in terms of other paradigms. The following are the typical properties of biological neural networks, as well as of most ANNs:

1. Neural networks are capable of *collective information processing and decision-making* (cf item 6 below); I reemphasize that even bacterial network structures are capable of choosing between alternative behavior strategies, e.g., by using quorum-sensing systems.
2. Neural networks can combine *serial processing* of information (the *input* → *hidden layer(s)* → *output* sequence in a perceptron)[56] and its *parallel processing* enabled by the coexistence of several units belonging to the same layer. In a similar fashion, several zooids can concomitantly respond to the same stimulus in a cnidarian modular network; several workers in an ant colony deal with the same operation and can change from one functional role to another, in terms of the *series–parallel* strategy mentioned at the end of the preceding section. Within the cerebral cortex, there are at least two parallel systems that process visual information: one system recognizes objects (the ventral stream: the "what" pathway); and the other deals with their arrangement in space and with movement (the dorsal stream: the "where"/"how" pathway). Language is analyzed by three different, although interconnected, networks inside the brain; memory also involves several information-processing systems (Kosslyn & Koenig, 1992). As for ANNs, combining serial and parallel information processing is considered an important practical advantage. "Since the neural networks are massively parallel in nature, they can perform computations faster and help find solutions in computation-intensive problems" (Balakrishnan & Weil, 1996, p.107). Importantly, information is partly delocalized in the brain, i.e., it can simultaneously be processed in a large number of brain structures.
3. A large number of neural networks can combine several functions. This is exemplified by the reticulate formation of the brain—a complex of "cells, cell aggregations, and fibres that form a network located in the brainstem (the medulla oblongata, the pons, and the midbrain). The reticulate formation is linked to all sense organs as well as to the locomotive and sensory areas of the cortex, thalamus,

[56] As I mentioned above, these feed-forward connections are found in many network types supplemented by feedback *output* → *input* links.

hypothalamus, and spinal cord. It controls the excitation level and the tone of various parts of the nervous system including the cortex and is involved in regulating the sleep-wake cycle, vegetative functions, and locomotive behaviors" (Dubynin et al., 2003, pp.56-57). It is still controversial whether different intellectual faculties including verbal, mathematical, spatial, and musical capacities share the same neurological basis (the *g* factor) that is evaluated during an IQ test. Nevertheless, this hypothesis is supported by the data that these capacities are closely interrelated, i.e., individuals with a high level of one of these capacities tend to perform similarly well with respect to other capacities (Gottfredson, 2001); the analogous parallel development of different intellectual abilities occurs in higher animals such as primates (Byrnes & Bates, 2007) and cetaceans, e.g., dolphins (Pearson & Shelton, 2010). These data can be accounted for by the involvement of the same *multifunctional* set(s) of neural networks of the lateral frontal cortex in different kinds of intellectual work.

4. Neural networks, including ANNs, are characterized by an *associative* mode of operation. This implies that neural networks can create the image of the whole object based on its fragments. Order is created from chaos due to the cooperation among a large number of network nodes. Different nodes may deal with different features of the objects to be recognized. As for the brain, not only do at least two different systems process visual data in parallel, but tactile information processing (object recognition using the sense of touch) involves the parallel functioning of two brain areas where the neurons are responsible for evaluating the macrogeometric properties (length, width) and the microgeometric peculiarities (surface structure) of objects (Bohlhalter et al., 2002). Brain networks, as well as many ANN types, can create a complete object image by generating details that have not been demonstrated to them. Moreover, they can distinguish the object from the background. This is the reason why visual illusions are possible: the brain can separate the object from the background in several different ways, e.g., recognizing either a vase or two human faces in the same picture. As for ANNs such as the Hopfield network, the object image represents the equilibrium state achieved by a set of nodes, with each node reflecting only a small part of the object. Object-recognizing ANNs, e.g., those developed for military purposes, face the same *object–background* dilemma as the human brain. Instead of recognizing tanks, an ANN may learn to recognize the landscape. In artificial *Cerebellar Model Articulation Controller* (*CMAC*) networks, the whole multidimensional space containing input information is subdivided into units (hypercubes), each of them corresponding to a special memory cell where the information on this unit is stored (see Balakrishnan & Weil, 1996). The output sums up the contents of all memory cells involved in processing the input object.
5. Neural networks are *adaptive*; they can change their structure in order to optimally solve problems. In neural networks, adaptivity is based upon the variability of the coefficients that influence their responses to signals. Coefficient (link weight) adjustment can be carried out by a teacher training the network. Alternatively, a network can learn by itself in a process similar to natural selection. It eliminates behavior that is maladaptive in a given situation.
6. Neural networks, as well as a large number of other networks (e.g., fish schools or insect societies), are often characterized by the comparatively simple behaviors of

their nodes (neurons or their analogs). A typical neuron is a binary system that can assume one of the two states digitally denoted as 1 (the active/excited state) and 0 (the inactive state). The 0 → 1 transition takes place if the summated excitatory stimuli received by the neuron exceed the threshold level. The highly complex behavior of the whole network that both recognizes images and makes decisions represents an emergent phenomenon. Only the relatively simple algorithms should be initially set, and network features such as link weights should be adequately adjusted. This often is the teacher's job. Even if a network "learns by itself," its successful functioning depends on the preparatory work done by a teacher/controller. Hopfield's recurrent networks converge to the correct image, provided that their characteristics (including feedback link weights) have been adequately set initially. This point is of relevance to the properties of network structures in general. Their seemingly spontaneously efficient behavior may actually depend on the involvement of an *external controlling agent*. At the molecular level of biological systems, of special interest are chaperones, i.e., molecules that control the assembly and folding of other biological molecules such as proteins. In some cases, a biological molecule includes a part whose function is to secure the correct arrangement of the other parts of the molecule. In human society, there are special organizations that mediate the relationships among networks as well as between networks and other types of social structures (see 1.8 above).

7. Parallel information processing by a large number of neural network nodes, as well as neural networks' capacity to readjust their structure (adaptivity, plasticity), contribute to the *enhanced reliability* of the networks. If a part of their nodes and links ceases to function, the whole network can be restructured, and some of the intact nodes and collateral connections can at least partially replace the impaired part of the network. This is an advantage of biological neural networks and ANNs over traditional computers where serial processing of information prevails over parallel processing. A number of brain functions, such as regulation of the sleep–wake rhythm, are performed by several structures in the brain. The issue to raise is whether, and to what extent, functional parts of the brain can replace their dysfunctional or missing parts. Such *compensation* is, in some cases, unfeasible[57], incomplete (*subcompensation*), or temporary (followed by *decompensation*, i.e., the relapse of the original problem caused, e.g., by cerebral trauma). Nonetheless, there are also brain lesions that can be completely compensated for by intact cortical neural networks, to the point of an excessive development of the originally lost capacities (*hypercompensation*). "Perception, movement, language, thought, and memory are all made possible by the serial and parallel interlinks of several brain regions, each with specific function. As a result, damage to a single area need not result in the loss of an entire faculty… Even if a behavior initially disappears, it may partially return as undamaged parts of the brain reorganize their linkages" (Kandel et al., 2000, p.15). Despite the functional specialization of neurons and their networks in the brain, some of its parts can be replaced to an extent, depending on the neural

[57] Such compensation is impossible or, at least, incomplete if the operation of the brain networks involved is based upon the AND principle (all the networks should function) rather than the OR principle (at least one of the networks involved should function). The AND principle is characteristic of a large number of psychologically important brain activities including the *social cognitive function (SCF)*.

networks' plasticity and the capacity of the intact parts to change their specialization for the purpose of compensating for the lesion[58].

The features listed above are characteristic of brain neural networks, as well as of ANNs. As mentioned above, other types of biological networks often display neural network-like behaviors such as parallel information transmission involving many network nodes and associative information processing ("creating order from chaos"); they are characterized by enhanced reliability owing to the overlapping functions of their nodes and the existence of collateral connections between them.

Quasi-neural features are also exhibited by a variety of network structures in human society, and some of them intentionally attempt to behave in conformity with the neural paradigm. For instance, business firms form networks (strategic alliances) aimed at developing distributed intelligence as an emergent phenomenon requiring concerted efforts of all the firms involved.

2.5.3. Networks and Hierarchies in Neural Systems

ANNs and biological neural networks include not only distributed network structures *sensu stricto*, but also hierarchies existing at several levels.

Network and hierarchical organization patterns are used in combination at different structural levels of the brain. This organ as a whole exerts hierarchical control over the operation of the whole body. This hierarchy, nevertheless, coexists with peripheral network structures that are influenced, but are not directly controlled by, the brain. I also reemphasize the importance of microbial networks that inhabit various niches, and especially the gastrointestinal tract, and release substances including neurochemicals that produce their effects on the human brain. This point was considered in section 1.8 above. Inside the brain, the network principle seems to prevail (Terekhin & Budilova, 1995).

Temporary hierarchical structures, like "project teams" formed by ants, are established for the purpose of doing an urgent task. Some groups of neurons temporarily play a dominant role in responding to an important external or internal stimulus or adequately behaving in a given situation. However, in this case, the brain combines hierarchical and network organization because a neuronal *network*, not an individual leader neuron, is at the top of the temporary, situation-dependent hierarchy that is involved in the current motivational or emotional state of an individual. The positive emotional state attained by satisfying an important need is associated with the activation of a brain network where the key role is played by the positive reinforcement center in the hypothalamus. In contrast, negative emotions caused by unsatisfied needs activate a different hierarchized network with the negative reinforcement center (another hypothalamus area) at the top. The two centers in the hypothalamus compete with one another. As already mentioned, this fact points to a contribution of *quasi-market interactions* to the functioning of cortical neural networks. However, both hierarchical and quasi-market interactions are generally less important than

[58] In a similar fashion, removing all specialists, e.g., all foragers from an anthill, does not stop the functioning of the whole system because a part of the other worker ants change their "profession" to compensate for the loss of the foragers (Zakharov, 1991).

network organization, particularly as a temporarily dominant subnetwork constantly communicates with other areas of the brain, including networks characterized by long-term excitation. This intrabrain "crosstalk" results in incorporating new brain structures into the expanding active brain network[59]. This pattern implicates the recruitment of originally less active neuron groups and is analogous to the "*active specialists + the pool of mobilizable generalists*" principle working in the eusocial structures of ants.

Permanent hierarchies in the brain are associated with the serial processing of information by brain neurons that finally result in the information's transmission to relatively few output cells. (A similar pattern is characteristic of many ANNs, such as perceptrons [Figure 2.8]). For instance, the system responsible for vision includes the dorsal and ventral stream, in which information from the primary visual cortex (V1) passes through the secondary (V2) and tertiary (V4, V5, etc.[60]) visual cortex. In both streams, neurons "at locations progressively farther anterior from V1 are selectively activated by increasingly complex visual images" (Nicholls et al., 2012). This implies that information from a large number of primary neurons is collected by a few secondary/tertiary neurons. Some of the neurons are activated by, e.g., human faces.

Many specialized brain *modules*, i.e., networks dealing with a specific function[61], include "subordinate" subnetworks that do simple routinized tasks, as well as "bosses", i.e., subnetworks that sum up the results of the "subordinates'" work. The top position in the hierarchy within the thalamus is apparently held by the dorsomedial thalamic nucleus that integrates information supplied by other hypothalamus structures; in turn, the nucleus is subordinate to the frontal cortex (Dubynin et al., 2003).

The *frontal lobes* of the brain perform particularly advanced functions linked to intentionality, purposefulness, and complex decision making, and they also serve the *social cognitive function (SCF)* that enables one to evaluate his/her own and another's social status and rank, while knowing what one must, may, and must not do in a particular social situation (Bechara, 2002). The SCF is closely linked to the capacity to understand others and to know what they can and what they cannot know (*theory of mind; TOM*). The frontal lobes are referred to as "the executive brain" that exercises leadership in the whole brain. "The frontal lobes are to the brain what a conductor is to an orchestra, a general to an army, the chief executive officer to a corporation" (Goldberg, 2001, p.2). The dominance of the frontal lobes is mitigated by intense communication with other brain structures that represent decentralized networks; they can significantly influence the decisions made by the frontal lobes. Moreover, different areas of the frontal lobes exert their leadership across different specific functions, similar to partial leaders of a hirama-type social network structure.

Since information finally concentrates on few neural network nodes, as exemplified by frontal neurons or perceptron output units, many neural networks can be considered as scale-

[59] The whole brain is an *isolationist*-type network (see 1.7.2.1) enclosed in the skull as an "external skeleton" (Bogdanov's *degression*). Inside the brain, some of the networks tend to approach the isolationist type (modules with precisely defined functions, e.g., auditory cortex cells responding to a specific sound pitch only), whereas others represent *interactive* structures tending to expand by recruiting neighboring brain neural networks.

[60] Some of the tertiary brain cortex structures are only involved in one of the two visual pathways.

[61] A brain module may represent a local structure, a group of adjacent brain structures, or a delocalized brain network that is responsible for a certain activity such as, e.g., memorizing information. Like a module (a zooid) in a cnidarian colony (see 2.2), it is a lower-level network structure which forms part of a higher-order network (the brain or, respectively, the cnidarian modular organism).

free Barabási-Albert networks (see 1.1) where the node degree distributions comply with the power law, and there are hubs that have a disproportionately large number of links. Nonetheless, it may also be expedient to apply the random (Erdö-Renyi) network concept to some neural networks[62].

Importantly, the frontal lobes' leadership is far from being authoritarian, and subordinate brain structures often perform their behavior without "asking their bosses for permission" (Dubynin et al., 2003, p.189). A large number of intracerebral information flow pathways bypass the frontal lobes.

2.5.4. The Fractal Properties of Neural Networks

Like a microbial colony, complex neural networks exemplified by the brain and advanced ANNs represent multilevel network structures, networks composed of networks. The whole brain network consists of a large number of smaller, partly autonomous, networks. Each network performs its functions and is subdivided into even smaller functional units that are composed of a limited number of neurons (like microcolonies and "rafts" inside bacterial colonies or biofilms).

One approach to the fractal properties of neural networks is based upon the *K set* concept. The term K set was suggested in the honor of Aharon Kachalsky, an early pioneer of neural dynamics (quoted from: Kozma & Puljic, 2013). In the brain (or its analog), K sets represent an intermediate level between individual neurons and large brain structures. These cell groups differ in their size; smaller K sets form part of larger K sets. A minimal K set (denoted as K0) consists of about 10,000 neurons; it is exemplified by vertical neuron columns that make up the primary visual cortex. Each column specializes in a particular task, such as firing in response to a specific visual stimulus (e.g., a vertical line). A K0 set per se tends to stay inactive unless disturbed, but it forms part of a larger KI set made of interacting K0 units; a KI set as a whole can maintain a non-zero activity level, owing to the constant interaction between the K0 components. In their turn, K1 sets combine to form a larger KII unit exemplified by the olfactory bulb or hippocampal regions. Due to the excitatory and inhibitory interactions between its KI components, a KII set displays much more complex activity patterns, including limit-cycle oscillations. KIII sets composed of KII sets include, in the brain, the olfactory system or the whole hippocampus where the main KII components are the CA1, CA2, and CA3 sections. "KIII sets generate broad-band, aperiodic/chaotic oscillations as background activity by combined negative and positive feedback among several KII populations with incommensurate frequencies" (Kozma & Puljic, 2013, p.103). Finally, global KIV units are made up of several KIII sets. An example of a KIV unit is the entire limbic system, which is directly involved in affective behavior and emotions. KIV "exhibits global phase transitions, which are the manifestations of hemisphere-wide

[62] Random networks can be created starting from a complete graph (where all possible links between nodes actually exist). By randomly removing, with a certain probability, some of these links, we can obtain random networks whose node degree distribution is approximated by the Gaussian curve. This method was actually used in an epileptic brain where some of the links between neurons can be microsurgically removed. It was predicted mathematically that, after deleting an adequate number of links, the resulting random network tends to attain a dormant equilibrium state in which its neurons do not synchronously fire and, therefore, do not cause epileptic seizures (Sakyte & Ragulskis, 2011).

cooperation through intermittent large-scale synchronization across the hemisphere" (Ibid.). Although the K0 through KIV units form a hierarchy in the brain, each of these components represents a distributed network structure.

2.5.5. The Matrix-Like Role of Glia

Neurons in cortical neural networks are supplemented by several types of glial cells (oligodendrocytes, Schwann cells, astrocytes, and others). They play most of the roles that are characteristic of the matrix of microbial colonies or biofilms (see 2.1). Glial cells (glia) perform *structural and integrative functions*: they "support neurons, providing the brain with structure. They also separate and sometimes insulate neuronal groups and synaptic connections from each other" (Kandel et al., 2000, p.20). Their *protective function* is exemplified by the involvement of glial cells in establishing the blood–brain barrier, allowing selective entry of chemical agents into the brain. They also perform a *communicative function.* Glial cells promote the communication between nervous cells. Oligodendrocytes and Schwann cells coat neuron axons with myelin layers, which drastically accelerate impulse transmission and, therefore, facilitate the synchronization of processes carried out by neural networks. Communication among the neurons and between their functional groups is facilitated by potassium buffering, interstitial volume control, and the maintenance of a low interstitial glutamate concentration. These functions are performed by astrocytes in the healthy brain, and the disruption of normal glial functions often occurs during the development of epilepsy (Wetherington et al., 2008).

In addition to these "quasi-matrix" functions, glial cells are directly involved in impulse transmission (see Nicholls et al., 2012), and many neurotransmitters are also called "gliatransmitters." Nonetheless, the "quasi-matrix" role is also of relevance.

Neural structures and their artificial analogs represent three-dimensional networks with partial/temporary leader cells (or leader subnetworks). In contrast to other network paradigms in biological systems, the neural paradigm predominantly implies local (synaptic contact-mediated) rather than distant coordinating interactions. The neural paradigm also places special emphasis on parallel information processing (while allowing for an important contribution of consecutive information processing), collective decision making, learning, and cognitive (mental) capacities, although these features are also displayed—to an extent—by biological networks that conform to other paradigms.

2.6. Egalitarian Paradigm

Three-dimensional network structures are characteristic of a number of primate species, including chimpanzees, bonobos (formerly called dwarf chimpanzees), Tonkin macaques, white-faced capuchins, and muriquis or woolly spider monkeys. Such structures are referred to as *egalitarian* in the literature (see, e.g., Butovskaya, 2002). Non-hierarchical relationships between individuals in them are associated with loyal (friendly) interactions, including mutual attraction (affiliation) and mutual aid (cooperation). These relationships manifest themselves in loyal behaviors (Figure 2.9, A) such as grooming, play behavior, greeting rituals, and reconciliation. They mitigate the hierarchical dominance–submission relationships that, nevertheless, exist in communities of, e.g., bonobos and, even more so, chimpanzees (de Waal, 1996, 1997; Deryagina & Butovskaya, 2004).

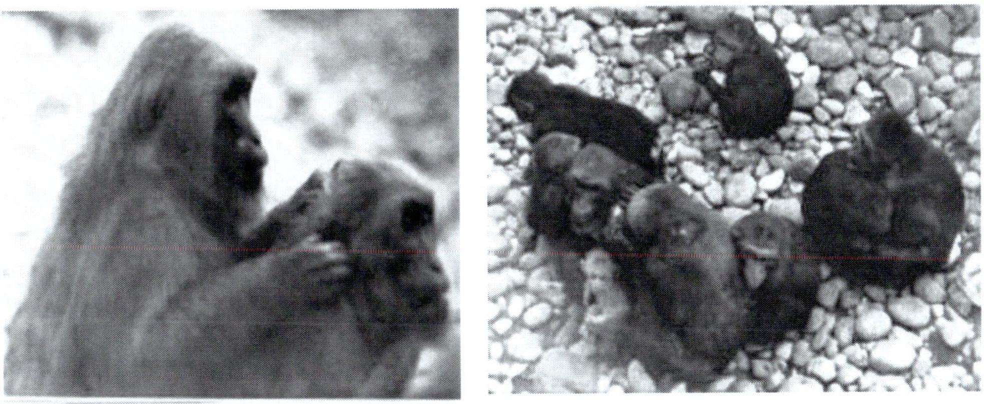

A B

Figure 2.9. Loyal behavior (grooming) in Tonkin macaques (A) and the formation of non-hierarchical groups ("clustering") in brown macaques (B). Both pictures are from Prof. Marina L. Butovskaya, with permission.

The term "egalitarian," although derived from the French word *egal* (equal), does not actually imply a complete equalization of social ranks in our close evolutionary relatives. The egalitarian paradigm of network organization is compatible with considerable hierarchization of the network structure. For instance, chimpanzee males "are ordered into a hierarchy so that one male usually emerges as the top ranking or alpha male"[63] (Goodall, 1996, p.243), although codominance by a male coalition is also possible (see below). However, even low-ranking individuals enjoy much personal freedom; high-ranking individuals are tolerant of them and acknowledge their "inalienable rights" such as the "right" to get one's share of meat or to quit one social group and join another (Butovskaya & Fainberg, 1993; Goodall, 1994; de Waal, 1996, 1997, 2006, 2008, 2009).

Like other network paradigms, the egalitarian paradigm applies to several different groups of biological species. Apart from some primates, cetaceans such as dolphins seem to comply with this paradigm (Lusseau, 2003; Lusseau & Newman, 2004; Pearson & Shelton, 2010).

Egalitarian networks tend to enhance the individuality of each of their members, which is in contrast to equipotential fish schools that tend to eliminate individual differences. The enhancement of the individuality of all network nodes implies that the development of their personalities is promoted and not suppressed.

Generally, in terms of psychology, the concept of *personality* means "that individuals display stability and consistency in their behaviour across time and situations, and that they differ from each other in the pattern of traits that constitute their personality" (Uher et al., 2008, p.100). Groups of each of the four main great apes (the bonobo, the chimpanzee, the mountain gorilla, and the orangutan) contain individuals with reliably different personal traits

[63] By contrast, chimpanzee *females* seem to be less hierarchical and form flatter networks when they establish long-lasting bonds with one another (Lehmann & Boesch, 2009).

that show cross-test persistency. Such traits are exemplified by different levels of aggressiveness to humans, arousability, anxiousness, competitiveness, curiosity, distractibility, and other individual features (Uher et al., 2008). Personality differences in rhesus macaques were evaluated on the basis of the following personality dimensions: sociability, boldness, excitability, and equability, and different dominance styles were revealed accordingly[64]. "Three of the four personality types in high-ranking males had a positive association with social power, suggesting that individuals of different personality receive signals of subordination for different reasons. Bold individuals likely receive signals because they are approaching individuals with greater frequency. Excitable individuals receive submissive signals from others because they are unpredictable; other group members may give signals of subordination by default to avoid unpredictable aggression. Equable individuals likely receive signals because they are respected and popular members of the group" (McCowan et al., 2011, e22350). Apart from dominance, personality traits influenced the intervention of rhesus individuals in settling conflicts between group members.

Studies on the network organization of primates are of special interest because primates include the chimpanzee and the bonobo—the closest evolutionary relatives of the human species. Suffice it to mention that 98.77% of the genome (the DNA) is common to chimpanzees and humans (Mikkelsen et al., 2005). Accordingly, in this section, we prepare the ground for the part of the book that deals with network structures in human society (Chapter three).

2.6.1. The Application of "Network-Science" Analytical Tools to Primate Networks

Many network structures considered above contain a very large number of nodes. They are exemplified by bacterial biofilms and cortical neural networks that may contain millions and even billions of bacterial cells and neurons (plus glia), respectively. Fish schools may include thousands of individuals[65]. Primate networks contain relatively few individuals, even though many of these networks display a multilevel structure. However, even the upper structural levels (e.g., megacoalitions in the chimpanzee; see below) typically contain several dozen or, maximally, about one hundred individuals. Therefore, the individual peculiarities of each node are of considerable importance for the functioning of the whole egalitarian network.

A similar phenomenon is characteristic of human networks. In small-sized networks, adding or removing a single member may drastically change the properties of the whole network and even result in its demise. By contrast, larger networks appear more homogeneous, and replacing thousands of their members with new people often presents no

[64] The existence of a number of alternative dominance/leadership styles that are correlated with different personality types raises the issue of whether several codominant individuals with different styles can coexist within one group of primates. This would result in a split leadership pattern that is similar to the *hirama* pattern (1.2.2) that includes several partial leaders. Indeed, the codominance of several high-ranking individuals is possible, and it is exemplified below (2.6.3.2) by the coexistence of two dominants in a single gorilla group (Deryagina & Butovskaya, 1992, p.65).

[65] However, such network structures consist of small-sized subnetworks (clusters, cliques) that are exemplified by small groups of fish only containing several individuals.

organizational problems; these networks are exemplified by giant international network structures.

Due to the relatively small number of nodes in many egalitarian networks, using some of the analytical tools developed with respect to networks *sensu lato* (1.1.2) presents difficulties. For instance, small-sized network structures are often difficult to unambiguously classify into random and scale-free networks.

Nonetheless, such "network science" concepts as node degree, clustering, and assortativity, seem to apply to egalitarian networks (Flack et al., 2006; Krause et al., 2009; Ramos-Fernández et al., 2009). Moreover, based on studies on primate egalitarian networks, it is evident that "network science" analytical tools, originally developed for networks in a broad sense, actually enable us to determine whether a system under study is a decentralized network *sensu stricto* or a centralized hierarchy.

In this context, of relevance is recent research on spider monkey (*Ateles geoffroyi*) groups. Using the centrality criteria discussed above (see 1.2.4.1), the researchers demonstrated that spider monkeys combine network and hierarchical organization. Their adult females form the high-ranking core subgroup, which, however, contains only random associations of individuals and, therefore, appears to represent a flat substructure per se (Ramoz-Fernández et al., 2009).

The "network science" approach revealed that the stability of the structure of the macaque, *Macaca nemestrina*—despite conflicts between its members—largely depends on the policing behavior of three high-ranking males that "receive a disproportionate number of subordinate signals called silent-bared teeth displays (SBT) from 45 mature (61 total) group members" (Flack et al., 2006, p.427). Removing the three high-ranking males resulted in destabilizing the group, which manifested itself in "smaller, less diverse, and less integrated… networks of social interactions" (grooming, play, and others) (Ibid., p.426). The whole structure appears to behave like a centralized hierarchy, although the dominance function is distributed among the several influential "hubs."

2.6.2. Primate Social Organization: Diverse Scenarios

Primates display diverse forms of interindividual interactions and social structures. The four main kinds of great apes (bonobos, chimpanzees, gorillas, and orangutans) are characterized by significantly different social patterns. Bonobos and chimpanzees establish three-dimensional networks dominated by females (bonobos) or more hierarchical males (chimpanzees). However, the degree of their hierarchization is limited because of the relatively egalitarian dominance style of the alpha individual. Gorilla groups have a dominant silverback male accompanied by a harem and, in some groups, nondominant males that are generally on friendly terms with the dominant silverback. Finally, orangutans are typically solitary; they only occasionally form aggregations, particularly if the pattern of resource distribution is patchy (Mitani et al., 1991).

Baboons, many macaques, and a large number of other monkeys establish rigid hierarchies, unless they lead a solitary lifestyle. Nevertheless, in addition to such hierarchies, loose non-hierarchical structures occur in their groups, which are particularly characteristic of associations of young individuals. In addition, not all monkey species behave in conformity with the hierarchical organizational pattern. To reiterate, Tonkin macaques, white-faced capuchins, and woolly spider monkeys display a mitigated hierarchy, which enables them to form relatively egalitarian structures characterized by advanced communication facilities. To an extent, these structures are similar to those of chimpanzees or bonobos.

Social organization, including the preponderance of either hierarchies or network relationships, widely varies even in closely related primate species, e.g., those belonging to the macaque genus.

"In despotic species like rhesus and Japanese macaques, rigid hierarchies co-occur with relatively high rates of intense aggression, unidirectional aggression down the hierarchy, strong nepotism and consistent operation of the 'youngest ascendancy rule' in rank acquisition (maturing females assume ranks immediately below those of their mothers and above any older sisters), and low conciliatory tendencies. At the other end of the spectrum, high tolerance is associated with weaker nepotism and less kin-bias in social relationships, low rates of intense aggression, less consistent operation of the youngest ascendancy rule, and high conciliatory tendencies" (Thierry, 2007, quoted according to: Watts, 2010, p.115).

The diversity of social organization scenarios in various nonhuman primates seems to be largely due to the different distribution patterns of the food and other resources they use (Ghiglieri, 1987). For example, "when feeding on clumped resources monopolizable by single individuals or by coalitions that include only some group members is important – i.e., when contest competition for food is important – and monopolization confers nutritional advantages, linear dominance hierarchies occur" (reviewed, Watts, 2010, p.114). In addition to the resource distribution pattern and other ecological conditions, the evolutionary prehistory of each primate species involved, as well as purely social factors, appears to exert an important influence on social organization.

Social factors are considered by Hemelrijk (2002) in terms of the "self-structuring approach" that is analogous to that used to algorithmically approximate the behavior of other biological network structures, particularly neural networks. Similar to a neural network that evolves from an initial state with preset link weight values to a final equilibrium state, groups of monkeys or apes are assumed to follow simplistic rules in their behavior towards one another.

"Model entities (i.e., individuals) are assigned initial 'dominance' values /i.e., the likelihood that an individual will dominate others in the group – O.A./… They then follow one of various alternative sets of rules that determine how they move and how they behave on encounters with others moving in the same space… Group cohesion also varies, along with the probability that more than two entities meet simultaneously and 'coalitions' form" (quoted from: Watts, 2010, p.116). As a result, a system may, in some cases, choose between several possible structural options regardless of food availability and other ecological constraints.

The apparent ease with which at least some nonhuman primates can switch from one social regime to another seems to be in stark contrast with the situation in civilized human societies where any serious structural change in the political system (i.e., dismantling a hierarchy and establishing a more egalitarian regime instead) typically requires a revolution.

The following part of this section concentrates on primate *egalitarian structures* based on several main principles:

- *Respect for individual freedoms* (particularly the freedom of choice);
- *Partial hierarchization* of the structure associated with some degree of respect for high-ranking network members and a significant contribution of dominance-

submission relationships; however, no network member can *permanently* exercize total control over the whole structure;

- *Loose links* between network members (typified by chimpanzee *fission–fusion* groups);
- *Loyal (cooperative)* interactions among individuals.

2.6.3. Behavior Coordination Factors in Primate Egalitarian Structures

Three-dimensional network structures occupy an intermediate position between flat (leaderless) networks and centralized hierarchies and, therefore, combine three types of behavior coordination factors: (1) *local interactions* that are typical of nodes in a decentralized network; (2) the *controlling behavior* of the dominant individual(s) to which other individuals submit (a hierarchy-specific mechanism); and (3) a *matrix* (in our terms) that predominantly represents a sufficiently rigid system of social norms and restrictions.

2.6.3.1. Local Interactions between Individuals

Local interactions between individuals resulting in copying each other's behavior, social learning, and coordinated action involve direct contact between neighbors and are typically based on intense communication. Such interactions involve interindividual loyal (affiliative) relationships. "The need for social communication may be stronger than the need for food. Primates kept in isolation experience severe stress and often die, even though one adequately takes care of them and provides them with proper food and abundant living space" (Butovskaya, 2013). Affiliation promotes coordination within egalitarian networks.

Primates and, in particular, the great apes are capable of effective social learning involving copying others' actions, resulting in non-hierarchical behavior coordination. Monkeys and apes display collective panic and mobbing behaviors, and they, like other gregarious animals, are characterized by the social facilitation of behavior. For example, the rate of food consumption increases in the presence of other members of the same group.

Coordinate behavior on the basis of local interactions can spread within an egalitarian network in terms of the relay mechanism (cf. the section on equipotential networks; 2.3), enabling efficient group-level activities such as cooperative hunting or territory defense without a rigid hierarchy (or with a mitigated, tolerant dominance pattern). Of interest in terms of local interactions-based coordination is the non-hierarchical synchronization of individual behavior in primate groups. It evokes the synchronization (entrainment) of oscillating nodes in systems investigated by Arenas et al. (2008). As emphasized in 1.6.2, initially asynchronous nodes of a network may become at least partly synchronized as a result of communication between them. Communication enables the transferring of emotions between individuals, which results in synchronizing the behavior of a whole primate group. Such synchronization manifests itself, for instance, in the excited behavior of a group of chimpanzees that have just received valuable food, such as meat. Even though the group includes a dominant individual (the alpha male), all group members are allowed to display their emotions without showing their submission to him and "without any apparent political inhibition" (Boehm, 1999, p.18).

Interestingly, groups of olive baboons (*Papio anubis*) engage in intergroup fighting along a "front line" that may be 100 m long. Their behaviors during attacks, retreats, and other maneuvers are coordinated without a central leader, i.e., non-hierarchically, in contrast to human warfare.

Chimpanzee males form killer parties raiding the territory of another (hostile) group (van der Dennen, 2011). It is an open question whether such raids are organized by high-ranking males functioning as "warleaders," or whether, which seems more likely, the efforts of killer parties are also coordinated non-hierarchically.

Among our closest evolutionary relatives, the bonobo is characterized by close friendly relationships among females (often implicating genito–genital rubbing; Parish, 1993) and the chimpanzee by coalitions and alliances among more hierarchical males (Ghiglieri, 1987), although there is evidence that chimpanzee females can also have distinct and sometimes long-lasting social bonds (Lehmann & Boesch, 2009). Much attention in the literature (Ghiglieri, 1987; de Waal, 1996, 2006, 2008; Nakamura, 2003; Lehmann & Boesch, 2009; Watts, 2010, and other works) has recently been given to *grooming* that includes "stroking the hair, picking the hair, removing things with hand(s) or lip(s), and scratching other individuals" (Nakamura, 2003, p.60). Grooming is widely used to ease social tensions and maintain socially important relationships, and grooming cliques ("directly connected configurations of grooming interactions at any given moment"; Ibid., p.59) are compared to human social gatherings.

More long-lived groups of several individuals are referred to as *coalitions*. They are characteristic of the chimpanzee (partly hierarchical, predominantly male coalitions) and bonobos (where more horizontal female coalitions are prominent) and also occur in other primates, e.g., olive baboons and brown macaques (Figure 2.9, B), as well as in dolphins and other cetaceans. Same-sex coalition members assist each other in obtaining valuable resources, including mating partners, despite the resistance of competitors. Affiliative relationships within coalitions are often associated with agonistic interactions with a third party. These relationships are characteristic of primates, where coalition members collaboratively direct aggression at joint targets (Harcourt, 1992). For example, "male baboons can sometimes forcibly take over consortships with estrous females from higher-ranking males, for example, but require coalition partners to do so" (Watts, 2010, p.111).

"One effect of coalitions is to help individuals to win contests they would otherwise lose… Many primate species… form coalitions, in which two or more individuals collaboratively direct aggression at joint targets" (Watts, 2010, p.110-111). Naturally, female coalitions also exist, and they appear to be widespread in bonobos.

Coalitions enable the members of egalitarian structures to make good use of their social links and status. They have direct relevance to biopolitics and particularly to its subfield, which is concerned with the evolutionary prehistory of *politics* in human society. Due to the sophisticated organization of the nervous system, i.e., the neural networks (see 2.5) and, accordingly, advanced cognitive capacity, apes and, to an extent, monkeys and cetaceans (e.g., dolphins), demonstrate *Machiavellian (political) intelligence*[66], i.e., "the decision making capacity that enables social animals to further their self-interests in situations that

[66] "Niccolo Machiavelli famously recommended politicians to use social manipulation for individual profit, hence the term 'Machiavellian intelligence' for the idea that social manipulation has been favoured in animal evolution" (Byrne & Bates, 2007, p. R715).

involve rivalry and quests for power and leadership" (Boehm, 1997, quoted from: Watts, 2010, p.127). Machiavellian intelligence primarily manifests itself in social manipulation, i.e., bluffing, appeasement, aggression, deference, and deceit, which implies knowing what others can and cannot know. As already mentioned, this capacity is called *theory of mind (TOM)*.

In terms of one of the currently widely accepted hypotheses on brain development, it is the complexity of social life with its intricate relationship network and the necessity to understand others that has accelerated the evolution of the brain in several groups of animals, including primates. It is pertinent that, in primates, the larger the typical social group, the bigger the whole brain and, in particular, the neocortex (Dunbar, 1992, 1998).

As emphasized above, various types of biological network structures possess neural network-like properties. Egalitarian networks are similar to neural networks in an important respect. Neural networks including ANNs (see 2.5 above) contain nodes that change their state in response to a change in the state of another node (an input neuron) and, in many networks, exert a feedback influence on the input neuron's state. In a similar fashion, individuals in monkeys or apes can respond to a change in the (psychological, emotional) state of another individual by changing their own state in accordance with this change.

"Perception of the emotional state of another automatically activates shared representations causing a matching emotional state in the observer. With increasing cognition, state-matching evolved into more complex forms, including concern for the other and perspective-taking" (de Waal, 2008, p.279). This process enables the development of *empathy*, i.e., understanding another's emotional state and, in animals with advanced cognitive capacity, another's problems and goals, as well as another's role in a social group (Povinelli et al., 1992; de Waal, 1996, 2006, 2008, 2009). A prerequisite for the development of empathy is the progressive evolution of the brain in primates.

Advanced cognitive capacity in primates enables true *imitation*, i.e., learning new skills, by assembling them from components demonstrated by others. This is particularly characteristic of apes (therefore, "aping") (Byrnes & Bates, 2010). Imitating your neighbors is likely to be a mechanism that promotes the spreading of cultural traditions in primate communities (see below).

Cognitive capacity and empathy are closely linked with social cognitive function (SCF), which enables an individual to adequately behave in a social situation and to follow social norms. The development of these capacities promotes reconciliation after a conflict. Reconciliation (de Waal & Roosmalen, 1979) takes a variety of forms, including direct affiliative interactions between former opponents (e.g., grooming) and the involvement of third parties engaging in affiliative behavior towards the victim (anthropomorphically termed "victim consolation," although this term is only considered psychologically valid with such intelligent animals as chimpanzees; Fraser & Aureli, 2008). Conciliatory behavior is considered to be more probable if the two opponents previously (before the conflict) supported each other and spent much time together (the valuable relationship hypothesis in the literature; reviewed, van der Dennen, 2011). Since conflicts pose the threat of disrupting the delicate balance inside a social group, special efforts to settle them also include, in primates, protecting the victim, soliciting help from others, and appeasing the aggressor; some other individuals actively function as mediators (Butovskaya, 2013).

At the beginning of this work (1.2.1), I briefly discussed the concept of *reciprocal altruism*, i.e., assisting those who are expected to assist you or have already helped you. The contribution of this mechanism to interindividual interactions in primates seems likely, particularly because their advanced brain enables individually recognizing others and keeping a record of one's previous relationships with each of them. However, there is evidence that individuals in groups of the brown capuchin *Cebus apella* engage in altruistic behaviors, benefiting other individuals without keeping record. This behavior is activated just by joint action with another individual, irrespective of whether the other has previously helped the altruist (Suchak & de Waal, 2012).

2.6.3.2. The Leader's Regulatory Role

Obviously, this hierarchy-specific coordination factor is of limited importance in an egalitarian three-dimensional network. To reiterate, high-ranking individuals display tolerant behavior towards subordinates in such networks. In the egalitarian groups of the chimpanzee and the bonobo, as well as in many primitive human societies, "group members were quite tolerant towards one another, they avoided aggression against group mates, and low-ranking individuals enjoyed the 'right to vote', the 'freedom of exploratory activities', ... and the 'property right'" (Butovskaya, 2002, p.49).

Even though the alpha male in a group of chimpanzees may bully other males that should display submissive behavior (pant–grunt greeting, crouching, bobbying, etc.), he seldom actually *controls* others' behavior, so that "every chimpanzee decides autonomously where to forage, and whether or not to join in a hunt or go on patrol" (Boehm, 1999, p.26). However, high-ranking individuals often engage in *communication* with others, and this communication enables us to consider them, under certain circumstances, as nurturing leaders. For instance, Chistopher Boehm (in the work cited above) observed the alpha male (Goblin) in a chimpanzee group encourage others to harass a dangerous python that finally escaped from the group-occupied area. Subordinate group members can approach a dominant male to "request reassurance by proffering a hand, and a dominant may respond by touching the nervous subordinate" (Boehm, 1999, p.24), thereby communicating a reassuring message to him.

The alpha male can behave as a nurturing leader and not a repressive dominant, even in the more hierarchical mountain *gorillas*. According to Dian Fossey (1983), the leader male in a gorilla group repeatedly helped group members escape from traps laid by poachers: he broke wire loops and set the victim free.

An important function of high-ranking individuals is their involvement in reconciliation following a conflict between group members. Frans de Waal (1982, p.124) described a quarrel between two chimpanzee individuals that "ended in biting and fighting. Numerous apes rushed up to the two warring females and joined in the fray. A huge knot of fighting, screaming apes rolled around in the sand, until Luit (the alpha male – O.A.) lept in and literally pulled them apart. He did not choose side in the conflict, like the others; instead anyone who continued to fight received a blow from him." In a similar manner, leaders in primitive human societies (e.g., peacemaking leaders that wear leopard skin) promoted the settling of intragroup conflicts.

The hierarchical component of egalitarian three-dimensional networks can be based on agonistic interactions involving aggression or threatening displays (an agonistic hierarchy; see 1.2) or the group's preferential attention (a hedonic hierarchy). The second option is characteristic, e.g., of chimpanzee networks where a male can attain a high social rank thanks to the support of a *coalition* (see 2.6.3.1) Forcefulness and assertiveness are not mandatory for the dominant status, although they may play a role. Nevertheless, relatively small and weak males can assume a high social rank if supported by others whom they actively groom (Popov, 2009; http://membrana.ru/particle/787).

Apes are characterized by a certain degree of Machiavellian intelligence (see above), and their "political roles" often change over time. In particular, chimpanzee "males seem intent on domination yet submit readily when it is necessary to do so" (Boehm, 1999, p.23). Moreover, there may be prolonged periods in which there is no stable hierarchy (for instance, a "subordinate male challenges a higher-ranking male and neither gives in"; Boehm, 1999, p.25). Alternatively, the dominant role can temporarily be played by a whole coalition instead of a single alpha male that may finally take over.

Not only do high-ranking individuals have limited power and respect some "rights and freedoms" of the lower-ranking ones; egalitarianism in a primate group is highlighted by the fact that "dominance roles become reversed in certain situations — as when a low-ranking male hunter possesses prey he himself has caught[67] or when a male has taken a female on an extended consortship and a higher ranking male tries to intrude" (Goodall, 1996, quoted from: Boehm, 1999, p.22).

In a hirama-like fashion, a gorilla group can be characterized by a split leadership pattern (*codominance*): "the leader's functions are distributed among several adult males. One of them coordinates group movements, and another guards the group at the resting site" (Deryagina & Butovskaya, 1992. p.65).

High-ranking chimpanzees establish "oligarchies" that collectively perform the functions of the group leaders. Coalitions are also established by subordinates that build up "organized opposition movements." In chimpanzees, "adult males are opportunistic in their relationships, making and breaking alliances for individual advantage as the relative power of each male waxes and wanes" (Pusey, 2001, quoted from: van der Dennen, 2011, p.62).

In human society, coalitions represented a powerful political force in various historical periods. Importantly, both ancient and modern politics involved establishing and maintaining coalitions including influential people. Octavian Augustus overpowered a strong rival, Antonius, and became the political leader in Rome because he established a sufficiently powerful coalition.

Mitigated hierarchies and coalitions involving grooming sessions are also characteristic of partly egalitarian structures established by meerkats in the deserts of South Africa. Their groups containing three to 50 individuals are characterized by dominance–submission relationships. The top of the hierarchy is occupied by the breeding pair of individuals that sires most offspring and dominates other adults and young individuals. Egalitarianism in a meerkat group is based on networks of grooming individuals. It was revealed that meerkats groom one another regardless of the dominance hierarchy; the more stable the hierarchy (the longer the same dominant pair remains on its pinnacle) and the larger the group size, the more evenly grooming-based relationships are distributed throughout the group. Importantly, loyal behaviors involving grooming prevail over agonistic interactions associated with hierarchical relationships, as well as over interactions based upon competition for food resources: "almost four times as many grooming interactions as either foraging competition or dominance interactions were recorded" (Madden et al., 2009).

[67] This situation is related to the "property right" of chimpanzee individuals that was briefly discussed above.

2.6.3.3. The Set of Social Norms (The Matrix)

Despite the apparently chaotic behavior of the members of some egalitarian networks (e.g., in a chimpanzee group), they behave in conformity with a system of norms and restrictions that should not be ignored, even by the alpha male. It is the matrix of social norms and restrictions that imposes limitations on the dominant's power.

He typically respects certain "rights" of other network members, e.g., the "property right" (the "concept of possession", Goodall, 1996); he often requests meat from an individual that has it, refraining from coercing this individual into sharing meat (Butovskaya, 2002, p.49). An adult male, normally superior to a female or a juvenile, can "beg, patiently, from a possessor who normally ranks below him in the hierarchy" (Goodall, 1996, p.250). Taking account of the whole list of such individual "rights," we can compare egalitarian network structures in apes to the most liberal democratic regimes of the modern-day world.

A number of scholars assume that *public opinion* exists, apart from *Homo sapiens,* in nonhuman primates. Interactions between two individuals (A and B) are the focus of attention of other group members that *gossip* about them. If an individual, irrespective of its rank, violates a social norm, the whole group may be involved in stopping the perpetrator.

"Jimoh, the current alpha male of the Yerkes Field Station group, once detected a secret mating between Socko, an adolescent male, and one of Jimoh's favorite females... Normally, the old male would merely chase off the culprit, but for some reason... he this time went full speed after Socko and did not give up. He chased him all around the enclosure... Before he could accomplish his aim, several females close to the scene began to 'woaow' bark. This indignant sound is used in protest against aggressors and intruders... The intensity of their calls quickly increased until literally everyone's voice was part of a deafening chorus... Once the protest had swelled to a chorus, Jimoh broke off his attack with a nervous grin on his face: he got the message. Had he failed to respond, there would no doubt have been concerted female action to end the disturbance" (de Waal, 1996, pp.91-92).

Functions analogous to the integrating matrix of cellular networks are performed by *(proto)culture* (Laland & Hoppitt, 2003; Whiten & Mesoudi, 2008), which is widely believed to exist in primates and other higher animals (e.g., dolphins) and includes nongenetically transmitted traditions, rituals, and technologies that spread due to social learning and imitation. Whiten (2010) mentions 40 behavior patterns in chimpanzees that seem to be group-specific and are not directly caused by hereditary or environmental factors. They are related to tool use, foraging techniques, social behavior, grooming methods, and courtship gambits. For example, different chimpanzee communities use different methods for fishing termites and they have group-specific greeting rituals, e.g., hand-clasp grooming in chimpanzees: "the two partners hold hands above their heads, grooming each other with their free hand" (de Waal, 1996).

"In the Goualougo Triangle, for instance, the local chimpanzees use a tool set, consisting of a stout puncturing stick and a slender probe, to exploit subterranean termite nests. It is assumed that the probing tools were already well established, since they are found in many chimpanzee populations, before the stout puncturing stick was invented. Another example from the same site is the brush-tipped termite probe, where the regular termite probe... undergoes an additional modification in which the tip is frayed, which makes it far more effective in gathering termites (which bite into the probe, and latch on more easily if the tip is frayed)" (reviewed, Pradhan et al., 2012, p.182).

The bonobo is also characterized by "protoculture" (Butovskaya, 2000), implying information acquisition by learning and accumulating social experience and incipient symbolic communication. The temporary leader of a moving group marks the route for other group members by biting tree branches in a special way. These primitive "culture artifacts" are the objectification of the *Follow this path* instruction.

As for cetaceans, e.g., dolphins, "there is evidence for traditional or cultural behavior in New Zealand dusky dolphins." For instance, "the seasonal movement between Admiralty Bay and Kaikoura is likely an example of inter-generational or oblique culture (i.e., transmission of knowledge between unrelated members of different generations, with younger males likely learning from older males)" (Pearson & Shelton, 2010, p.345).

Such primitive protocultural traditions presumably promote individuals' identification with a social network and can potentially be associated with xenophobia and conflicts with the members of other networks formed by representatives of the same biological species (e.g., of other chimpanzee groups).

2.6.4. Fission–Fusion Structures: The Fractal Properties of Egalitarian Structures

Many egalitarian network structures are characterized by very loose links between individuals (nodes). Chimpanzees form small temporary networks called *fission–fusion groups* (Le Hellaye et al., 2010).

"Chimpanzees comprising a community recognize one another and may travel in peaceful association, although they do not move about as a socially cohesive unit. Rather, individuals form and re-form in constantly changing temporary associations" (Goodall, 1996, p.243). The size of such temporary groups varies depending on the external conditions. Particularly large fission–fusion groups were observed during the dry season characterized by abundant fruit. Fission–fusion groups often consist of chimpanzee males or males plus estral (sexually active) females; inactive females prefer a solitary lifestyle, and breastfeeding females form their own fission–fusion groups (Le Hellaye et al., 2010).

Fission–fusion groups also occur in cetaceans, ungulates, elephants, and bats (Patriquin et al., 2010; Kelley et al., 2011). The formation of such temporary small groups is considered in the literature in terms of a trade-off between benefits caused by group formation (collective defense of the group area, easy access to mates, intragroup information transmission promoting protocultural traditions, protection from predators) and costs associated with enhanced competition for limited resources (Lehmann & Boesch, 2004).

Egalitarian networks are multilevel structures. Several temporary fission–fusion groups form larger, more stable network structures that can be called *megacoalitions* (a term suggested by me – O.A.). Ape megacoalitions contain up to 100 individuals (Deryagina & Butovskaya, 2004). Chimpanzee individuals primarily identify with such large networks, which can result in hostile behaviors towards representatives of a different megacoalition (Le Hellaye et al., 2010) that are "dechimpized" (van der Dennen, 2011), i.e., treated as if they belonged to a different species. Reconciliation after a conflict typically occurs if the former opponents belong to the same social group (i.e., in our terms, to the same megacoalition or a subnetwork inside it). "There is no – *never* – reconciliation after episodes of intergroup violence in chimpanzees" (Adang, 1999, quoted from: van der Dennen, 2011, p.62).

The larger the entire network (the megacoalition), the more clear-cut are the internal boundaries between small fission–fusion groups (Le Hellaye et al., 2010), i.e., the more manifest is the megacoalition's community structure (cliquishness). The network appears to consist of loosely connected subnetworks. These subnetworks themselves may be multilevel structures. For example, they include grooming cliques mentioned above that consist of 2–4 males or 5 or more females (Nakamura, 2003); several grooming cliques form *grooming clusters* "in which 10 or more individuals groom in the same session and the membership changes frequently" (van Lawick-Goodall, 1968, quoted from: Nakamura, 2003, p.60).

In cetaceans such as bottlenose dolphins, individuals form two- or even three-level alliances (Connor et al., 2006). Dolphins maintain social ties by engaging in affiliative behaviors including swimming in contact with others, synchronously rising to the water surface, and touching each other with their fins (Sakai et al., 2006). In killer whales, small pods form higher-order aggregations counting up to 100 individuals. Such aggregations are compared to social clubs where killer whales establish and maintain social ties; an additional goal of such "clubs" (like many clubs in human society) is promoting sexual interactions (Filatova et al., 2008).

Hence, the egalitarian paradigm exemplified by the three-dimensional network structures of some primate species is characterized by mitigated hierarchies with tolerant high-ranking individuals that function in conformity with a set of social norms—the matrix*—that restrict their power and secure the "rights" of all network nodes, including low-ranking individuals. The hierarchy is also mitigated by a system of decentralized horizontal interindividual interactions involving empathy, affiliation, cooperation, and social cognitive function. These interactions manifest themselves in grooming, play, greeting rituals, food sharing, and postconflict reconciliation. Many egalitarian structures are multilevel systems (megacoalitions); they often include small-sized temporary fission–fusion groups.*

2.7. Diversity of Network Paradigms in Biosystems

This chapter is focused on some of the main organizational patterns (paradigms) of network structures in biological systems. It should be reemphasized that each paradigm is not necessarily confined to a specific group of biological species. Some paradigms apply to a relatively wide variety of species. In addition, social structures of many species can be described in terms of more than one paradigm. This is exemplified by social insects: although predominantly considered from the viewpoint of the eusocial paradigm they, to an extent, fit the modular paradigm emphasizing the quasiorganismic features of their social structures. Bacterial biofilms can also be envisaged as analogs of modular organisms, as emphasized by James Shapiro (1988). All six paradigms are potentially applicable to human society (Chapter three), as they act as attractors to which real-life network structures tend under certain circumstances. The paradigms are compared in Table 2, given below.

The following addresses the issues regarding in what way and to what extent various network structures in biology conform to the general features of networks (*sensu stricto*), as set forth at the beginning of this work in section 1.2.

2.7.1. Decentralization

Decentralization is characteristic of all the variants noted above (paradigms) of network organization in biological systems, but they can be classified into two subgroups depending upon the degree of decentralization:

- *Flat (leaderless) networks* such as most cellular (microbial) network structures, fish shoals/schools, and modular organisms of cnidarians (except for colonies with leader zooids, exemplified by pennatularians)
- *Three-dimensional networks* with partial leaders (pacemakers). They include the eusocial structures of ants containing functional groups (clans) with leaders, the egalitarian structures of primates with high-ranking individuals, and neural networks with dominant subnetworks. However, partial leaders are embedded in higher-order networks. The leaders of ant clans are the members of larger decentralized networks (columns and pleiads, Zakharov, 1991).

The question to be raised is what features are typical of each of the two subgroups of network structures in biological systems.

Most of the flat networks considered above consist of relatively *uniform* nodes (cells, individuals). Cells in most microbial networks are only insignificantly differentiated in functional terms, except for some special systems. As emphasized above, equipotential fish schools tend to minimize individual differences between the fish they include. Modular structures of cnidarians may consist of uniform nodes (zooids). Alternatively, the nodes may significantly differ in morphological and functional terms. However, many modular structures are made up not directly of zooids, but of their uniform complexes of zooids, cormidia, as repetitive functional units.

By contrast, many three-dimensional network structures are *heterogeneous*. In typical eusocial systems, individuals belonging to different castes or different specialized groups reveal significant morphological and functional differences. Neurons are subdivided into different morphological types; they are functionally differentiated. Neural networks include cells with manifest individual peculiarities, e.g., “mirror neurons” in the cortex (Dubynin et al., 2003). Egalitarian networks tend to enhance the individuality of their members that are granted individual “rights” (Deryagina & Butovskaya, 2004). Presumably, individual differences between network nodes promote their stratification in terms of social ranks.

Some three-dimensional networks implement the *active specialists plus the mobilizable pool of generalists* pattern. It is typified by network structures that are formed by social insects where active specialists form hierarchical structures (clans) embedded in higher-order decentralized networks. It is also used by neural networks. While the brain is focused upon a specific task, a few subnetworks in it are predominantly activated (as evidenced by fMRI scans). These active specialists can be supported by supplementary mobilizable neural networks that otherwise exhibit low activity. A similar pattern is characteristic of human social networks of the hirama type; it is potentially applicable to creative groups of artists, small-sized businesses, service clubs, and so on.

2.7.2. (Quasi-)Organismic Properties of Biological Networks

The flat networks discussed above demonstrate (quasi)organismic features, although they manifest themselves to a different extent in different structures. Microbial network structures are envisaged as multicellular organisms (Shapiro, 1988, 1995, 1998; Ben-Jacob, 1998, 2003: Ben-Jacob et al., 2004) or, at least, as integrated microbial systems, "cities of microbes" (Watnick & Kolter, 2000; Nikolaev & Plakunov, 2007; Karatan & Watnick, 2009). The quasi-organismic properties of equipotential networks such as fish schools are less manifest; however, some important functions (e.g., protection against predation) are performed in them at the collective—rather than individual—level, and efficient coordination of individual behavior enables a fish shoal/school to function as a coherent entity that has a distinct geometric shape. A modular network structure, such as a cnidarian colony, represents a single organism (Marfenin, 1993, 2002, 2009). To an extent, flat network structures of microbial colonies and fish shoals/schools are comparable to modular organisms composed of a large number of repetitive uniform units, i.e., cells and fish individuals, respectively.

Generally, quasi-organismic properties are less characteristic of three-dimensional networks. Highly individualized fission–fusion egalitarian structures, e.g., of primates, seem to virtually lack such properties. Although a part of the animal (human) organism, the nervous system and the neural networks it comprises do not form an organism per se; besides, some neurons are characterized by conspicuous individual features.

Eusocial structures seem to stand out as three-dimensional networks that, nevertheless, possess manifest quasi-organismic properties—the reason behind considering them *superorganisms* (Wheeler, 1928; Chauvin, 1963; Hölldobler & Wilson, 1990, 2009, 2010; Kipyatkov, 1991), although this idea is still open to debate (Zakharov, 1991, 2005, 2009); moreover, their "nodes" (ants, bees, termites) demonstrate important individual features and often compete, and not only cooperate, with one another.

Chauvin and a number of other scientists (cited above) envisaged the whole ant family (the pleiad, Zakharov, 1991) as a "superorganism." If we only take into consideration the level immediately below the family level, we can consider the pleiad a flat, not a three-dimensional, structure: its nodes (columns) form no hierarchy and, moreover, are quite uniform. The system is hierarchical and heterogeneous only at the still lower level of a single clan that includes a leader (to reiterate, several clans form one column, Zakharov, 1991, 2005, 2009). At the higher levels of a column and a whole pleiad, an analog of a modular organism (consisting of homogeneous repetitive units) emerges. It can be termed a modular *superorganism,* to use the term "superorganism" preferred by Wheeler (1928), Chauvin (1963) and, more recently, by Hölldobler and Wilson (1990, 2009, 2010). However, the repetitive units are composed of heterogeneous parts (clans) that are hierarchical structures. In sum, hierarchical structures lacking quasiorganismic properties per se can combine to form more organism-like network structures (e.g., an ant family).

2.7.3. The Main Coordination Mechanisms in Biological Networks

I reemphasize that, apart from a lack of a centralized hierarchy, network structures *sensu stricto* are characterized by cooperative interactions of their nodes. A prerequisite for efficient cooperation among network nodes is the coordination of their behavior. The three main coordination factors (local interactions, distant interactions, and the matrix or its functional analog) outlined above (1.6.4) are used in varying proportions by various biological network structures (Table 2).

Despite an indesputable contribution of cell–cell contacts, cellular systems rely upon distant chemical regulators that coordinate metabolic activities and cell movement within the system against the background of the extracellular biopolymer matrix that performs a number of integrating functions, including maintaining the spatial organization of these network structures. Equipotential networks are largely coordinated using a relay system of information transmission. Zooids in cnidarian modular structures are characterized by predominantly competitive local interactions (Marfenin, 2002, 2009), and the whole structure would be similar to a quasi-market system, except for the integrative influence of the coenosarc (a matrix analog). As a result, the system follows the *competition for cooperation* principle that is characteristic of network structures. In three-dimensional networks, network-specific local[68] and distant coordination factors are supplemented by those typical of hierarchies, including the *collective imitation of the leader's behavior* and, to a limited extent, *control exercised by the leader*.

In Table 2, the line dealing with *specific organizational principles or scenarios* is focused on the prominent features of each of the network paradigms considered in this Chapter. As for the eusocial and the egalitarian paradigm, the respective cell in the Table contains two organizational principles that are combined in these networks, but they can be used independently in model structures that are created in technical systems and in human society (in terms of social technologies).

2.7.4. The Matrix and Its Functional Analogs

Apart from their nodes and links, networks contain an additional component performing structural, protective, communicative, and other network-level functions. This component is exemplified by the biopolymer *matrix* of cellular systems; analogous functions are performed by glia[69] in neural networks and by the coenosarc (together with the perisarc) of the modular organisms of cnidarians, ascidians, and bryozoans. In eusocial structures, matrix analogs include not only material structures (anthills, beehives, termite mounds, and the burrows of mole rats), but also an immaterial component, a system of social norms. This immaterial component is the only matrix analog in fish shoals/schools and in the egalitarian network structures of, e.g., primates or cetaceans where it comprises the social group-consolidating set of behavioral regulations.

---

68 In neural networks, local interactions are mainly based on impulse transfer across synapses, in eusocial networks on contact communication during which the antennas of two individuals touch, and in egalitarian networks on the whole system of face-to-face interactions between, e.g., two chimpanzees.

69 According to recent data, glial cells are also directly involved in impulse transmission in neural networks.

Table 2. Paradigms of network organization in selected biological systems

Paradigm	Cellular	Modular	Equipotential	Eusocial	Neural	Egalitarian
Predominant organizational pattern	Flat[70]			Three-dimensional		
Predominant mechanism of behavior coordination	Distant interaction	Local interaction	Local interaction enabling relay transmission of information[71]	Complex of several coordination factors	Local interaction	Complex of several coordination factors
Functional specialization of network nodes	Low	Different	Low	High	Different	Low
Node individuality	Insignificant			Different		Significant
Integration level	Different	High				Low
Level structure pattern	Flat network consisting of flat networks			Alternate network and hierarchical levels	Three-dimensional network made up of three-dimensional networks	
Specific organizational principles or scenarios	Long-range exchange of chemical signals among uniform nodes	Integrity of the whole network is secured by a matrix analog (the coenosarc in cnidarians)	Minimized individual differences promote uniform behaviors coordinated by relay mechanisms plus distant coordination factors	1. Nodes are subdivided into functional castes. 2. Active specialists interact with a pool of mobilizable generalists	Nodes are capable of learning/ training and information processing	1. Node individuality is enhanced. 2. Nodes form fission–fusion groups.
Representative biological systems	Colonies/biofilms of microorganisms, cell cultures	Colonial cnidarians, bryozoans, and ascidians	Shoaling/ schooling fish, cephalopodes	Ants, termites, bees, and other social insects; naked mole rats	Animal or human nervous systems	Apes (chimpanzees, bonobos), monkeys (capuchins, muriquis), cetaceans (e.g., dolphins)

[70] Some cellular systems, e.g., those with complex developmental patterns exemplified by myxomycets may be three-dimensional (with temporary pacemakers).

[71] However, distant communication based on chemical signals or electromagnetic fields also seems to contribute to the coordination of behaviors of individuals in equipotential structures.

2.7.5. The Algorithmization of the Behavior of Biological Network Structures

Behavioral regulations in biological systems can be approximated using algorithm-based model systems applied, e.g., to the collective behavior of fish schools and ant families (see 2.3.8 and 2.4.10 above). Simple algorithms prescribing the behavior of a node vis-à-vis its neighbors enable us to model complex behavior and to account for the formation of spatial patterns in a large number of biological systems relatively successfully. This opens up new opportunities in terms of research concerning the behavior of various biological networks. Besides, models using algorithmized principles of biological network structures have recently been used in the field of technology. For instance, "ants, honeybees, and other social insects" have been employed to develop technologically applicable "models of self-organizing systems" (Hölldobler & Wilson, 2009, pp.53-54).

This approach is also applicable to the behavior of the human species in certain situations. Groups of humans in life-threatening situations, similar to what is evident in fish shoals, often obey primitive behavior rules exemplified by the following: (1) look for the shortest path leading to a safe place; (2) avoid colliding with your neighbors; and (3) move in the same direction as your neighbors (Helbing et al., 2000). This results in a sufficiently complex pattern of the whole crowd's behavior. Unfortunately, such collective behavior often blocks the crowd's movement (Sumpter, 2006).

Algorithmization is particularly widely used in studies with neural networks and their artificial analogs. Importantly, other kinds of network structures in biological systems are, to an extent, quasi-neural, i.e., they demonstrate many of the features that are characteristic of neural networks and ANNs, including parallel information processing by a large number of network nodes, adaptability (ability to learn/train), associativity (creating order from chaos), and increased reliability of the whole network due to the concomitant operation of several subnetworks performing the same function.

It should be emphasized that algorithmization in many systems results in oversimplifying the actual behavior, particularly if the algorithm is only based upon local interaction between neighboring nodes. Importantly, many network structures in biological systems also use long-range rapid communication channels enabling synchronizing node behavior within the whole network or its large parts. They are exemplified by quorum-sensing systems in microorganisms. Analogous long-range communication channels based on chemo- or electroreception are likely to function in fish shoals/schools (Pavlov & Kasumyan, 2002, 2003). There is evidence that behavior is efficiently synchronized in groups of mammals that occupy a large area (Zaitsev, 1992). Many human network structures tend to synchronize their members' behavior even in the absence of a central leader, and this subject will be revisited in several sections of Chapter three.

2.7.6. Multilevel Biological Network Structures

The networks considered in this Chapter are multilevel systems. In most of them, each of the structural parts (subnetworks) is structurally similar to the whole network, i.e., such a network possesses fractal properties. *Flat networks* (of microbes, cnidarians, or fish) *are made up of flat subnetworks*, which may consist of still smaller "subsubnetworks". As for neural and egalitarian networks, these *three-dimensional networks are composed of three-dimensional subnetworks*, and so forth… In eusocial structures, the pattern is more complex: *Eusocial networks consist of hierarchies* that, in their turn, can be made up of networks composed of hierarchies (see 1.8.6 and 2.4.6 above). I reemphasize that such "layer-cake" structures are characterized by a repetitive interlevel pattern that makes the entire system self-similar (although the pattern is more complex than, e.g., in flat networks composed of flat subnetworks).

As pointed out in 1.6.1, some biological networks, e.g., microbial structures represent *temporally self-similar/fractal systems* because the first stage of their development includes second-order stages (substages) that are remarkably similar to the later stages (Smirnov et al., 1982).

2.7.7. Network Structures as Ambivalent Systems

Network structures *sensu stricto* constantly interact with structures of other kinds (hierarchies and [quasi-]markets); these structures can interconvert (see 1.7). These interactions and transformations, as well as the existence of several different network paradigms can be interpreted in terms of the concept of *ambivalence* used by researchers dealing with animal or human behavior.

The existentialist philosopher Nikolai Berdyaev emphasized that Russia and the Russian soul are self-contradictory: they are torn apart by *antinomies* (pairs of mutually exclusive categories). In a similar fashion, the behavior and social structures of both animals and humans can be envisaged as the result of a trade-off between opposing tendencies (Table 3).

The tendency to *establish dominance hierarchies* is in opposition to the tendency to *equalize social ranks*, and the whole gamut of interactions between networks and hierarchies is set in the stress field created by this pair of opposing tendencies.

The predisposition of both human and animal individuals and groups for *cooperation* coexists with their predisposition for *competition*, and this is the stress field in which network–(quasi-)market interactions are situated.

The tendency of individuals to stay together (*affiliation*) coexists with the opposite tendency to stay alone (*isolation*). Depending on the position of the balance point between these tendencies, networks can be dense or loose (see 1.2). For example, egalitarian networks are often very loose (fission–fusion groups).

The tendency towards *activity* struggles with that towards *passivity* (inertia), and networks with active specialists and a pool of mobilizable generalists exist in the resulting stress field. To reemphasize, such ambivalence is characteristic of both animal and human individuals and social structures.

In sum, each network paradigm is not necessarily limited to a specific group of biological species. Some paradigms apply to a relatively wide variety of species. The social structures of many species can be described in terms of more than one paradigm. All six paradigms are potentially applicable to human society as attractors to which real-life network structures tend under certain circumstances. The paradigms are compared in Table 2. The behaviors and social structures of both animals and humans can be envisaged as the result of a trade-

off between opposing tendencies; individuals and their networks exist in stress fields that are molded by these tendencies (Table 3).

Table 3. Opposing behavioral tendencies and organization of biological systems

Pairs of tendencies	Implications for network organization
Hierarchy formation—hierarchy elimination	Hierarchy ↔ network interconversion
Tendency towards cooperation—tendency towards competition	Network ↔ quasi-market interconversion
Tendency towards affiliation—tendency towards isolation	Formation of dense or loose networks
Tendency towards activity—tendency towards passivity	Networks with active specialists and a pool of mobilizable generalists exist in the resulting stress field
Tendency towards information acquisition (learning), plasticity, and self-reorganization—tendency towards conservative, routinized behavior	Networks become more quasi-neural as the tendency towards information acquisition increases
Unitary organization—modular organization	Depending on the balance point, parts of a network vary in their capacity to exist autonomously as "reduced copies" of the whole network

Chapter III

Network Structures in Human Society

Not only animals but also humans form hierarchical, network, and (quasi-)market structures, often in combination. However, unlike most other biological species, humans can intentionally *create* social structures of various kinds in accordance with their interests and goals, and they do not merely *live* in them.[72] As for the biological network paradigms discussed in Chapter two, humans can combine these structures and supplement them with uniquely human organizational patterns.

It should be emphasized that all paradigms used in this work for describing networks in biological systems have actually been suggested by the very humans who investigated them (not by animals themselves, naturally). Therefore the paradigms, of necessity, reflect not only the real organizational pattern of the biological systems involved, but also the scientific views and personal preferences of these biologists. This is reflective of the biologists' "tacit awareness" (a term used by Michael Polanyi, 1966) and their intuitive understanding that influence seemingly factual statements.

Network paradigms are not necessarily applied to a single group of biological systems alone. Although originally developed in studies with one particular group of species, some paradigms can be extrapolated—with varying degrees of success—to a large number of other biological systems. To reiterate, this is exemplified by the modular paradigm that, apart from cnidarians, bryozoans, and ascidians, has actually been applied to systems of social insects where it corresponds to the *superorganism* concept.

Since paradigms applied to biological networks are developed by humans, they are inevitably *anthropomorphic* to an extent; i.e., intuitively, we believe that nonhuman organisms are similar to humans in terms of their behavior. This anthropomorphism is not as straightforward as that observed in Aristotle's writings or in Bernard Mandeville's *The Fable of the Bees: or, Private Vices, Public Benefits* (Mandeville, 1988 [1732]). Nonetheless, anthropomorphism is a typical feature of the human mind and, therefore, it finds its way into modern research pertaining to the behavior of various biological systems. Both the advantages and disadvantages of the anthropomorphic approach were discussed extensively

[72]Unfortunately, humans do not always make good use of their creativity and SCF; they often prefer to *live* in preexisting social structures without trying to creatively optimize them.

with respect to animal behavior (Lorenz, 1966; Dewsbury, 1978; Panov, 2001; Oleskin, 2012).

The anthropomorphic approach that is implicit in the network paradigms considered above actually *promotes* their application to human networks, which is the main subject of this Chapter. These paradigms that have been designed by *humans* and applied to other creatures are to be used in this work—in combination with a number of concepts developed in the social sciences and the humanities—to describe the functioning of *human* networks.

Interestingly, such an approach has repeatedly been used by recent political movements that are exemplified by anti- and alterglobalists, including the Zapatistas of the South of Mexico. These groups set up their network structures in conformity with the "a logic as old as the hills and the forests, an eco-logic, a bio-logic, the profound logic of life" (Networks, 2003). They try to make good use of our knowledge about the organization and behavior of biological systems. The Zapatistas compare the decentralized political network structures that were established by them to neural networks and, to an even larger extent, ant societies. "The ants can teach us that by working locally and continually sharing our local stories globally, by connecting everything and creating a plethora of feedback loops, we don't need to – indeed cannot - 'organize' the global network, it will regulate itself, swarm-like, lifelike, if we develop the right structures and conditions" (Ibid).

The impact of the discussion on the relative importance of "biological" paradigms and purely human elements for network structures in human society can be illustrated with the following example. A large number of non-governmental network organizations, including those dealing with environmental issues (see 3.9.1), contain a hierarchical core substructure and a more loosely organized peripheral network (Davydova et al., 2008). The relative importance of the core and the peripheral network varies from organization to organization, and the typical scenarios are as follows:

- The hierarchical core (e.g., containing the central body with the president) clearly dominates the periphery. This is actually a rigid centralized bureaucracy, and the powerless submissive peripheral network is merely a disguise;
- The core is less hierarchical (partly decentralized), and the peripheral network is as influential as the core. This "dynamic equilibrium" situation is characteristic of a large number of international organizations including the Biopolitics International Organization and the International Eco-Ethics Union that includes interconnected networks called "local chapters".
- The network dominates the core; its hierarchy is only a smokescreen behind which the networks carry out their plans. This scenario was described in the futuristic book *Netocracy...* (Bard & Söderqvist, 2002), which predicted the advent of a network society.

Even though such scenarios cannot be reduced to network paradigms used in biology, modified elements of some of these paradigms are incorporated within them, either intentionally or unintentionally. Of relevance is the *neural* network paradigm (2.5). Brain neural networks or their analogs often contain active task-oriented subnets where the activities are supplemented by the operation of the less active network periphery. Some human social networks, e.g., those set up for educational purposes (3.6), reveal striking similarities to neural networks. The *hierarchical core + network periphery* scenario mentioned above also evokes the *active specialists plus a pool of mobilizable generalists* principle that is characteristic of some eusocial systems (2.5).

In the following, two kinds of human social networks are considered: (1) network structures that spontaneously arise in human society and (2) networks intentionally created in terms of social technology with the purpose of carrying out specific tasks.

Networks in human society can be *formal* or *informal.* Formal networks are based on official agreements between their members. In contrast, informal network structures were denoted in the early work by Chrisholm (1989) as "informal organizations" that lack an official basis, although they have roles and tasks that are "set not by any single authority but by components /i.e., network members and their groups – O.A./ themselves".

Like other kinds of *sensu stricto* networks, network structures in human society can be flat or three-dimensional, dense or sparse, and homo- or heterogeneous (see 1.2 above).

In Chapter one (1.5.2) and in several sections of Chapter two, it was emphasized that integrated network structures with advanced coordination mechanisms can be considered as organisms (modular structures such as those of colonial cnidarians as well as, in terms of Shapiro's and Ben-Jacob's concepts, microbial colonies/biofilms) or, in somewhat looser terms, as superorganisms exemplified by the eusocial systems of ants, bees, or termites. The issue concerning the applicability of the term *organism* to human society has already been mentioned in this work. Generally, analogs of biological organisms or, at least organism-like systems, can be found among human network structures that are based on a rigorous set of ideological principles and behavioral norms that constitute their matrix, the matrix—the "impersonal leader"—provides guidelines for all of the network's members. Such modular organism-like networks include some sectarian and politically subversive networks. These are considered below in the example of *al-Qaeda'*s network structure (3.2.8).

However, in the author's opinion, a more important role in modern-day society should be played by network structures that prefer communication to control and that enhance, rather than suppress, their members' individualities. To an extent, such human networks are analogous to the egalitarian structures formed by some primate species (2.6). To reiterate, the term *organism* hardly applies to such structures. It will seldom be used in this chapter. There is the term *superorganism*, which refers to an integrated system made up of components that represent organisms per se. However, this term has certain negative social connotations because it is widely used with respect to the eusocial systems of ants and other insects (Wheeler, 1928; Chauvin, 1963; Hölldobler & Wilson, 1990, 2009, 2010; Kipyatkov, 1991). Should human network structures also include several different castes? It seems likely that some other concepts and terms originally used in the life sciences hold greater potential value with regard to human social networks. For example, the neural paradigm that emphasizes learning, distributed information processing, and creative work is of considerable interest in this respect.

Humans can combine and creatively modify network organization paradigms that work in the biological world; they supplement them with uniquely human elements.

3.1. Role of Network Structures in Primitive Society. Political System Formation as the Hierarchization of Network Structures

The transition from ape to human social structures was made during the long period of human evolution. This section is concerned with the organizational principles of primitive human societies. The relevant data include (1) ethnographic findings concerning still extant primitive societies that are influenced by civilization and that often try to retain their identity by emphasizing their opposition to it (i.e., using the *isolationist* strategy, which was discussed in 1.7.2.1.) and (2) archaeological data, e.g., excavated funerary ornaments.

Researchers have revealed a broad spectrum of primitive societies that differ with respect to the relative importance of hierarchies and horizontal structures. It seems likely that, in prehistoric times, "even adjacent populations of early hominids could have different types of social systems ranging from rigid hierarchical… to egalitarian structures. This diversity was largely due to ecological, social, and historical factors as well as specific traditions of the populations involved" (Butovskaya & Fainberg, 1993, p.216), but *egalitarianism*, nevertheless, seemed to be a widespread social pattern.

3.1. Egalitarian Organizational Pattern in Primitive Societies

According to the data available in the literature, a large number of primitive societies were characterized by a tendency to equalize social ranks. Many anthropologists believe that such societies consisted of relatively small groups (one estimate suggests that there were about 25 people, Meyer, 1996) that engaged in gathering, scavenging, and/or hunting. Conventionally described as "hunter-gatherers," these were cooperation-promoting, low-density networks that give an individual a chance to migrate and to stay isolated. "Hunting and gathering societies are as close as humans have ever come to constructing social patterns compatible with their primate and genetic legacy - a composite of both strong and weak ties, an egalitarian ethic, and a sense of community resting on cooperation and exchange among several interrelated but self-sufficient families that are free to disperse or come together depending on individual preference and available resources" (Maryansky and Turner, 1992: p.112). Egalitarian principles were, to a significant degree, implemented in a large number of societies that practiced primitive agriculture (Boehm, 1999), particularly if they used the "immediate return" system, where they did not store large amounts of food for a long period of time (Woodburn, 1982).

In egalitarian societies, "people were guided by a love of personal freedom". They used a system of disciplinary measures in order to "curb individuals who showed signs of wanting to dominate their fellows" (Boehm, 1999, p.65). A similar system where individual dominance is prevented is characteristic of some modern sectarian movements including Mennonites, who also implement egalitarian principles (Wilson & Sober, 1994).

However, egalitarianism in hunter–gatherer bands did not imply a complete lack of social hierarchy. Egalitarian hunter–gatherer groups were influenced by at least two competing tendencies: the propensity to establish hierarchies, and the tendency to abolish them (cf. Table

3)[73]. Primitive societies already include vestiges of the future political hierarchies that were typical of early civilizations (see below). Therefore, the applicability of the term "egalitarianism" to primitive hunter–gatherer bands has been questioned in the literature (Ames, 2010).

Many primitive social structures apparently represented three-dimensional networks. There were status differences between members of human groups and between whole groups that formed parts of the higher-order structures. This is consistent with archaeological data concerning inhumations with special funerary gifts provided for privileged people that were characteristic of hunter–gatherer bands in the Baikal Region (Siberia) in the Early Bronze Age (Weber & Goriunova, 2013) and—much earlier—of Upper Paleolithic people that lived at Saint-Germain-la-Riviere in France (Vanhaeren & d'Errico, 2005). People with a high status ostentatiously wore their ornaments and demonstrated other prestigious items (*costly signaling*), which serves as an analogy to modern-day conspicuous consumption.

There is evidence (Summers, 2005) that many primitive human groups were much more egalitarian than chimpanzee or bonobo groups. Similar to nonhuman primates, primitive humans established mitigated and changeable dominance hierarchies, and they formed coalitions/alliances that facilitated their acquisition of resources or their ability to attain a higher social status. However, one important difference was that humans communicated using verbal language and they developed an advanced system of ethical norms[74], which acted as a strong immaterial *matrix*.

Implementing these egalitarian norms despite the attempts of "upstarts" to achieve their dominance requires "considerable effort" (Boehm, 1999). People in hunter–gatherer bands, as well as some primitive horticulturists (who also lived in egalitarian societies), made this effort because they were aware that, generally, "egalitarian societies are vulnerable to takeovers" (Ibid., p.11)[75]. Actually, the transition from primitive to hierarchized civilized society may involve such a takeover—"a political transformation from the reversed hierarchy … to an orthodox one which is dominated by a powerful male (or female), or by a power coalition" (Ibid., pp.10-11). This transformation conforms with the hierarchization scenario that is based on the emergence of strong warleaders (item#3 in 3.1.2 below).

In conformity with these verbally expressed norms, people developed egalitarian attitudes and showed respect for the individual rights of each group member. High-ranking group members were expected to display tolerant, and often supportive, behavior towards other group members. Plausibly, advancements in stone tool technology, which resulted in the development of effective projectile weapons (Bingham, 1999), promoted the spreading of egalitarian behavioral norms because these tools made it much easier to punish an exploitative leader (as well as cheats and free-riders). The instructions of such leaders were disobeyed. Other group members could even kill them, making good use of the projectile weapons. "The advent of projectile weapons allowed many more individuals to

---

[73] Richerson & Boyd (2001) attributed the tendency to form hierarchies to our primate heritage and the predisposition for reciprocal cooperation, which is essential for establishing network structures, to "tribal instincts" evolving in primitive human society. This, in their opinion, accounted for our ambivalent attitude towards leadership. However, both tendencies seem to be much more ancient than the species *H. sapiens* (Chapter two).

[74] Evolutionary precursors of such norms apparently exist in social groups of some nonhuman primates; they are exemplified by "public opinion" in chimpanzees (de Waal, 1996, 2008, 2009).

[75] Boehm (1999) stressed that a similar vulnerability to takeovers is characteristic of ancient and modern democracies.

simultaneously attack a single individual, increasing the firepower brought to bear on the transgressor, distributing the risk of retaliation among more punishers" (Summers, 2005, p.109).

However, the more widely used strategy of getting rid of an upstart or a cheater was by ostracizing him throughout the entire group. Generally, *ostracism* appears to be an effective collective strategy for dealing with potential upstarts, and it is also a method that is used to punish those who try to violate other social norms. For example, one of the norms implemented by egalitarian societies is sharing, and a band of Utku Eskimos avoided speaking with a female (Niqi) who failed to share food with others and displayed other kinds of socially inacceptable behavior (Briggs, 1970).

This power of the weak over a single strong "alpha-male" was referred to as a "reversed hierarchy" by Boehm (1993, 1999, 2004); this is a structure in which "the basic flow of power must be *reversed* definitively" (Boehm, 1999, p.10, italicized by the author cited).

As found in chimpanzee groups (van der Dennen, 2011), primitive social structures are characterized by negative attitudes towards outgroups. For instance, neighboring groups of extant primitive Nambiquara Indians in Brazil are typically hostile, rather than friendly, to one another (see Panov, 2001).

Hierarchical relationships in primitive hunter–gatherer bands were limited, and these bands were largely characterized by *heterarchy*[76] (see 1.2.2 above) due to the following factors:

- *Leadership was split.* In many hunter–gatherer bands[77], leaders were temporary, enjoyed relatively few privileges, and their authority was limited to specific areas of activity. In bands of extant nomadic Ona foragers (Terra del Fuegos, South America), "one man might seem leader today and another man tomorrow, according to whoever was eager to embark upon some enterprise" (Bridges, 1948; quoted from: Boehm, 1999, p.62). In many hunter–gatherer bands, the shaman dealt only with magic and related spiritual matters, and had little authority otherwise. A number of primitive tribes included "headmen". The functions of these leaders were confined to settling arguments and representing their group during tribal meetings (Maryanski & Turner, 1992). Some leader functions were performed by people who were particularly skillful in certain activities like hunting or fishing; naturally, their authority was only limited to these activities. The split-leadership principle is being revitalized in present-day society in terms of various network structures (e.g., it is used in a *hirama*) (see 1.2.2). Overall, "useful leadership roles could develop without subverting the system" (Boehm, 1999, p.10).
- *Several individuals were located on the same rung of the hierarchical ladder*, so that they actually entered into horizontal (network-type) relationships with one another. Archaeological data on cemeteries (e.g., at Saint-Germain-la-Riviere) revealed that funerary ornaments did not reflect individual status or role; the ornaments were different because the deceased belonged to different (more or less privileged) groups.

[76] Heterarchy does not necessarily require complete equality of individuals, but it implies a lack of rigid centralized hierarchy that is referred to as *homoarchy* (Bondarenko, 2004).

[77] *Complex hunters* predominantly living in areas with abundant resources can have permanent and powerful group leaders; however, the linkage between social hierarchy and resource abundance is doubted by some anthropologists.

For instance, "the ostentatious wearing of ornaments made up of numerous exotic objects obtained by exchange often characterizes individuals belonging to a dominant social group" (Vanhaeren & dÉrrico, 2005, p.118).

- *Different structural levels of primitive societies were characterized by different hierarchies.* In still extant primitive groups of Tsimane (Amazonia) in Bolivia (von Rueden et al., 2008), dominance hierarchies at the lowest level (relationships between two individuals) are largely determined by the physical features of these individuals ("might is right"; the hierarchy is agonistic, see 1.3.2 above); at the level of a larger group, agonistic hierarchies are partly replaced by hedonic hierarchies based upon the coalitionary support of other group members. At the level of the whole primitive society, the hierarchy is predominantly hedonic because it is based upon prestige and respect.
- To reiterate, *the reversal of the dominance hierarchy* that occurs under certain circumstances in groups of apes (see 2.6.3.2 above) becomes a permanent phenomenon in those primitive human hunter–gatherer or agricultural societies where the tendency towards egalitarianism is particularly strong. In these societies, "the pyramid of power is turned upside down, with a politically united rank and file decisively dominating the alpha-male types" (Boehm, 1999, p.65).

Like egalitarian network structures of chimpanzees, many primitive social structures consisted of fission–fusion groups. "Large group size is a costly strategy in energetic terms, forcing foragers to engage in what is a high mobility strategy in the long term" (Grove, 2009, p.232), i.e., to split into smaller groups and get together again, depending on the ecological conditions (food availability, predation pressure, neighbors' attacks). A minimum group counted several dozens of people, although the exact number varied from location to location and from period to period.

These groups formed the parts of larger entities, including relatively coherent social structures. At the uppermost structural level, there were ethnolinguistic communities that counted 500 to 1,500 members (Richerson & Boyd, 1998, 2004). Even though such communities could be quite loose in structure, they promoted mutual aid and cooperation among their members. These social structures were stabilized (e.g., in Australian original societies) by marriages between representatives of different bands within the community. To some extent, such communities served as prototypical political units, precursors of more organized political systems. In addition to gene exchange promoted by intra-community marriages, they were also cemented by cultural factors, including unitary cultural traditions, a common language, and community-specific symbols and rites that enabled the people to distinguish "us" from "others".

Within such large communities, safety nets (Whallon, 2006) were established. They were stable network links connecting spatially separated "clusters" (local groups); these connections that were rooted in reciprocal altruism helped people survive when local food resources ran short. The network links also promoted information exchange between network clusters. Information exchange could be associated with, e.g., religious rituals and ceremonies.

This fractal system is analogous not only to the fission–fusion groups of apes (that form large, loose "megacoalitions"; see 2.6.4) but, to an extent, to modular network structures composed of repetitive units (subnetworks) that were considered previously, mainly in the example of colonial cnidarians (2.2).

Manifest fractal properties are also characteristic of some modern social and political structures. They are exemplified by the political structural units (*cantons*) that form part of Switzerland; each canton has its constitutions, legislatures, and courts; these units are further subdivided into autonomous subunits (*Gemeinden*), which will be discussed in 3.3.3 below. This modular paradigm can be used creatively in terms of social technologies that envisage the establishment of network structures in present-day societies. Unfortunately, these fractal properties are also typical of some criminal and terrorist networks, including those of *al-Qaeda.*

People in primitive bands also implemented the scenario in which relatively few active specialists interact with a pool of mobilizable generalists. An analogous scenario was considered in Chapter two in relation to a number of biological systems, including the *eusocial* structures of ants. Partial leaders with limited authority (exemplified by the skilled hunters and shamans mentioned above) constituted the core of active functional groups, and these groups were supported—to an extent—by normally less active generalists. When a group of primitive Indians in South America ran short of food, the group leader, who had almost no privileges, formed a "project team" that included himself and one of the female members of the group. They took baskets and collected edible grasshoppers, while the "less active pool of generalists" was lying on the grass and waiting for the meal (according to: Panov, 2001).

In contrast to eusocial systems, individuals in primitive human society were not subdivided into castes (caste systems were characteristic of certain types of civilized societies). However, the age-related transition from one set of professional duties to another was documented in a number of primitive societies, as typified by the Migili society of Nigeria (Barkow, 1982); it contained three age subgroups whose members carried out their respective duties of helpers, warriors, and specialists in religious rituals. To reiterate, a similar transition was described in eusocial systems (e.g., honeybee workers undergo at least four age-dependent stages; Johnson, 2010).

In sum, primitive human societies combined hierarchical and network organization. The ratio between the two patterns varied from society to society; however, networks prevailed over hierarchies in a large number of primitive societies. In organizational terms, such networks were sufficiently similar to the egalitarian structures of nonhuman primates. Nevertheless, they exhibited uniquely human features, such as verbal communication, and developed ethical norms that could make their organization even more egalitarian.

In Chapter one, it was emphasized that (quasi-)market structures are a third organizational option, in addition to hierarchies and networks. Such predominantly competitive quasi-market structures were of limited importance in primitive human societies, even though primitive hunter–gatherers engaged in exchanging goods. It is due to such exchanges that primitive hunter–gatherers tried to assume a high social status, despite the pressures of egalitarianism (see above), by obtaining exotic shells, deer teeth, and other prestigious items that were ostentatiously worn for "costly signaling". Nonetheless, it is debatable whether and to what extent such interactions associated with goods exchange can be considered quasi-market relationships.

It seems likely that goods (e.g., arrowheads and shells) in many primitive societies could be exchanged as gifts rather than as commodities, and that the whole exchange exemplifies *reciprocal altruism* (i.e., helping your helper), as was mentioned at the beginning of Chapter one. Such an exchange could, accordingly, emphasize friendly, loyal relationships between the individuals and groups involved. In this case, the goods to be exchanged did not necessarily have an equivalent value. It is still considered improper if the giver of a gift receives an exactly equivalent item (in terms of price) in return. Exchanging goods or services without demanding their exact equivalence in terms of price is characteristic of modern network structures, including clandestine networks typified by present-day Russian *blat* network structures (see 3.2.7). Generally, this sets networks *sensu stricto* apart from markets and their analogs, which tend to exchange items that are precisely equivalent.

Apart from reciprocal altruism, egalitarian relationships between individuals and groups in primitive human society involved kin altruism (i.e., the *support your kin* principle) and group-level sanctions against cheats, free riders, and repressive dominants.

3.1.2. Hierarchization of Egalitarian Society: The Formation of Political Systems

A challenging issue is concerned with the transition from primitive egalitarianism to rigid hierarchies, which was conducted during the Neolithic Age. As pointed out by Knauft (1991), the pathway of human social evolution was U-shaped. Our closest evolutionary relatives (the chimpanzee and the bonobo) and, in all likelihood, the human–chimpanzee–bonobo last common ancestor (LCA), formed hierarchies. They are at least partly replaced by more egalitarian structures in hunter–gatherer societies. Advanced agriculture is associated with the revitalization of hierarchies that are characteristic of chiefdoms and, more so, of civilized political systems.

However, it seems likely that the U-shaped trajectory began before the emergence of the LCA because chimpanzee/bonobo hierarchies are typically less rigid than the hierarchies of most lower primates; dominant individuals in ape hierarchies are often tolerant of their subordinates; split leadership is also possible in ape groups. Therefore, as emphasized in 2.6, the term *egalitarianism* is also applied in the literature to the groups of great apes (and to some other primates).

More rigid hierarchies were revived "around 10,000 years ago" when "increasingly complex social entities such as villages, tribes, and chiefdoms began to appear, followed by kingdoms, empires, and states… The evolution of these societies was associated with increasing despotism, in terms of control of resources and differential reproduction" (quoted from: Summers, 2005, p.107). The development of political hierarchies is considered as a process involving a variety of factors (Summers, 2005) such as population pressure necessitating the control of land, sedentism, transition from hunting and gathering to agriculture, the implementation of large-scale projects (e.g., constructing irrigation dams), and ideological control (see 3.1.2.4 below). As far as the potential driving forces of the hierarchization of egalitarian society are concerned, two possible mechanisms discussed in the literature (Summers, 2005; Rosenberg, 2009; Hooper et al., 2010) seem to be of relevance (cf. 1.7.1):

- *A violent takeover* by an aggressive dominant or a power-seeking group, resulting in the establishment of their coercive power in the originally egalitarian group
- *Decisions made voluntarily* by a primitive group's members, who consider establishing a hierarchy, "as a solution to the key challenges of life in social groups, such as conflict over resources, coordination failures, and free-riding in cooperative relationships" (Hooper et al., 2010, p.633).

The second option is considered more feasible in many primitive groups (Rosenberg, 2009) that gradually increase in size. This makes free rider identification and collective work organization more difficult, as long as there is no central leader (Summers, 2005)—a challenge also faced by present-day networks tending to expand. In addition, the group leader is expected to settle internal conflicts and promote reconciliation, and this evolution-based primary leader function is also performed by alpha males in groups of nonhuman primates as well as by "leopard skin-wearing leaders" in some African primitive tribes.

The two options noted above roughly correspond to the agonistic and hedonic strategy of hierarchy formation. As pointed out in 1.3.2, there are also several mixed strategies. Certain types of mixed strategies are typical of *aggrandizers*, i.e., aggressively ambitious individuals that could exist in hunter–gatherer bands (Hayden, 1995). As we know from history, aggrandizers tried to persuade their potential subordinates into considering them as group leaders using both social skills and physical prowess. One weak point in the idea that aggrandizers promote hierarchization is that "aggrandizer theories tend to ignore subdominance and subordination as active social strategies" (Ames, 2010, p.23). Actually, the option based on voluntary decisions implies that not only the potential leader but also the potential subordinates make efforts to establish a hierarchical structure.

Towards the beginning of this section, we mentioned *kin altruism, reciprocal altruism,* and *sanctions against free riders*. It seems likely that these behavioral tendencies continued to play a role during the development of political hierarchies.

1. Kin altruism could be redirected at all members of an alliance consisting of several family-sized groups: during a meeting of several formerly independent kin groups (or extended family groups), their representatives collectively made the decision to worship the same mythical ancestor or totem (often an animal, a plant, or a celestial body). This facilitated their interactivity in terms of trade, collective work, and religious ceremonies. Finally, the groups merged into one tribal unit, whose members were considered *relatives*. People in a particular kin group traditionally respected its senior members (the father, the big brother, and so on), and they also obeyed these members. In the new large unit, the notion of a "senior member" signified not only the age of the specific member or their role in the family, but also their status in the developing power structure.
2. The reciprocity principle is apparently characteristic of relationships between chimpanzees and other evolutionary relatives of our species. As for primitive humans, they developed a "prestige-based economy" involving the exchange of valuable symbolical gifts between individuals or whole groups. Such gifts became "costly signals" of one's high social status. Reciprocity-based relationships could become increasingly skewed, as some people inside a group/band/community accumulated large amounts of goods. Such rich people ("Big Men") could hold a

monopoly on valuable resources and become usurpers who exploited other group members and finally claimed their own political power. This scenario is actually a variation on the theme of "hierarchization via quasi-market formation", which is discussed in 1.7.1.

3. Sanctions against social norm violators were imposed on free riders by all group members in primitive society. It is conceivable that further evolution of human society resulted in the formation of a special privileged caste that specializes in punitive actions. Apart from punishing free riders in their own group/community, they could be involved in disciplining or subjugating outgroups. The leaders of these warriors could become increasingly affluent as a result of conquering their enemies and collecting the booty. They gradually gained power in the developing potestary systems (chiefdoms) and, subsequently, primitive political systems.

For example, a centralized state emerged in South Africa in the 19th century in the area formerly occupied by a number of tribes of primitive cattle breeders and agriculturists. In 1816, Shaka, the 29-year-old leader of the Zulu clan, conquered a large area (ca. 20,000 km^2) with the help of warriors armed with spears and shields (Corning, 2003b).

The scenario is consistent with the *conflict theory* suggested by Carneiro (1970). He believed that the state emerged as a result of competition among communities or villages, where the weaker competitors were engulfed by the stronger ones.

Scenario #3 initially resulted in the establishment of a military democratic system, where political decisions were made during warriors' meetings. Subsequently, the military democracy was superseded by a military hierarchy, implying that there was the unlimited power of the warleader, who later became the monarch. The scenario was apparently carried out in a number of Gothic tribes that established feudal kingdoms in the early Middle Ages.

Scenarios #1–3 are not mutually exclusive. For example, a senior tribe member can also be a rich Big Man while also being a warleader.

It should be noted that neither any single evolution-based behavioral predisposition, nor their combination, are sufficient enough to account for the generation of complex hierarchical social and political systems. Drawing on Roger Masters (1989), Losco (2001, p.175) stressed that the state is not "merely an extension of human sociality. Evolutionary theory only extends sociality via kin selection to a number of related others and via reciprocal altruism only to a small number of individuals with whom we have the occasion for face to face contact. Additional environmental and ecological factors must be considered in accounting for the emergence of the state as a device for extended social cooperation".

For example, reciprocal altruism supplemented with free rider sanctioning is sufficiently effective only in small-sized groups, not large societies (Richerson and Boyd, 1998). The establishment of advanced political systems, therefore, implicates the operation of factors working at the group level, including both the selection of advantageous biological traits (gene selection) and cultural features that were called *culturgenes* (Lumsden & Wilson, 1981) or *memes* (Dawkins, 1976). This ongoing gene–culture coevolution apparently proceeded with an increasing contribution of cultural factors.

Additional scenarios that prioritize cultural factors are also suggested in the literature, including the following: In primitive society, a shaman did not have much authority, and shared his special knowledge with others during initiation rites. The transformation of primitive society into more advanced political systems may be associated with the formation of a special priest caste which has the exclusive right to perform religious functions, which can be used to acquire dominance in the developing political system. This scenario occurred in some Australian Aboriginal societies (Artemieva, 2002). The tribal leader in such a system combines wordly and clerical power, playing the role of an intermediary between people and supernatural beings.

However, invoking culture as an evolutionary factor with respect to political systems presents serious difficulties. Importantly, egalitarian primitive societies developed ethical norms conforming to their social organization and lifestyle (termed the *Bauplan* of primitive society). Imposing a new hierarchical system on this society meant radically changing those egalitarian ethical norms. According to Rosenberg (2009), such *Bauplan* changes could only be made on the group level; they could not result from an individual's decision: a deviant individual would be punished as a social norm violator. A sizable group of dissidents could decide to violently change the social organization pattern, making others submit to the new hierarchical system. However, the more feasible scenario is that a weak social group, which is ousted from their territory by an enemy, seeks protection from a stronger, friendly group. If the stronger allies allow the weak group to stay with them, the weak group's members will subsequently, of necessity, assume a subordinate status to their mighty protectors. "In any such disadvantaged and dependent situation, a refugee group would be highly susceptible to subtle manipulation, if not subtle coercion, by their hosts to the relative economic and reproductive advantage of the hosting group" (Rosenberg, 2009, p.38). The result could be the establishment of a rank hierarchy and, ultimately, of a developed political system.

The social system that is characterized by a strong leader at the top of its hierarchy (a chiefdom) still remained modular, like a primitive hunter–gatherer community, because it consisted of autonomous units such as villages. The crucial difference between a chiefdom and a primitive community was that the leader status was permanent and legitimate in a chiefdom. The leader enjoyed a number of privileges including the right to wear special garments and the ability to hold power symbols in his hands. As the leader's power and the territory increased as a result of conquering neighboring communities, large-scale potestary systems and primal city–states emerged (e.g., in Mesopotamia and Central and South America in the fourth–third millennium B.C.). These early city–states were characterized by the formation of several social classes/strata and the differentiation of labor. They included districts of potters, coppersmiths, and so on, and were surrounded by villages inhabited by free farmers. Their political life was based on three central elements: the palace (residence of the monarch), the sanctuary, and the city assembly.

As a result, city–states combined hierarchical and non-hierarchical structures. However, the hierarchy gradually overpowered egalitarianism, particularly as a city–state developed into a despotic empire, due to conquering adjacent city–states. This "Asian" state establishment scenario was characteristic (apart from the situation that arose in the Middle East) of African, American, and the Oceanian states.

Interestingly, our species' evolution-molded capacity for forming coalitions/alliances manifested itself, under despotic regimes, in the establishment of strategic alliances among political elite members for the dual purpose of holding large populations of subordinates under control, while preventing their competitors (other elites) from gaining political power (Summers, 2005).

During the course of the development of state-level political systems, modular primitive societies are replaced by layer cake-like stratified societies. These systems include a special social stratum that is professionally involved in state government, and which is referred to as *primary bureaucracy*.

Full-blown stratified societies are typified by Asian-type despotic monarchies, some of which were characterized by caste systems. Each social caste is usually endogamous, i.e., inter-caste marriages are prohibited. Each caste performs specific social functions. Traditional Indian society consisted of four *varnas*: the *Brahmins*, which were vedic priests and learned men; the *Kshatriya*, including secular elite and warleaders; the *Vaishyas*, which were comprised of agriculturists, craftsmen, and merchants; and the *Shudras*, including laborers and service providers. Each varna was subdivided into several castes. Of relevance is the ancient comparison between caste societies and eusocial systems that also include several castes (reproductive individuals, workers, and soldiers), with each caste subdivided into several subgroups with specialized functions; one of the obvious differences is, nonetheless, that some castes in insect eusocial systems cannot reproduce per se.

Special emphasis should be placed upon the alternative Antique scenario of political evolution. In a number of cities (*polises*) of Ancient Greece, a relatively egalitarian unstratified[78] society evolved that was characterized by a lack of a coercive apparatus (Berent, 2002, p.223). Despite the large number of participants (up to 22,000 citizens took part in political meetings in Athens), the polis was characterized by predominantly horizontal, largely personalized, relationships. Any important political question was resolved during a meeting that featured the participation of the people at large, and some of these individuals behaved as temporary discussion leaders. As a result, the conventional translation of polis as a "city–state" invites confusion with the increasingly hierarchical structures of modern nation–states (cf. Strauss, 1964). Preferably, such a city–state, e.g., in Athens, could be called a "stateless polis".

Presumably, civilized society can come into being without establishing a state that contains characteristic features such as a powerful elite with a coercive apparatus and considerable social stratification. This issue can be formulated in more general terms: *is the establishment of rigid hierarchies resulting in state formation the only feasible pathway leading to civilized human society* (Bondarenko, 2004)*?* I reemphasize that heterarchy is characteristic of a large number of still-extant primitive societies. Of relevance is the fact that Mayan Indians resisted the centralized administration system, preferring decentralized political power (Cook, 2004). Do these heterachical systems mark the beginning of another pathway that leads to the development of social structures with a low degree of hierarchization and the predominance of egalitarian and cooperative relationships between individuals, i.e., to networks *sensu stricto* that occurred in various historical periods and that are in existence in the modern-day world?

The evolution of a complex political system can be considered in terms of Peter Corning's synergism theory. "Progressive evolution, whether at the microscopic level of the one-celled creature or at the macroscopic level of transnational political organizations, embodies cybernetic principles coupled with functional synergism. Human political systems are not a thing apart but a variation on one of the most primordial evolutionary themes. The purposes served by our political systems spring from our basic needs as biological creatures, and the patterns of organization that we have evolved are successful only insofar as they

[78] Nonetheless, a democratic Greek polis could not exist without slaves as an exploited class; in addition, only *male* citizens were endowed with political rights.

embody the same basic principles that govern lower levels of biological evolution" (Corning, 1983, pp.396-397). Corning regards the integration of formerly autonomous systems with the formation of a more complex system as the central pathway of progressive evolution. It works both with biological systems, as exemplified by the formation of multicellular organisms, and with human society. Interestingly, Corning regards the formation of increasingly complex political systems as the development of *superorganisms*.

However, this process is counterbalanced by *devolution*, which results in the fragmentation of complex systems, their simplification, and even their complete destruction. A similar process that occurs in various kinds of systems was denoted by the Russian scholar, Alexander Bogdanov (1980 [1921]), as *disjunction*; the integration process was called *conjunction*. The devolution (disjunction) of political systems can result in their reversion to primitive organizational patterns. This was the case with Greece at the beginning of the first millennium B.C., after the destruction of Mycenian political systems by Doric invaders.

In the political arena, integration and devolution processes are carried out in parallel. The collapse of the Soviet Union, the fragmentation of Yugoslavia and Czechoslovakia, and the separation of East Timor from Indonesia were accompanied by the progressive development of the European Union and of the alliance of the Pacific Rim states.

Nevertheless, it is integration/conjunction that seem to play the predominant role. A scenario envisaged by a number of philosophers and futurologists is based on the merging of all states into one political global "superorganism of superorganisms". It is to be hoped that, like the EU, this super-superorganism would evolve as a result of a gradual cooperative process. The union of political systems should be based upon mutually beneficial interactions and not result in establishing a political regime that would benefit only a part of the international community, i.e., the more affluent countries and transnational companies (Corning, 2004, p. 212). In other words, the global political system should represent a horizontal network structure with split leadership, rather than a rigid hierarchy with a central leader at its pinnacle.

The industrialization age followed by the modern postindustrial period has replaced stratified societies with more complex systems that are characterized by the greater social mobility of individuals and groups (an "equal opportunity society"). The boundaries between social strata are no longer impenetrable, and forces attempting to reestablish an egalitarian society—on a new cultural and technological basis—are at work in the present-day world.

Primitive societies have used a number of different organizational patterns. However, many of these societies were based upon non-hierarchical cooperative relationships among hunter–gatherers that formed small-sized, closely-knit, autonomous social modular units. During the course of the development of political systems, egalitarian modular primitive societies developed into stratified systems. Industrialization, followed by the modern postindustrial period, has replaced stratified societies with more complex systems that are characterized by the enhanced social mobility of individuals and groups.

3.2. Network Structures in Civilized Society

Despite the transition from the relatively non-hierarchical primitive society to hierarchical political regimes, non-hierarchical cooperative network structures have coexisted with hierarchical structures during the whole course of humankind's history. Moreover, a new impetus has recently been given to their development.

3.2.1. Egalitarian Religious and Humanitarian Organizations/Movements

The term "egalitarian" is, to an extent, ambiguous. According to Fried (1967), egalitarianism means that "leadership is informal and leaders are chosen for their abilities in certain circumstances and their leadership does not carry over. In such societies individuals also have equal access to the means of production; reciprocity and generosity are valued" (quoted from: Ames, 2010, p.24). However, it should be stressed that the term "egalitarianism" can refer to a whole spectrum of social structures; it seems "to conflate an array of social practices from weak, unstable or stable rank orders with no repression of prestige competition to active repression of competition and a strong egalitarian ethos" (Ibid., p.265).

In the following section, the interpretation described above of the term *egalitarian* (which was applied to primitive society) will be used. This interpretation of the term lays stress on the following points: (1) the individuality of each member of the network is emphasized; (2) each member enjoys what, in civilized society, is referred to as "inalienable rights"; (3) the structure has a vertical dimension, but high-ranking individuals are expected to behave tolerantly towards other network members and to respect their "rights"[79]; and (4) the dominant's power is restricted by strong coalitions/alliances of their subordinates, to the point of "reversing the hierarchy" (Boehm, 1993, 1999). In modern human society, egalitarianism in these terms seems to conform to the *equal opportunity* principle, emphasizing individual differences and chances to actuate one's potential assets.

In contrast, the *equal outcome* principle leveling everybody's income and living standard is to be considered a human analog of the *equipotential* paradigm; it is characteristic of fish schools (see 2.3) that downplay the role of individuality.

According to Alexander Berkman (1929, pp.164-165), "equality does not mean an equal amount but equal opportunity... Do not make the mistake of identifying equality in liberty with the forced equality of the convict camp… It does not mean that every one must eat, drink, or wear the same things, do the same work, or live in the same manner. Far from it: the very reverse in fact... Individual needs and tastes differ, as appetites differ. It is equal opportunity to satisfy them that constitutes true equality... Far from levelling, such equality opens the door for the greatest possible variety of activity and development. For human character is diverse".

The author cited was an anarchist; another Russian anarchist, the prominent scholar Pyotr Kropotkin (1972 [1902]), held views that were also in conformity with the interpretation of egalitarianism described above. In contrast to Darwin, Kropotkin emphasized cooperation, and not competition, as the main driving force of biological evolution. Kropotkin also applied this idea to the evolution of human society. He preached a new type of social organization based on voluntary work, anarchism, and complex interactions among "diverse groups and

---

[79] Such high-ranking individuals are respected rather than obeyed, and their rank/status is temporary rather than permanent.

federations of all sizes and ranks". This concept was corroborated by facts concerning animals that are capable of "unconscious mutual aid". Kropotkin was convinced that humans are sociable by virtue of their nature and, therefore, should form a wide variety of voluntary societies and associations (which people could join in accordance with their individual capacities and preferences), rather than coercive political systems. Kropotkin considered Greek *polises* as important precursors of the stateless system.

The principles of egalitarianism were obeyed by the Early Christian church (in the third and fourth century A.D.), which still lacked a developed hierarchy and complex institutions. The network was united by the goal of spreading Christ's Word in the world; its members placed special emphasis on the following passage from the Gospel: "There is neither Jew nor Greek, slave nor free, male nor female, for you are all one in Jesus Christ" (Galatians 3:28, Bible, 2011). In later historical periods, egalitarian network structures were established by heretics (Cathars) and sectarians, as well as by various socialist and communist movements. Kropotkin's followers were among the settlers in Palestine that established the first non-hierarchical *kibbutzim* (see 3.2.4). Egalitarian principles can be implemented at two different levels:

- *Goals and objectives* pursued by an organization
- *Organizational patterns.*

Among modern Christian organizations/movements, the *Christians for Biblical Equality (CBE)* organization preaches, in terms of its "core values", that "believers have equal authority and equal responsibility to exercise their gifts without regard to gender, ethnicity or class and without the limits of culturally-defined roles" (CBE International, 2014, http://www.cbeinternational.org). Nonetheless, the CBE structure includes a hierarchically organized center, the *Leadership*, that is headed by the President, and that is assisted by the Board of Directors and Board of References. However, the apparent mismatch between the professed egalitarian goals and the hierarchical structure becomes less important if we take account of the network's organizational pattern of the sufficiently influential periphery with its local CBE *Chapters.* Generally, the organizational design of the CBE apparently fits a relatively widespread pattern in human society that was briefly described above (Chapter 3, introductory text) as the *core + network periphery* scenario; it evokes the "insect" principle of combining active teams of specialists and a less active pool of mobilizable generalists. In this case, the headquarters, with its hierarchical structure, actually represents a collective analog of what is called an *external leader* in a hirama network (see 1.2.2). The hierarchical center, rather than controlling the network structure, advertizes and promotes it in society. On the whole, such an organization is a three-dimensional network of the *interactive* subtype (see 1.7.2.2) that aims to expand in society and recruit new people.

In organizational terms, network structures are still more prominent in Hutterite colonies in North America; Hutterites believe in religious values, pacifism, self-sufficiency, and a democratic (though obviously unfeminist) decision-making method involving all male colony members (Esau, 2004, p. 10). Each colony contains 10 to 20 families and has three high-level leaders (the Minister/President, the Secretary/Business Boss, and the Assistant Minister also working as the German teacher). There are a number of other bosses, reminiscent of the *hirama* pattern, with its several partial leaders dealing with the same pool of *hirama* members. Even if we ignore the split leadership pattern that exists at the colony level and regard the Minister/President as an at least nominal central leader, we should take account of the fact that several colonies form part of a higher-order structure, a *branch* (such as *Dariusleut, Lehrerleut,* or *Schmiedeleut*), with branch-level bishop committees. Within a branch, the President of each Hutterite colony is only considered a partial leader, and the whole multilevel structure of Hutterites bears some structural similarity to eusocial systems (e.g., that of ants), where relatively hierarchical worker teams (clans) are embedded in higher-order horizontal network structures (columns and pleiads; Zakharov, 1991, 2005)[80].

In North America, sectarian non-hierarchical structures were also established by Russian *dukhobors* and by West-European Perfectionist communes such as "Oneida". A wide variety of Moslem religious organizations also represented decentralized cooperative structures, i.e., networks *sensu stricto*.

Three-dimensional network structures that implement egalitarian principles are characteristic of a large number of modern organizations/movements pursuing social, political, and/or humanitarian purposes. They include human rights organizations, movements aimed to promote more humane warfare methods, e.g., the *International Campaign to Ban Landmines*, structures formed by activists in the field of bioethics and ecological ethics, e.g., the *International Eco-Ethics Union*, and various political protest and revolutionary organizations exemplified by the *Red Brigades* in Italy in the 1960s–1980s. In most of these structures, egalitarian principles are obeyed not only at the level of individual relationships between their members, but also in terms of interactivity between their structural parts (substructures, clusters). This allows for the decentralization of the management of these organizations. The *International Campaign to Ban Landmines (ICBL)* network is run by a group of six non-governmental organizations such as *Afghan Landmine Survivors' Organization, Action on Armed Violence (UK), Handicap International,* and others (ICBL, 2014; http://www.icbl.org/ index.php).

Not only does the degree of decentralization significantly vary from organization to organization, but it may also drastically change during the lifetime of a single organization. This was demonstrated in a study concerning the aforementioned *Red Brigades* in Italy, which were gradually hierarchized during the course of their evolution (Alexeenkova, 2005). It should also be reemphasized that networks and hierarchical substructures can permanently coexist within one organization, as exemplified by *Christians for Biblical Equality* which, as mentioned above, includes a hierarchical center and a network periphery.

3.2.2. Peasant *obschinas* in Russia

During the course of many centuries, the specific cultural traditions of Slaves promoted the formation of weakly hierarchical structures called *obschinas*, particularly among peasants. Although network structures are often considered an originally Western phenomenon, similar Slavonic structures were spontaneously established several centuries ago.

An *obschina* was a local community usually enclosed by a fence. Typically, it was a village with about 40 or 50 houses, to a maximum of 100 houses. An *obschina* cherished archaic values

[80] Once the number of the members of a Hutterite colony reaches a particular upper limit, the collective decision is made to branch off a daughter colony. In a similar organizational fashion, an ant family (pleiad) buds off daughter families (sociotomy) that combine with the parental family to form a colony which, in turn, forms part of a stable non-hierarchical secondary federation (Zakharov, 1991, 2005, see 1.8.6 and 2.4.5-2.4.6).

that were characteristic of the pre-state historical period, as well as ancient forms of social bonds. All issues were resolved during a meeting (*skhod*) involving the heads of all peasant families/households, who actually performed the functions of network-type partial leaders. Both the geographical and the social space of an *obschina* were limited. An *obschina* was based upon the autarchy (self-government) principle. It preferred decentralized decision-making procedures (Oleinik, 2003).

In terms of my classification, an *obschina* represents an *isolationist*, rather than an *interactive*, network structure. Besides, the individual features of *obschina* members were often less important than the collective identity of an *obschina*. It was not uncommon that the whole collective unanimously voted for a particular *obschina*-level decision. To an extent, an *obschina* was comparable to an *equipotential* network typified by a leaderless fish shoal or school (2.3.2).

In contrast, the spirit of communalism, as well as other *obschina*-like ("fish school-like"), features are less characteristic of more individualized, less close-knit, and more pragmatic non-hierarchical cooperative structures that have been developing in the present-day Western world. For instance, several businesses form a network aimed at jointly suppressing a strong competitor, which is reminiscent of an ape coalition trying to overthrow the alpha male. Goods suppliers and consumers form a non-hierarchical alliance in order to minimize the costs they would otherwise incur if they entered into classical market relations (see 3.7. below).

3.2.3. Swiss Communes (Gemeinden)

Horizontal ties were also of considerable importance in medieval urban and rural communes. Of relevance are the Swiss communes (German: *Gemeinden*; French: *communes*; Italian: *comuni*), also known as municipalities. The communes used to be "non-hierarchical, undivided, that is, non-feudal" communities (Steinberg, 1976). In present-day Switzerland, these communes still hold town meetings, as the legislature; councils are headed by a mayor who serves as the executive. In an Alpine environment, communes have typically represented peasants' or townsmen's associations. Currently, Switzerland includes over 2,000 communes (2,352 according to Répertoire officiel des communes de Suisse, Mutations, 2013), in part due to an extremely heterogeneous language, as well as to the geophysical environment; indeed, the Swiss still identify with their communes, which remain a key element of political and social life. "Typically, there is no 'Swiss citizenship' as such. A Swiss person is in the first place a citizen of his *Gemeinde* and as such a citizen of his canton and hence automatically a Swiss citizen" (Steinberg, 1976).

Throughout Swiss history, communes possessed much political power, so that the resulting political system was "bottom-heavy". The communes joined together to form cantonal leagues which, in turn, collectively formed a federal body (German: *Eidgenössische Tagsatzung*; French: *diète fédérale*; Italian: *dieta federale*) that, before 1848, represented only "a formal assembly of ambassadors with no power to coerce" its members (Steinberg, 1976, p.21). Even though the situation changed after the adoption of a constitution that made Switzerland a federal state in 1848, cantons and the individual communes inside them have retained much sovereignty.

The communal organizational pattern has promoted the development of Swiss "microcapitalism", a decentralized network of small-scale enterprises manufacturing high-quality products such as watches.

The Swiss in their communes pay special respect to their legendary spiritual leader, who is held responsible for their traditions. This ideal "behavioral coordination factor" is embodied by Wilhelm Tell, the legendary "hardy mountaneer who defies the Habsburgs... Its <the legend's> truth is the truth of a communal tradition by which Swiss defined and made precise their public values" (Steinberg, 1976, p.19). At least in one case, the role of the spiritual leader was played by a real person. In 1481, a civil war between urban and rural cantons was prevented by the aristocratic hermit, Niklaus von Flüe (afterwards known as the legendary Brother Klaus). He persuaded the townsmen and peasants to reach a peace agreement.

Medieval communes and their higher-order networks also arose in a number of other countries, including the Russian Novgorod Republic, which held the meeting of citizens (*veche*) as the supreme decision-making body.

3.2.4. Israeli Kibbutzim

Kibbutzim (the plural of *kibbutz*) were established as "communities characterized by collective economic production, direct participatory democracy, and communal childrearing" (Snarey et al., 1985, p.152). However, a majority of them have recently undergone serious economic and organizational changes, calling into question the viability—under modern-day conditions—of this type of network in human society.

The first *kibbutz* (more precisely, *kvutza*, i.e., a small-sized collective agricultural settlement) was founded in 1909 near the southern end of the Sea of Galilee by followers of A.D. Gordon's doctrine. Gordon idealized physical labor (especially farming), cooperation, and mutual aid as means through which to improve the individual and to achieve self-realization (Political Dictionary of the State of Israel, 1993). The first larger, full-scale *kibbutz* called Eid Harod was established in 1921. Many of the early *kibbutzim* considered Gordon as one of their spiritual leaders. For some *kibbutzim* formed by Russian emigrees, Kropotkin also played the part of a legendary spiritual authority. Emphasizing the *fractal* properties of networks, *kibbutzim*, from the beginning of their history, formed socially and politically influential higher-order structures such as the *United Kibbutzim Movement*.

Kibbutzim were originally characterized by the following main distinctive features:

- Communal ownership of the land, tools, buildings, and other related elements.
- "Medical treatment, education, food, housing, clothing, and cultural activities... offered to all *kibbutz* members according to their needs" (Leviatan, 1982, p.143).
- A collective lifestyle: *kibbutz* members used to work and socialize together.
- The principle of absolute equality in terms of income: the *kibbutz* "shared equally all earned income, independent of an individual member's occupation, skills or work effort" (Ruffle & Sosis, 2006, p.148).
- Decision making by the general assembly of *kibbutznikim* (*kibbutz* members) or, at the level of a plant, by the workers' assembly, and "the rank and file workers who are *lowest* on the hierarchical ladder are also the *highest* in normatively accorded influence in their roles as members in the workers' assembly (where they form a majority)" (Ibid.).

- The rotation of important social roles, including leader positions (executive secretaries, treasurers, economic coordinators, and so on). Ex-office holders still retain, for some time, their influence and prestige; those who will hold important offices in the near future already enjoy an elevated social status. Overall, an increased percentage of *kibbutz* members are involved in managing the *kibbutz*. Traditionally, there were no social status-dependent privileges, and a leader's position was considered a supplementary duty, in addition to his or her main social role as a worker (Rosner, 1975).
- Split leadership in *hirama*-like fashion (see 1.2.2 above). A top position in one area of activity is associated with a low-level position in other areas. For instance, those "who have high offices in the domain of work would not hold high offices in the domain of social organization and vice versa" (Leviatan, 1982, p.150).

Despite the professed egalitarian ideology, *kibbutz* members actually did have social ranks—informal and formal. Many *kibbutzim* dealt with industry (apart from agriculture), and their plants/factories had a hierarchy. The hierarchy was relatively flat, with three to five hierarchical levels in the ten plants studied by Rosner (1971, quoted according to: Leviatan, 1982). Similar to primitive hunter–gatherers, *kibbutznikim* used effective methods to mitigate the hierarchy, turning it into a heterarchy by creating an "ambiguity in the order among levels of hierarchy in terms of their influence and power" (Leviatan, 1982, p.147). Apart from establishing heterarchy—to the point of creating a "reversed hierarchy" (Boehm, 1999)—the egalitarian paradigm implies enhancing the individuality and the freedom of all members of the network involved. The *kibbutz'* attitude to the individuality of its members was ambiguous.

The professed aim of education on the *kibbutz* was to promote personality development and the actualization of individual potentialities. Many *kibbutzim* succeeded in attaining these goals, as a large number of *kibbutz* members made important achievements in academia, business, music, and the military (Gavron, 2000, p.166). The educational system and the social environment of *kibbutzim* also proved beneficial in terms of moral education: a longitudinal study revealed that *kibbutz*-born or *kibbutz*-educated young people tend to reach higher levels in terms of moral development than their urban contemporaries; this was attributed to the influence of "the principles of justice, collective education, and self-government" on the *kibbutz* (Snarey et al., 1985).

However, along with enhancing individuality, a large number of *kibbutzim* tended to downplay its importance in favor of the spirit of communality. "Among the main principles in the social organization of the kibbutz is the accent placed on the communal over the individual, two expressions of which are identical housing and the collective raising and educating of children" (Meir, 1990, p.189). According to the results of a special study (Lobel & Agami-Rozenblat, 1993), the prevalence of the communal over the individual manifested itself in the decreased social self-esteem of *kibbutz* members, compared to that of urban Israeli citizens. As for *kibbutz* females that are generally more sensitive to the influence of the social milieu, they are also characterized by reduced levels of academic aspirations and academic self-esteem.

The *kibbutz* was in a stress field generated by two different paradigms—the individuality-enhancing *egalitarian* and the individuality-eliminating *equipotential* (see 2.3 above) paradigms. The latter is typified by the organization of a leaderless fish shoal or school that is preferentially composed of individuals of similar size, with similar levels of bravery, and similar dietary preferences (Croft et al., 2005, 2008).

Interestingly, even modified present-day *kibbutzim* that have abandoned some of their original ideological principles are highly assortative as far as accepting/rejecting potential newcomers are concerned. Even accepting new people as *nonmembers* staying in "expansion neighborhoods" who cannot use a *kibbutz'* collective property is based upon the principle of "looking for people like us". Typical *kibbutz* members, and even recently invited nonmembers, should all be secular, Jewish, and petit-bourgeois; the newcomers should be relatively young, accept the social lifestyle of the community, and share the dream of living in a single-family house (Charney & Palgi, 2013).

It should be emphasized that most traditional and modernized (renewing) *kibbutzim* are gated communities that use explicit and implicit sorting and admittance criteria and, therefore, belong to the *isolationist* type of networks (see 1.7.2.1 above). The whole community of members decides who will become a new member (a multistage procedure: a newcomer assumes the status of a candidate after a trial period of about 1 year; after another 1–2 years, the decision is made whether the candidate is to become a *kibbutz* member; see Charney & Palgi, 2013). However, some *kibbutzim* represent a mixed network type. Although they stringently select new members, they are actively involved in the social and political life of the whole nation. Starting from the first half of the 20th century, they have been building a spreading higher-order network. This is characteristic of *interactive* networks.

A decentralized cooperative network structure is cemented by a material or immaterial matrix. An ideological matrix is typical of Israeli *kibbutzim*. Although there are a small number of religious *kibbutzim,* the majority of them are united by socialist principles (not necessarily of the Marxist type), "work ethic", and frequently by Zionism. Importantly, the *kibbutz* mentality, in addition to the "sense of community", is based upon entrepreneurship and social elitism (Ben-Rafael, 1997).

The term "social elitism" in this context does not mean special privileges or prominent political positions, although some *kibbutz* members became Israeli parliament members. The main point is that *kibbutznikim* consider themselves responsible for an important social and political mission. In the first half of the 20th century, they helped establish the state of Israel. *Kibbutz* settlements on the periphery of Israel were involved in determining and protecting its frontiers; they greatly contributed to the Zionist self-image.

Hence, the *kibbutzim* found themselves in a yet another stress field created by the struggle between (1) biological paradigms of network organization (the egalitarian and, to some extent, the equipotential paradigm) and (2) uniquely human features of spirituality. Even predominantly atheist *kibbutzim* were aware of their special spiritual mission and, therefore, represented an immaterial social entity comparable to a church congregation as a spiritual structure.

Recent serious economic, ideological, and organizational changes in most *kibbutzim* have resulted in a transition from a predominant network structure to a more complex pattern also incorporating hierarchical and market structures. These changes include:

- *Extensively using hired labor.* This promotes not only the marketization but also the hierarchization of the whole structure because hired workers (including Arabs and, more recently, people from China and Thailand) do not have members' rights and are obviously inferior in social rank to them[81]. While creating a mediating interface between the core *kibbutz* and the capitalist market and increasing a *kibbutz's* economic success in the short term, this practice tends to undermine the *kibbutz's* main organizational principles in the long run. On the assumption that the market wage-rate for hired labor is lower than the wage paid (in an egalitarian manner) to *kibbutznikim,* hired labor will be more economical than communal "self-labor", as is typical of *kibbutznikim. Kibbutz* members may ultimately be displaced from all job positions except for those where hired workers are not admitted (Satt & Ginzburg, 1992). Some *kibbutznikim* do not agree to work alongside hired workers and quit such positions.

- *Introducing a differential wage system.* A majority—about three-quarters of the total number as of 2012 (Charney & Palgi, 2013)—of *kibbutzim* (renewing *kibbutzim, kibbutzim mithadesh*) pay their members different wages/salaries for doing different jobs; the rest (*kibbutzim shitufi*) still follow original egalitarian principles in terms of income.
- *Partially replacing communal ownership with private ownership.* This was exemplified above by the system of introducing "nonmembers" that have their own private houses in "expansion neighborhoods". The area used by the *kibbutz* is then broken up into the part belonging to the *kibbutznikim* proper and the part where the nonmembers reside. These individuals all live according to capitalist principles and represent a mediating structure between the *kibbutz* as a non-hierarchical nonmarket network structure (*sensu stricto*) and the market relations-dominated capitalist environment. Such mediating structures that promote constructive interactions between different types of social structures were discussed in general terms in 1.8.3–1.8.4 above.

These recent developments pose the threat that the *kibbutz* as a specific type of social network will ultimately become extinct. However, it seems more likely that most *kibbutzim* will just find a new balance point in the multidimensional network-hierarchy-market continuum. The *kibbutz* becomes more adaptable and flexible in the modern-day dynamic world. The currently increasing interest in at least some of the extant *kibbutzim* in Israel gives us hope that the original communal spirit and cooperative lifestyle of the *kibbutz* movement can also exist, albeit in a modified form, in the 21st century. In conformity with this lifestyle, even renewing *kibbutzim* still guarantee their members a certain minimum income, irrespective of what each individual earns personally (the safety net principle). As pointed out by Igal Charney and Michal Palgi (2013, pp.49-50), "privatization processes and adherence to the market on the one hand, and the intrinsic rural-like and quasi-communal characteristics of the kibbutz on the other, attracted a growing number of people to join the kibbutz as nonmembers" that form a tandem with *kibbutz* members within the framework of the renewing network-type community.

---

[81] Apart from the hired workforce, modern-day *kibbutzim* increasingly rely on volunteer work. Obviously, volunteers and *kibbutz* members have different social status and rank.

3.2.5. Modern Communes

A typical example is provided by the *Twin Oaks Intentional Community* (Washington, D.C.). The community is situated in rural central Virginia and is made up of around 85 adult members and 15 children. According to its website, "since the community's beginning in 1967," its way of life has reflected the "values of cooperation, sharing, nonviolence, equality, and ecology". Importantly, the community does not have a central leader and is based on a "form of democracy with responsibility shared among various managers, planners, and committees" (Twin Oaks, 2014; http://www.twinoaks.org). *Twin Oaks* is characterized by communal ownership and communal satisfaction of the members' needs. Special emphasis is placed on the principle of social status equality (no personal titles are used in addressing specific people). Like a *kibbutz*, this community rotates its leadership and also some of its professional roles. Everybody is free to choose the job (or combination of jobs) preferred for the coming week. Every professional is expected to share specialized knowledge with others. Thus, the commune actually uses the principle of changing social roles, which has some biosocial implications. Individuals in biosocial systems do not always have fixed roles (see 2.4.3). As an analogous example, individuals in many ant societies can change their "professions" if necessary: a scout can become a forager, a nurse, and so on; these role changes can also be age-dependent.

3.2.6. Voluntary "Service" Organizations

This subsection deals with organizations that often combine network and hierarchical principles: a large number of them includes an official hierarchical center and the network-type periphery. Typically, however, the periphery is much more important than the hierarchical headquarters, which only provides general guidelines and is necessary for legal purposes.

Upon entering most American cities and towns—and, frequently, in Western Europe—one sees signs indicating the time and place for the weekly meeting of *Kiwanis, Rotary,* or other "service" clubs. These clubs all have hierarchically organized headquarters with a specified location that can be found on the map, a director general, a board, and an official document stating the main principles of the club, as well as the club's goals. The activities of these networked organizations are actually coordinated at the level of autonomous local clubs. Usually composed primarily of local business and professional people, these networked clubs combine sociability with fundraising and other programs that respond to social needs and deal with humanitarian projects.

For example, the following are the "six permanent Objects of Kiwanis International", which were approved at the 1924 Kiwanis International Convention in Denver, Colorado:

- "To give primacy to the human and spiritual rather than to the material values of life.
- To encourage the daily living of the Golden Rule in all human relationships.
- To promote the adoption and the application of higher social, business and professional standards.

- To develop, by precept and example, a more intelligent, aggressive and serviceable citizenship.
- To provide, through Kiwanis clubs, a practical means to form enduring friendships, to render altruistic service and to build better communities.
- To cooperate in creating and maintaining that sound public opinion and high idealism which make possible the increase of righteousness, justice, patriotism and goodwill" (Kiwanis, 2013; http://www.kiwanis.org).

While sometimes associated with regional institutions, such as the *Shriner Hospitals for Crippled Children*, many of these organizations emphasize informal local activities and concerns. Alexis de Tocqueville, who studied the 19th century US social and political system, noted that such voluntary associations have been ubiquitous in the United States. In addition to secular groups, many of these networks emerge around religious institutions. These predominantly horizontal, voluntaristic networks have long been viewed as one of the essential keys to the vigor of free representative or constitutional government in the West.

Similar organizational patterns giving local peripheral branches a chance to develop their own projects based on network decision-making procedures are characteristic of a large number of business networks such as *cooperatives* (*co-ops*) that are discussed in 3.7.2.3 (the subsection on small-sized business networks).

3.2.7. Informal (Shadow)[82] Networks: Blat

In various historical periods and parts of the world, spontaneously developing informal networks (that do not become official organizations) enable people to engage in beneficial interactions with other people "in the right places", overcoming bureaucratic hurdles and legal restrictions. Some of these shadow structures are hierarchized, and this is typified by *clientelles* (see 3.3)

In Russian society, less hierarchical[83] shadow structures are more widely spread. They are called *blat* networks. People within *blat* networks exchange favors; their relationships are not based on the strict equivalence principle and often do not involve money (Ledeneva, 1998), in an analogy to primitive society. Valuable resources that are otherwise difficult to obtain are exchanged between partners who establish personal contacts with one another. The partners may have an equal status within the *blat* network (a horizontal exchange pattern) or their interactions may be somewhat hierarchical. Predominantly, this is a split hierarchy (a heterarchy) because each influential node in the network plays the dominant role only with respect to one specific kind of valuable resource. For example, a *blat* member has the power to decide who will become a Moscow State University student or will get a draft deferment. To an extent, *blat* structures in Russia (and similar shadow networks elsewhere) are comparable to the egalitarian structures of primitive hunter–gatherers and even to those of apes (chimpanzees and bonobos; see 2.6).

---

[82] In virtual networks such as *Facebook*, the term "shadow networks" is used in a different, although somewhat similar, meaning: it refers to relationships that are never professed online.

[83] The degrees of hierachization vary from network to network within the higher-order general *blat* structure in Russia.

3.2.8. Dark Networks. Al-Qaeda

Network organization is a potentially powerful social technology that, like genetic engineering or nuclear technology, can be used not only for good, but also for evil purposes. The term *dark networks* actually refers to mixed social structures combining both networked and hierarchical organizational patterns (Raab & Milward, 2003)[84]. However, network organization plays a dominant role in many "dark networks". They include criminal structures trafficking in drugs, youth gangs and, most importantly, terrorist networks. Fighting dark networks presents serious difficulties due to the following problems (Sparrow, 1991; Krebs, 2002; McBride & Hewitt, 2013): (1) we usually possess incomplete information concerning the network's structure, including the key nodes (hubs) and links and, therefore, have to face dilemmas such as "whether to remove the node with the most observed connections among those whose connections were monitored (the sure thing) or whether to arrest another node for whom some but possibly not all connections are observed (the risky choice)" (McBride & Hewitt, 2013); (2) dark networks have fuzzy boundaries, so that it is difficult to decide who to include and who not to include (Sparrow, 1991); and (3) dark networks are dynamic structures that constantly change their organizational pattern, adapting to the changeable environment. Criminal and terrorist structures can efficiently learn/train: "organisations learn when their participants acquire, analyze, and act on information and experience, changing existing practices or creating new ones" (Kenney et al., 2013, p.740).

Special studies revealed that covert networks whose members risk exposure with uniform probability attain maximum resilience if they are structured

- as a complete graph (i.e., a network where all possible links between nodes are present, provided that the exposure probability is sufficiently low), or
- as a hierarchy with a single central element, which is exemplified by a star graph (Figure 1.5, A in Chapter one), at a higher exposure probability (Lindelauf et al., 2009).

Due to their capacity to collectively learn/train and to adapt their organizational pattern to changing conditions, dark network are, to a large extent, comparable to *neural networks* in biological systems (2.5).

The decentralized organizational pattern of many dark networks is one of the reasons why a war on terrorism is often extremely difficult, particularly if it is waged by *a hierarchically organized army*; the experience of the US army during the recent campaigns in Afghanistan and Iraq seems to support this statement. A terrorist network contains several partial leaders; moreover, if these leaders are removed, new replacement leaders soon emerge.

Terrorist network structures can be subdivided into local and delocalized (potentially global) networks.

1. *Local networks* "typically emerge within specific countries to eliminate governments that are perceived to be inimical, on religious, political or other grounds" (Rathbone & Rowley, 2002, p.6). Such local organizations are exemplified by the *La Cosa Nostra*, *Hamas* and *Hezbollah*, *Baader-Meinhof*, and *The Shining Path*. Until recently, terrorists often

---

[84] Several dark networks pursuing similar goals can compete with one another, forming a higher-order quasi-market structure.

established hierarchical structures. They included the ideological goals-setting central headquarters that could be linked to a legal political party. A militant organization was subordinate to the headquarters. Typical examples include the *Irish Republican Army* and the organization of Basque separatists. In the political arena, they are represented by the *Sinn Fein* and the *Herri Batasuna* party, respectively.

Presently, these political movements are increasingly being restructured according to network organization principles. The originally hierarchical Palestinian *Fatah* has been supplemented by horizontally integrated networked *al-Aqsa Martyrs' Brigades* that can establish links with global terrorist networks.

2. The network organization pattern is particularly conspicuous in *delocalized dark organizations* that are predominantly united by their ideological matrix. A typical example is provided by *al-Qaeda.* This network is aimed at carrying out the Salafi[85] *jihad* mission. It considers Allah its mystical guru[86] and pursues the ultimate goal of establishing a Moslem fundamentalist state (a *califat*) stretching from Morocco to the Philippines. From the very beginning of its organizational inception, the ideological matrix of *al-Qaeda* included the goals of (1) overcoming the Judeo–Christian influence around the globe that allegedly poses a threat to the Muslim world (*ummah*) with its sanctuaries, such as the *al-Aqsa* Mosque in Jerusalem and (2) destroying the USA as the "head of the snake" (Sageman, 2004, 2008).

It is well known that *al-Qaeda* committed a large number of terrorist acts, starting from the bombing of the Golden Mihor hotel in Aden, 1992, and of the World Trade Center, New York, 1993 (Wander, 2008). The history of *al-Qaeda* demonstrates that the matrix of a social network contains not only ideological, religious, or philosophical tenets, but also a system of loyal interindividual relationships.

As for the *al-Qaeda* network, these originally informal relationships based upon friendship, kinship, and discipleship still play a very important role in motivating the new people to participate in the *jihad* movement, and this "social capital" is secondarily reinforced by ideological concepts and values (Sageman, 2004, 2008). The strong matrix with a social and an ideological component encourages the new members to get involved, even in a completely leaderless *jihad* movement (Sageman, 2008), whose organizational pattern resembles that of *equipotential* fish schools/shoals (2.3.2).

From the structural viewpoint, the *al-Qaeda* global network that incorporates elements of a number of pre-existing Salafi structures including, e.g., the Islamic Liberation Organization and the Egyptian Islamic Jihad, represents a multilevel network composed of groups and organizations, most of which are also characterized by network organization (Kahler, 2009a, b).

Indisputably, *al-Qaeda* originally had a strong hierarchical structural component headed by the influential leaders Osama bin-Laden, Ayn al-Zawahiri[87], Jamal al-Fadl, and others who proclaimed its formation at a meeting in Peshawar (Pakistan) in August, 1988: "the creation of the group brought together extraordinary Saudi wealth, the expertise of a lifetime Egyptian militant, and a philosophical foundation for jihad from a Cairo intellectual" (Wander, 2008). *Al-Qaeda* included the Central Staff with a Shura that contained the Military, Money/Business, Law, and the Islamic Study Committees.

---

[85]The word *Salafi* is derived from *salaf* (Arabic: ancient, ancestral; related to the prophet Mohammed).

[86] Presently, the legendary guru's role can be played by Osama bin Laden, who was assassinated on May 1, 2011.

[87] Ayn al-Zawahiri was in charge of his own group, *al-Jihad,* which officially merged with *al-Qaeda* in June, 2001; the merger strengthened the decentralized network component present in the structure of *al-Qaeda.*

More recently, the hierarchical headquarters have become just one of the several major clusters of the global *jihad* network. Apart from Central Staff, such hubs include the Core Arabs, the Maghreba Arabs, and the South East Asian cluster (Sageman, 2004). Most of these clusters represent three-dimensional networks rather than hierarchical structures, although they contained sufficiently influential leaders. One of the clusters, *Jemaah Islamiyah* representing South East Asia, was distinguished by its rigid hierarchy with an emir on its pinnacle. The cluster was subdivided into four regional subdivisions. Each of them was headed by an emir and a council. Because of the hierarchical rather than networked organization of the South East Asian cluster, it was comparatively easy to disrupt.

Overall, *al-Qaeda*'s network structure exhibits a high degree of clustering, which is an analogy to many biological networks including fish schools/shoals. *Al-Qaeda* used to consist of partly autonomous clusters where the nodes are connected by a large number of intensely functioning links. In contrast, different clusters are only linked by a limited number of edges. It was estimated that a synchronous targeted attack on 5%–15% of the members of this type of network could disrupt it (Sageman, 2004).

The vulnerability of *al-Qaeda* was used by its enemies, who concentrated on eliminating the relatively few nodes connecting its major clusters. These individuals—without whom the network falls apart—can exercise much *bargaining power*, which limits the power of cluster emirs and confines the emirs to partial leadership. Each of the intercluster nodes directly contacts field commanders responsible for terrorist operations (Sageman, 2004). Each cluster is composed of smaller subclusters (cliques) that are close-knit groups of friends that trust one another.

The importance of the network's structural component considerably increased during the course of *al-Qaeda*'s development. After 1996, the Central Staff was not directly implicated in conducting terrorist operations, and the three other major clusters involved in carrying them out were connected to Central Staff representatives via special messengers (Sageman, 2004). The successful War on Terror waged by the USA in the wake of 9/11 effectively neutralized *al-Qaeda*'s central hierarchy, which lost its sanctuaries, funding, and communication channels; a large number of leaders were also eliminated. The network's survival depended on its subsequent decentralization and the establishment of autonomous, self-financing subgroups that were characterized by fuzzy boundaries; *al-Qaeda* "has become, in effect, a trademark or a brand name" (W. J. Strope, comments to: Feiser, 2004) that binds these subgroups together, along with a powerful ideological matrix and a religious value system, as well as the images of the currently otherworldly gurus such as Osama bin Laden. One of these autonomous affiliated subgroups committed a number of terrorist acts in Iraq; the subgroup's vulnerable hierarchical component was eliminated as its leader, Abu Musab al-Zarqawi, was killed on June 8, 2006 in a US military operation (Wander, 2008).

Using the Internet further promoted the decentralization of *al-Qaeda* while also providing new opportunities for information exchange, the dissemination of the Salafi *jihad* ideology, and even the mobilization of *al-Qaeda* members and adherents (including the adoption of the flashmob technique).

The process of "networkization" of an originally more hierarchical structure was investigated by Kenney et al. (2013) using a number of "network science" criteria (see

1.2.4.1). *Al-Muhajiroun*, a radical organization close to *al-Qaeda* in terms of its ideology[88], set itself the goal of establishing an Islamic state in Britain. Its founder and leader (*emir*), Omar Bakri, left Britain for Lebanon and was forbidden from returning to Britain following the suicidal bombing of the public transport network of Britain on July 7, 2005. The organization itself had been banned earlier (in 2004); subsequently, the two successor organizations, *al-Ghurabaa* and the *Saved Sect*, were outlawed as well.

The former *emir*'s closeness centrality decreased, as evidenced by the fact that the short paths between him and the other nodes became longer; for example, he was directly linked (path length = 1) to 19 members before the 7/7 bombing; to 13 members following the terrorist act; and to only 7 members after the banning of the successor organizations (Kenney et al., 2013, p.742). Nonetheless, the *al-Muhajiroun* network downsized but survived. The former central leader, Bakri, still played a prominent role, although as a partial network-type leader. To reiterate, three-dimensional networks, in contrast to rigid centralized hierarchies, are characterized by discordant centrality values with different centrality criteria.

While losing points on the closeness centrality scale (and also, obviously, on the degree centrality scale, because Bakri had fewer direct links to other nodes after the two setbacks occurred), Bakri still ranked high in terms of betweenness centrality (Kenney et al., 2013, Table 2). This indicates his continued influence in terms of communication within the organization that preferred a network structure to a hierarchy. "Bakri's leadership has changed from direct oversight of al-Muhajiroun to symbolic, geographically removed leadership" (Ibid., p. 743), including online lectures and visits of the network's members to Lebanon.

"While Bakri remains highly respected by his students, several al-Muhajiroun veterans based in Britain have essentially replaced their mentor as day-to-day emirs of the evolving network. When al-Ghurabaa and the Saved Sect were banned by the British government, it was these AM[89] veterans that created 'Ahlus Sunnah wal Jamaah,' an Internet discussion forum, and Islam4UK, a platform for conducting the group's political protests and Islamic 'roadshows,' largely without Bakri's direct involvement" (Ibid.).

Interestingly, Omar Bakri even gained in importance in terms of *eigenvector centrality*. According to this criterion, "a given node is considered central to the network to the extent that its neighbors are central" (Ibid., p.745). This upward trend is attributable to the fact that the geographically distant partial network leader lost many contacts with low-centrality rank-and-file members, and his direct partners were AM elite members. Kenney et al. (2013) make the prediction that networks with high eigenvector centrality should be more adaptable to hostile environments; the ability of these networks to learn and adjust, in a neural network-like fashion, should be emphasized.

Despite the uniqueness of human society and its network structures, dark networks, including *al-Qaeda* and related organizations, reveal striking organizational similarities to some types of networks in the biological realm. Dark networks combine elements that are comparable to those characteristic of several paradigms of biological networks. Their neural network-like learning/training capacity was stressed above.

---

[88] *Al-Muhajiroun*'s relationship with *al-Qaeda* was questionable. Although Osama bin Laden ranked second on the betweenness scale concerning *al-Muhajiroun,* this fact was attributed to "newspaper artifacts" by the authors of the article cited (Kenney et al., 2013).

[89] AM is short for *al-Muhajiroun.*

The network paradigm that does not seem to apply to most dark networks is the *egalitarian* paradigm, which encourages tolerant, individuality-enhancing attitudes of (partial) leaders towards other network members, and this is typical of many primitive hunter–gatherer groups.

Dark networks exemplified by *al-Qaeda* obviously do not pay much attention to their members' individual freedoms, rights, and personal preferences. On the contrary, they are ready to sacrifice many individuals, as is necessary, for the successful operation of the whole network or any of its clusters; *al-Qaeda*'s members are expected to obey its fundamentalist principles (the "impersonal leader"), as well as the organization's relatively short-lived earthly leaders that act on his behalf. *Al-Qaeda* is largely based upon the controlling action of leaders inside small subclusters that directly execute terrorist attacks, and the organization is characterized by a modular organizational pattern. Therefore, the global *jihad* network is, to an extent, similar to a cnidarian colony that conforms to the *modular* paradigm (2.2) of biological networks.

Removing the nodes responsible for intercluster links is analogous to severing the stolon (stalk) that connect individual zooids within the body of the modular cnidarian organism.

While the unneeded polyps in the colony are eliminated (Marfenin, 2002), and dysfunctional cells in an organism undergo apoptosis (programmed cell death), *al-Qaeda* and other similar dark networks consider their members "cannon fodder" to be used, e.g., to fight Americans in Afghanistan or to commit a terrorist act. Importantly, the ideological matrix that incorporates the goal of purifying and renewing Islam and overthrowing "apostate" local governments places special emphasis on self-sacrificial heroic deeds and, therefore, pays no attention to the value of a *human life* (Soloviev, 2004, http://www.intertrends.ru/five/006.htm, emphasis added—*O.A.*).

Terrorist dark networks also exhibit features that are comparable to those of *eusocial* networks in biological systems:

- Eusocial three-dimensional networks exemplified by ant families are characterized by alternate networked and hierarchical structural levels (2.4); at the lowest level, there are hierarchical "project teams" (clans) whose foremen represent partial leaders embedded in the higher-order network structure (a column). In a similar fashion, clusters of *al-Qaeda* form task-oriented teams. A largely hierarchically organized team hijacked planes and crashed them into the Pentagon and WTO towers on September 11, 2001. The dominance of the hierarchy over the *sensu stricto* network is confirmed by the results of analyzing the terrorist acts in terms of the network centrality criteria. It was revealed that the one of the hijackers, Mohammed Atta, was characterized by the highest degree and closeness centrality values; in addition, he ranked second in terms of betweenness centrality (Krebs, 2002). I reemphasize that the convergence of the three centrality-related values attests to the strictly hierarchical organizational pattern of the hijacker team. Therefore, it cannot be called a three-dimensional network. Nonetheless, the project team leader represents just one of the several hubs at the higher level of the whole *al-Qaeda* cluster (a "column", to use an insect-specific term) that included a number of overlapping teams that were tasked with specific missions (Krebs, 2008, http://orgnet.com/tnet.html).
- Eusocial systems include several castes. Likewise, many dark networks consist of the elite and the rank-and-file members that are often considered "cannon fodder". In the

case of *al-Muhajiroun*, the high eigenvector centrality of the former *emir*, Bakri, indicated that he predominantly dealt with other elite members in the organization, and that his departure additionally emphasized that the network included at least two partly isolated "castes".

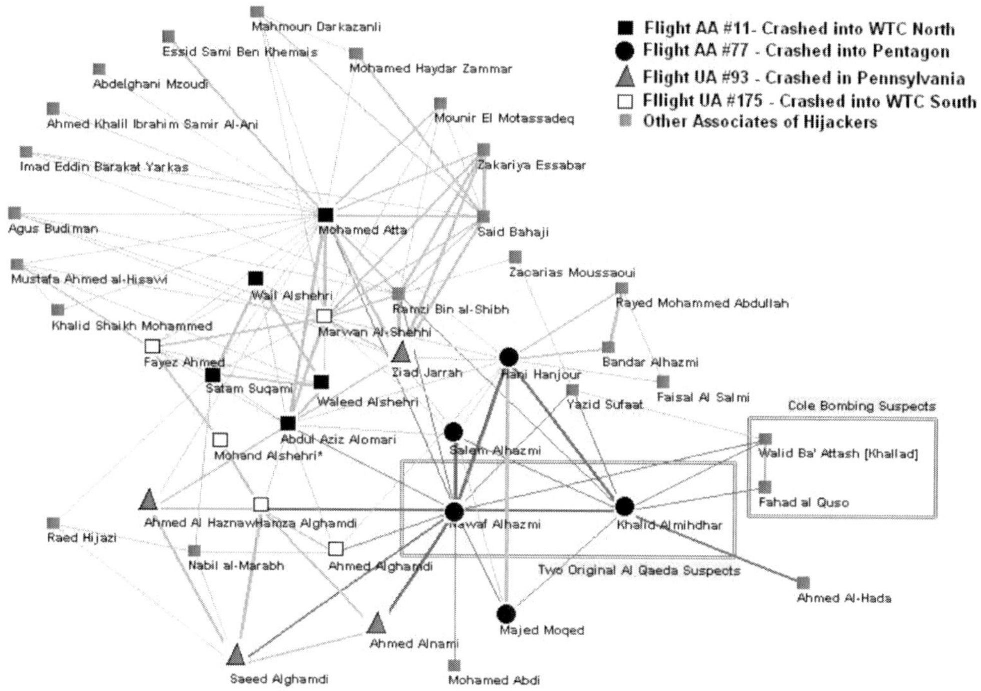

Figure 3.1. The structure of the team involved in hijacking planes and crashing them into the Pentagon and the WTC towers on September 11, 2001. The boxes contain the links that were the first to be discovered by the investigators. The picture demonstrates the leading and coordinating role of Mohamed Atta during each of the terror acts. However, the hierarchical team under him was part of a higher-order decentralized network structure, so that the whole system was based upon the "network composed of hierarchies" principle, in an analogy to the eusocial structures of ants (see 2.4). Reproduced with permission from the following website: *Connecting the Dots. Tracking Two Identified Terrorists.* Krebs, V. E. http://orgnet.com/tnet.html. Copyright © 2001 - 2008, Valdis Krebs. Note: the diagram was slightly modified to make its black and white version easy to read .

- In the example of ant families, the coexistence of teams of active specialists and of the pool of passive generalists in eusocial systems was demonstrated. In a similar fashion, terrorist networks, e.g., *al-Qaeda*, do not solely include the active members that are directly involved in carrying out important missions. Their activities and, in particular, their "percolation" in society, are facilitated by the presence of "passive supporters". These "are normal people who do not need to express their position explicitly. They just do not oppose a jihadist act in case they could" (Miralles Canals, 2009, p.903). In a similar way, many people in Chechnya let militant separatists hide from the federal Russian army in their settlements (*auls*).
- The *stigmergy* principle ("the results achieved by working ants stimulate further work at the same site", see 2.4.8) is widely used by human networks (Lewis, 2013),

including dark networks typified by groups affiliated with *al-Qaeda*. There are data that their terrorist acts tend to become increasingly destructive, claiming more and more human lives. One of the reasons behind this is that *al-Qaeda* creates a competitive environment where different subnetworks (clusters, cliques, or terrorist cells) vie for prestige and status. "Al Qaeda may be portrayed as a contest organizer… Each group observes the results of previous attacks perpetrated by other groups. Consequently, each group maximizes its efforts by launching attacks more destructive" (Caruso & Schneider, 2013). Therefore, *al-Qaeda,* a polystructural system, obviously includes a quasi-market (competitive) component, which is, nonetheless, embedded in a network structure-dominated environment (cf. a similar situation in an ant family, see 2.4.4 and 2.4.8).

In conclusion, it should be emphasized that effective counter-terrorist structures should be isomorphic; i.e., structurally similar to the networks they aim to disrupt. To reiterate, fighting a networked organization presents serious difficulties to a hierarchically organized police department or army unit. According to Krebs (2002, p.51, with reference to Ronfeldt & Arquilla, 2001, http://www.firstmonday.org/ojs/index.php/fm/article/ viewArticle /889), who investigated the 9/11 hijacker network, "to win this fight against terrorism it appears that the good guys have to build a better information and knowledge sharing network than the bad guys".

As far as network structures in modern human society are concerned, of special note are informal youth groups. In principle, they may be based upon different organizational patterns which reminds us, once again, of our ambivalent evolutionary legacy in conjunction with our contradictory behavioral trends (see Table 3 in 2.7). Some of these informal groups establish rigid hierarchies which is, unfortunately, a widespread trend among criminal groups. Other groups, termed "clubs", prefer horizontal relationships between their members. Typical examples include Russian hippies or the art group *Mit'ki (митьки)*. A stranger could rely on the group members' helpful and friendly behavior (Gromov & Kuzin, 1990).

3.2.9. General Features of Network Structures in Human Society

As was described, network structures can be characterized as coherent, cooperative, and predominantly non-hierarchical systems. They are structurally similar to primitive hunter–gatherer bands and even ape groups, especially when discussing the networks that place special emphasis on egalitarian principles. The idea that these principles are consistent with the biological side of human nature, which still exists in our genes, is consistent with the fact that networks are spontaneously formed in human groups and organizations, regardless of the organizational pattern favored by the boss of that organization. If people, e.g., staff members of one company, interact for a sufficiently long time, the sociological laws of small groups come into play: The collective is restructured based on informal relationships that are not stipulated in formal instructions and regulations. For such restructuring to take place, it is sufficient that people have breakfast together. Free "social breakfasts" are provided by some companies for their staff. In this situation, people are encouraged to establish informal groups of psychologically compatible individuals that have similar views and principles, and who are friendly to one another and ready to cooperate regularly in terms of business as well as in everyday life. Obviously, such an informal group constitutes the embryonic stage of a network structure characterized by spontaneous partial leaders and predominantly horizontal relations between members.

Some general features of network structures in human society outlined in Chapter one (1.2.2) should be reemphasized at this point, while taking into account the examples discussed in this section.

- Network structures are based upon horizontal and cooperative interactions between individuals or groups, which supplement the inevitable existence of dominance-submission relationships not only in egalitarian hunter–gatherer bands, but also in civilized society.
- Many network structures use the split leadership principle. As noted above, the Twin Oaks community had as many partial leaders as they did social or business functions. Each of these partial leaders holds limited power and offers a specialized area of competence.
- Network structures pay special attention to the psychological atmosphere and interpersonal relationships in the group, which may require introducing additional social roles that are concerned with these matters. They are exemplified by the shaman in hunter–gatherer societies, the partial psychological leader in a hirama, and the moderator in modern-day networked clubs.
- Many networks use a system of overlapping/rotatable social (and professional) roles, thus making it possible for group members to change or combine their occupations.
- The long-term viability and efficiency of a network usually implies the existence of a system of ideas and values, as well as social behavior norms (the matrix).

Importantly, network structures do not exist in isolation. They have to interact with the hierarchical and (quasi-)market structures of human society. Modern society is marked by the predominance of bureaucracies, i.e., hierarchies based upon formal organizational principles.

Despite the rise in hierarchical structures, network structures were of considerable importance during the entire history of human society. These structures are steadily gaining importance in the present-day world. A wide variety of network structures emerged in different historical periods and in different parts of the world. Network organizations are characteristic of such social or political structures as the early Christian church, the Swiss commune/Gemeinde, the Russian peasant obschina, the Israeli kibbutz, American communes, clandestine nonmarket "favor exchange" structures (e.g., blat networks in Russia), and criminal and terrorist "dark" structures typified by al-Qaeda.

3.3. Bureaucracy. Nonbureaucratic Organizations

Social structures in the traditional stratified hierarchical society were predominantly characterized by the following features:

- The activities of human social groups (collectives) were primarily aimed at attaining the goals personally set by the leader (the Pharao, the liege, the landlord, or somebody appointed by them).
- Labor distribution depended on the leader's personal decision.
- The hierarchical stepladder was established according to the leader's decision.
- The leader also made all other major organizational, structural, and financial decisions.
- The leader's authority was sanctified by cultural tradition; to an extent, it could also depend on the leader's personal charisma.
- No organizational regulations and rules were solidified in legal terms; personal relationships between the subordinates and their leader were of paramount importance.

Indisputably, such personal leader–subordinate relationships still exist in shadow hierarchies such as in clientelles that were established, e.g., within the framework of the Sicilian *mafia (Cosa Nostra)*. Such an informal hierarchy has an influential person at its top; this informal leader controls important resources and can provide support to "clients". The clients are expected, in return, to be loyal and submissive to the "patron" (reviewed, Oleinik, 2003).

3.3.1. Bureaucracies

Bureaucratic structures already existed in the state apparatus and the army of the Ancient Rome, for instance. However, bureaucracy started playing the dominant role as industrial society took shape in the West.

Bureaucracy[90] is based upon a centralized hierarchy, a narrow specialization of all of the members of a bureaucratic organization, and the dependence of relations between them on the members' official status and formal regulations rather than on sentiments. According to Max Weber (1956), the development of bureaucracy marked the "rationalization" period in the history of industrial, political, scientific, and religious organizations. They were compared to machines (clockwork, cars) and their people to machine parts, with each part performing a specific function. The following is a list of the main distinctive features of bureaucracy (Weber, 1965; partly quoted according to: Meulemann, 2008):

- The work of the members of a bureaucratic organization pursues the goals set by official documents, e.g., contracts or the organization's regulations.
- Functions are distributed among various organization parts and workers on a clear legal basis involving narrow specialization, a carefully defined division of labor, secure jobs and salaries, and promotions that vary according to seniority and merit.
- There is a rigid unambiguous hierarchical structure, or a chain of command: officials are subject to the authority of their superiors.

[90] Note: This subsection presents, in an abridged and modified version, part of the materials discussed in more detail in my work that was previously published by *Nova Science Publishers, Inc.* (Oleskin, 2012).

- Authority is impersonal and is vested in rules that govern official business.
- Legal documents (regulations, statutes, constitutions) stipulate behavior norms and decision-making rules, depending on the member's official rank, position, and status, not on informal relationships.

Towards the end of the 20th century, human society, at least in the Western (First World) countries, entered the postindustrial stage of its evolution. The future of bureaucracy became dubious. Despite the worldwide spreading of bureaucratic structures and their indubitable advantages—such as their efficiency and easy manageability in a stable environment, as characterized by predominantly routine technologies and planning methods—bureaucracies faced increasingly serious problems in a more dynamic and unpredictable world. The whole organization and its main parts tend to become closed in on themselves; the links between them become far too weak. Bureaucracies also tend to formalize too many aspects of human life in conformity with official regulations.

By narrowing down the members' personal responsibilities to their workplace functions, a bureaucratic organization decreases their interest in carrying out their duties, which is in contrast to the methods used by primitive hunter–gatherers. Eskimo hunters, unlike a bureaucratic clerk, will never complain: "Hunting seals is a sheer waste of time… I need a more interesting and meaningful job" (Bernhard & Glantz, 1992).

The disadvantages of bureaucracy cause inadequate and slow decision-making in a changing, unpredictable situation. The efficiency of bureaucratic organizations could be increased by subdividing them into smaller groups of people "who interact on a regular basis, thus maximizing reciprocal cooperation" (Axelrod, 1984, p.131). This organizational change actually meant taking a step towards *nonbureaucratic* organizations that became a reality in the second half of the 20th century.

3.3.2. Nonbureaucratic Organizations

Nonbureaucratic organizations mitigate hierarchies, aiming to make good use of personal relationships with informal leadership. Such organizations are particularly advantageous in the circumstances where bureaucracies lose their efficiency. The transition of the computer industry in the USA and in some other countries to nonbureaucratic structures was primarily due to the necessity to cope with the highly dynamic commercial and technological environment. These structures could more successfully solve complex and unclearly formulated (fuzzy) problems requiring creative teamwork rather than narrow specialization, obedience to the boss, and compliance with the organization's formal rules and regulations. Of much potential value is the idea that was extensively used in Japan after World War II, where each creative team member, if necessary, should be ready to switch from one area of activity to another. This obviously promotes the coordination and integration of various types of creative work. These features of nonbureaucratic structures facilitate solving interdisciplinary problems or carrying out commercial projects.

Typical nonbureaucratic structures in the modern-day world are exemplified by (1) *matrix* structures that are set up if the organization carries out several projects; they thus include several project leaders in addition to the general boss, and (2) *adhocratic* structures that are established to cope with unexpected serious problems. They minimize formal hierarchical relationships and function as a single team that collectively carries out challenging tasks and distributes the risks and rewards involved among all of their members.

Other kinds of nonbureaucratic structures, such as participatory and entrepreneurial structures, have been designed as well (see Mescon et al., 1986).

Nonetheless, bureaucracy still possesses advantages over alternative organizational options in many situations. It still "can be seen as a rational tool for executing the commands of elected leaders" (Olson, 2005, p.5). An important advantage of bureaucracy—unless it becomes "over-bureaucratic", i.e., "rules are followed too slavishly", or it becomes corrupt—is that it "is based on the rule of law, due process, codes of appropriate behavior, and a system of rationally debatable reasons". Currently, bureaucracies are "part of a repertoire of overlapping, supplementary, and competing forms co-existing in contemporary democracies, and so are market-organization and network-organization" (Ibid., p.26). It is the ever important bureaucratic hierarchies of the state apparatus that should establish mutually beneficial and productive relationships with nongovernmental network structures that form part of civil society; this is vitally important in terms of present-day politics in a large number of countries including Russia.

As for nonbureaucratic structures, they can develop in compliance with two alternative scenarios:

- They can transform into *(quasi-)market structures*. For instance, organizations that are concerned with public policy can convert into a kind of "supermarket" that delivers "a wide variety of public services" (Ibid., p.9).
- They can evolve into *networks*—a process that is of paramount importance in terms of this work.

Bureaucracies that are widespread in industrial society are based upon a centralized hierarchy, a narrow specialization of all members of a bureaucratic organization, and the dependence of relations between them on their official status and formal regulations. The current postindustrial historical period is marked by the spreading of nonbureaucratic organizations with mitigated and split hierarchies that evolve either into (quasi-)market structures or, alternatively, into sensu stricto *networks.*

3.4. Network Structures in Terms of Social Technologies

This section is predominantly concerned with *social technologies based upon network principles*, i.e., with intentionally created (not spontaneously emerging) network structures that function in various areas of society. Their characteristic features and practical potential is the subject of a large number of recent works (see, e.g., Baker, 1994; Castells, 1996, 2000, 2001, 2004, 2009; Oleskin & Masters, 1997; Chuchkevich, 1999; Adler, 2001; Churakov, 2001; Arshinov, 2003; Sergeev & Sergeev, 2003; Oleskin & Kirovskaya, 2005; Oleskin, 2007a,b; 2013b; Meulemann, 2008; Kahler, 2009a, b). In addition to research on networks, practical steps are made to establish them in modern-day society. Network organizational principles were used by Alexander Krel' and others who established the *AntEra Association* (AntEra, 2013, http://celenie.ru/index_eng.html), which was aimed at treating people suffering from rheumatism in Russia. Apart from health care workers, the network structure included patients and their relatives.

Social technologies can be used to create a variety of different kinds of networks. Of relevance to the fractal properties of networks is the fact that network structures may be small (the size of a primitive hunter–gatherer band) or large. With large networks, the multilevel organization of network structures becomes obvious: a network structure consists, in turn, of smaller network structures. In addition to general *sensu stricto* network characteristics (decentralization, cooperative internode interactions, and the matrix including the network's goals, values, and social norms), intentionally created networks are marked by a number of specific features including the following:

- a *wide specialization* of its members: instead of narrowly focused specialists, there are generalists who cope with the whole task (mission, role) carried out by the network, tending to work as a single team;
- efforts to *promote informal personal relationships* among network members (individuals or groups);
- a *management style* focusing not only on attaining the goals pursued by a network structure, but also on satisfying the needs of the people involved in this network;
- a policy *enabling the members of a social network to overcome loneliness* and even to play the role of a partial leader and to develop the feeling of belonging to a coherent social group (a team of enthusiasts), which actively supports each of its members while respecting their individuality.

Hence, in sociological terms, decentralized non-hierarchical cooperative networks tend to evolve into *primary groups* (*Gemeinschaften* in the words of Ferdinand Tönnies (1988), a classical German sociologist). In his view, a *Gemeinschaft* is similar to a family in that the interpersonal relations inside it resemble parent–child, husband–wife, or brother–sister relations. Human relationships are based upon reciprocal attachment and support, as well as the feeling of belonging to the group and self-identification with it. Any personality in a *Gemeinschaft* is regarded as unique, and the whole group is significantly changed with the appearance or disappearance of a single member. In contrast, a *Gesellschaft* (also termed "secondary group" or "society") envisions people only as incumbents of partial, and often temporary, social roles. While in a *Gemeinschaft*, people live together; in a *Gesellschaft*, they just coexist as temporary "fellow travellers", business partners, service clients, and so on (Tönnies, 1988).

Bureaucracies belong to secondary groups (*Gesellschaften*) where people interact in conformity with official regulations because they are "coworkers and not friends". A different situation is typical of a networked organization. Formal interactions are supplemented by informal relationships involving personal likes and dislikes, sentiments, and informal leadership that is typically partial and situation-dependent. Even networks only pursuing the goal of doing a practical job tend to evolve into quasi-communes. This point received the attention of the developers of the *hirama* project (see Oleskin, 1996, 2012; Oleskin & Masters, 1997; and 1.2.2 above), who added to the list of partial leader positions (1) a psychological leader playing the role of a conflict manager and, if necessary, a psychiatrist, and (2) an "external" leader, one of whose duties are to organize leisure activities for *hirama* members. From the psychological viewpoint, even virtual interactions among online social network users, e.g., on the *Facebook* or *MySpace* websites, promote mutual trust, self-diclosure, multi-aspected communication, and even intimacy, provided that the virtual network structures robustly function for a sufficiently long time.

A comparison of network structures and more traditional social hierarchies typified by bureaucracies reveals that a centralized hierarchy can outperform a network in terms of "efficiency criteria such as speed, message count, and frugal use of resources... The network lacked a central coordinating mechanism and spent more time negotiating procedures" (van Alstyne, 1997).

However, it is important to reemphasize the creative potential of network structures, particularly with regard to inchoate projects and innovative ideas. Both the relatively slow tempo of a network and its creative potential are demonstrated below in the example of networked and hierarchical student teams in a classroom (3.6.1).

Innovative ideas generated in hierarchies are "more likely to be discarded on the grounds that the central person was too busy, that the innovation was too bothersome to implement, or that current practices required no improvement" (van Alstyne, 1997). Moreover, "persons in the network structure are apt to value their participation more and be much happier with their experience. The one exception is the central person in the hierarchy who was generally quite happy" (Ibid.).

Since communication among network members is not limited to business, and since it tends to become informal and deals with various aspects of each member's life, this promotes a multifaceted approach to the tasks faced by the network. In a similar fashion, neural networks in the brain try more than one strategy to deal with a problem, which often implies that the active neural subnetwork dealing with this problem activates a number of other subnetworks within the brain. This results in incorporating new brain structures into the expanding active brain network, as pointed out in 2.5.2. In network structures made up of scientists, an analogous strategy promotes an integrative, interdisciplinary approach to the subject of their research and facilitates innovative thinking strategies (see 3.5). This involves even those who are seemingly incompetent in the area of research and those who originally represent the "pool of passive generalists". This organizational pattern has been repeatedly discussed in this work with respect to a variety of network structures.

In contrast to bureaucracies characterized by a narrow focus, communication within a network is aimed at overcoming interdisciplinary boundaries. This is one of the reasons why networks hold much value in terms of coping with interdisciplinary issues that require the involvement of specialists in many different areas of research. Such issues include interdisciplinary scientific research (3.5), innovative educational methods (3.6), and networked business strategies (3.7). Networks also set themselves the goals of struggling against terrorism[91], overcoming ethnic conflicts, coping with the ecological crisis, and carrying out various humanitarian missions such as helping the homeless, the aged, and HIV-infected people (see the network structure in Figure 3.2).

The structural complexity of networks with its numerous links between nodes actually may give it an advantage in solving complex, interdisciplinary, and "fuzzy" problems; it is

[91] Unfortunately, terrorists set up their own "dark" networks; this point is discussed above (3.2.8).

known that the complexity of the problem that needs to be dealt with should be commensurate with the complexity of the system that is to solve it. The comparison between a network structure and the thinking brain is to be reemphasized at this point. Like the brain, a networked problem-solving team generates the solution at the level of the whole collective of its members. Each member can be considered analogous to a neuron.

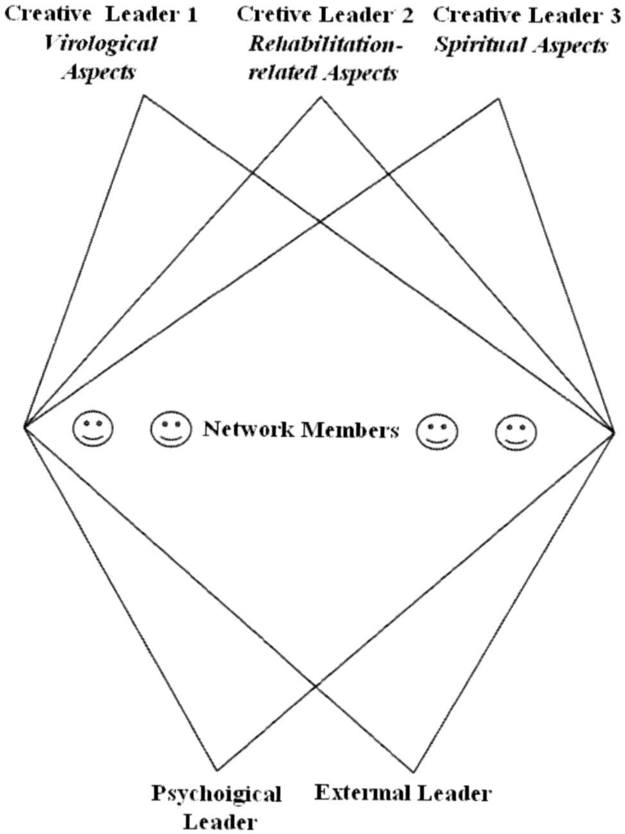

Figure 3.2. A *hirama*-type network structure in human society aimed at aiding HIV-infected patients, with special attention paid to the virological, rehabilitation-promoting, and spiritual aspects of the mission.

A paradoxical situation often arises in which no *single network member possesses the whole information* used by the network. The information concerning the problem to be solved is stored "everywhere and nowhere" in the network, in an analogy to the sensory integration system (SIS) operating at the level of a whole fish shoal or school (see 2.3.7 above). This is both an advantage and a disadvantage of a network. Despite promoting collective problem-solving, collective information storage predisposes the network's members to make wrong individual decisions based on incomplete data because nobody can personally make use of the whole relevant information concentrated by the network (van Alstyne, 1997). This also concerns technical networks exemplified by the Internet. "It is a curious fact that although the Internet is a man-made and carefully engineered network we don't know exactly what the structure is, since it was built by many different groups of people with only limited knowledge of each other's actions and little centralized control" (Newman, 2012, p.3).

In some social projects, networks are envisaged as parts of more complex structures that also incorporate hierarchical and/or (quasi-)market components. They are exemplified by the *Kinoviy* project (mentioned in 1.2.7) that is aimed at revitalizing archaic prestate modular chiefdom-like structures denoted as "social cells". They combine hierarchical and network organization features. In particular, the *krug soglasiya* (consensus-promoting commission) that performs some administrative functions in such a "social cell" is internally structured according to network principles (see Lebedev, 2007, p. 152).

Network structures established in terms of social technology can be applied to a variety of cultural, scientific, educational, and political projects. Their successful operation is facilitated by their organizational features including decentralization, wide specialization of their members, the development and creative use of informal interpersonal and intergroup interactions, and the presence of a uniting goal, as well as social norms and values (the matrix). Although hierarchies can perform better than networks in terms of efficiency, networks are more creative and are more likely to generate innovative ideas at the level of the whole structure; this is analogous to a neural network. Collective information storage and processing, despite their obvious advantages, pose the risk that network members may make wrong decisions and draw wrong conclusions if they have access to only a part of the whole network-stored information.

3.5. Interdisciplinary Scientific Research: Role of Network Structures

As far as interdisciplinary scientific (fundamental and applied) research is concerned, *sensu stricto* network structures often have advantages over traditional bureaucratic structures. Bureaucracies exemplified by research institutes that, in many countries (e.g., the USA and Russia), belong to the respective Academy of Sciences typically have a more narrow focus in terms of their areas of research. Network organization is particularly advantageous if a discipline or a research trend is still in its infancy and its practical results are not obvious.

A network was spontaneously established by a group of Russian microbiologists (Kirovskaya, 2004; Kirovskaya & Oleskin, 2003, 2004; Oleskin & Kirovskaya, 2005, 2007). During the second half of the 20th century, an informal association of enthusiasts took shape within the microbiological community in the Soviet Union.

These enthusiastic microbiologists belonged to different formal institutions, e.g., Moscow State University, the Institute of Microbiology of the Academy of Sciences of the Soviet Union, the Institute of Epidemiology and Microbiology of the Academy of Medical Sciences of the Soviet Union[92], and others. They specialized in different fields of microbiology or other life sciences (Figure 3.3). However, all the scientists took an interest in the same subject, which can be denoted as "social microbiology". They were fascinated by the social life of microorganisms, the collective behavior of their cells, the information exchange among them,

[92] These two institutions were called the Institute of Microbiology of the Russian Academy of Sciences and the Institute of Epidemiology and Microbiology of the Russian Academy of Medical Sciences after the demise of the Soviet Union in 1991.

and the microbial populations as coherent, organism-like systems. This common denominator of the enthusiasts' theoretical approach to microbes can be termed "the population- and communication-centered paradigm in microbiology". In essence, they developed a borderline field between microbiology and biopolitics.

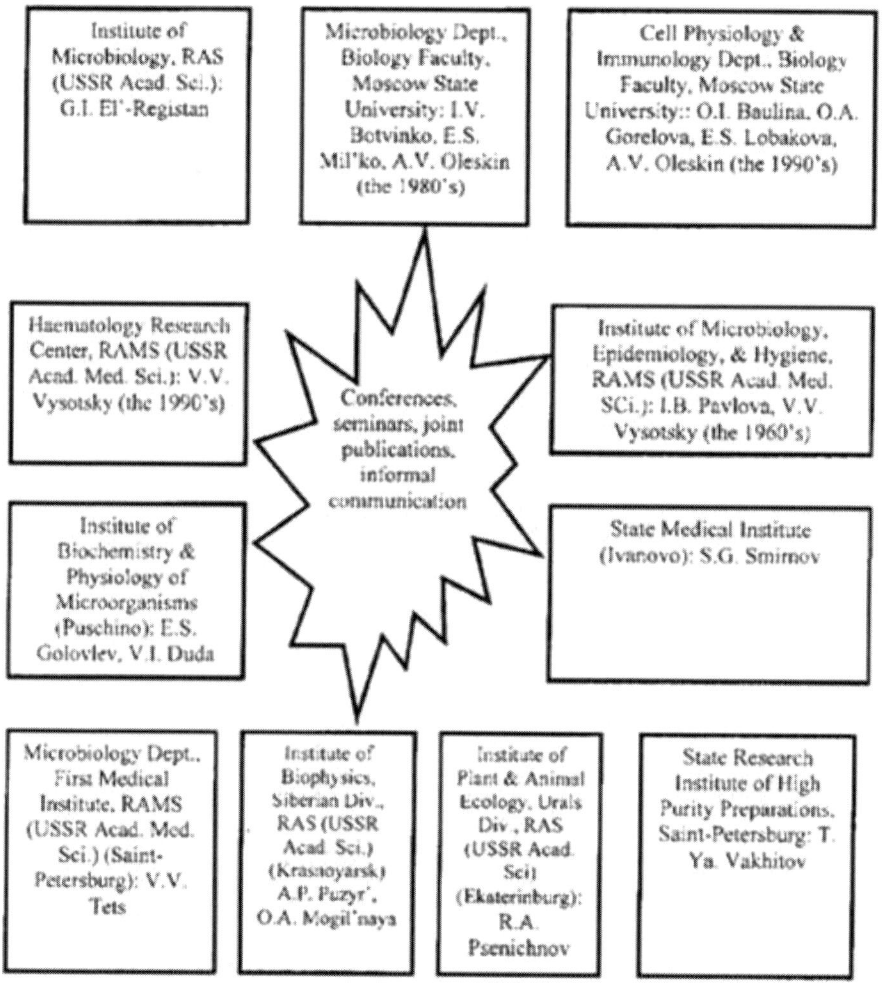

Figure 3.3. A spontaneous network structure within the Russian scientific community in the late 20th century (the 1960s–1990s). The network structure focused on the organization of microbial populations and on the communication between microbial cells. Based on the following works: (Kirovskaya, 2004; Kirovskaya & Oleskin, 2003, 2004; Oleskin & Kirovskaya, 2005, 2007). RAS, Russian Academy of Sciences; RAMS, Russian Academy of Medical Sciences. Note: 1. The period spent by a scientist at a particular institution is given in parentheses after the scientist's name; 2. Some institutions changed their names during the period in question; 3. Some scientists changed their position and affiliation during the same period.

Owing to their common views, the adherents of this paradigm established business relationships with one another. In the absence of a central bureaucratic leader, these relationships were decentralized and non-hierarchical. Despite their different backgrounds and institutions, the enthusiasts held meetings and conferences, conducted joint research initiatives, and published their results. The independently working scientists V. V. Vysotsky,

P. L. Zaslavkaya, A. V. Mashkovtseva, O. I. Baulina, and others established an informal network in the mid-'80s when they all participated in a microbiological conference at the town of Ivanovo. It was during this conference that they found out that they held similar views on the microbial population as a system composed of diverse individuals—each of them making its own contribution to the well-being of the whole system. The microbiologists decided to work together for the benefit of the whole network that was cemented by the new paradigm. For example, they produced the article "Polymorphism as a Developmental Trend in Populations of Prokaryotic Organisms" (Vysotsky et al., 1991). Several years elapsed after the fruitful meeting at Ivanovo, and the network group discovered the article "Bacteria as Multicellular Organisms" by the American scientist James Shapiro (1988) who held quite similar views.

This spontaneous network structure in the Soviet (Russian) microbiological community, unlike a traditional hierarchical school of thought, was not founded by a single eminent scientist and supported by a group of his disciples. Instead, the network functioned as an *invisible college*: information was constantly exchanged between autonomous groups or individual scientists that were united by a common microbiological paradigm. Eminent scientists were among the members of the network structure. Nicolai D. Yerusalimsky, Full Member of the Academy of Sciences of the Soviet Union, was the central leader and the founder of a hierarchically organized school of thought, and he was an inspiring partial leader in the network. The hierarchy under his leadership dealt with microbial growth kinetics and industrial microbiology; developments in these areas were only the "tip of the iceberg" of Yerusalimsky's scientific ideas. His research and educational activities also promoted the development of the network; of paramount importance in these terms was his concept of the *microbial culture as a coherent, organism-like system.* The dissemination of Yerusalimsky's concepts in the Russian scientific community contributed to the development of new theories and methods by a large number of independently working Russian microbiologists.

It was the common microbiological paradigm—the *matrix*—that cemented the network in the Russian microbiological community. Networked, rather than hierarchical, organization significantly influenced their research strategy and the results obtained; creative dialogues between people with different backgrounds significantly broadened their horizons and encouraged them to develop ideas and concepts of interest in terms of general biology, science as a whole, and even philosophy. This was demonstrated in a series of works by Tatiana A. Kirovskaya and myself (Kirovskaya, 2004; Kirovskaya & Oleskin, 2003, 2004; Oleskin & Kirovskaya, 2005). The following features of the research conducted by the members of the network within the Russian microbiological community were related to the influence of the network structure on scientific creativity:

1. *They put forward important hypotheses related to areas far beyond the original scope of their research.* A large number of researchers in the field of the population organization and communication of microorganisms produced articles, theoretical papers, and monographs on general microbiological issues, including the book *Low Molecular Weight Microbial Autoregulators* (Khokhlov, 1988), the review *Distant Informational Interactions in Bacteria* (Nikolaev, 2000), and the seminal theoretical

work *Bacterial Exopolysaccharides* (Botvinko, 1985)[93]. Indisputably, research on microbial communications was related to ecosystem theory and global studies. In addition, Yerusalimsky emphasized the relationship between his research on the development stages of microbial populations and the general biological concept of the *life cycle* of an organism, which was advanced, among other scientists, by Nicolai P. Krenke and Alexey N. Severtsov.

2. *Their scientific developments had significant philosophical implications.* Golovlev (1999) aptly called Yerusalimsky's theory "the philosophy of the microbial population". Smirnov (2004) pointed out that his studies on the developmental stages of the microbial population were encouraged by his philosophical views concerning biological time.
3. Remarkably, *the network composed of microbiologists revealed structural similarities (isomorphism) to networks formed by microbial populations.* Like microorganisms with their functionally differentiated subclusters in decentralized biofilms, microbiologists with their different professional backgrounds and different, although overlapping, areas of research formed a heterogeneous network with a large number of "active centers". Both kinds of non-hierarchical structures were coherent and cooperative[94] systems cemented by the matrix, which was comprised of a biopolymer substance in microbial biofilms, and a set of ideas and methods along with the sense of belonging and enthusiasm in the networked community of the adherents of the population-and-communication-centered paradigm in microbiology.
4. *The network of enthusiastic microbiologists was capable of generating important scientific ideas that were "ahead of time"*: some of these ideas were similar to those put forward several decades later by the mainstream scientific community. The network was capable of the *anticipatory reflection* of the *future achievements* of the global microbiological community.

The term *anticipatory reflection*, originally applied by Smirnov et al. (1982) to the lag phase, whose sub-stages foreshadow the subsequent phases of the development of a microbial population, should also be used with respect to the network structure composed of microbiologists. Anticipatory reflection is related to the heterochrony of a network (see section 1.5) where, to reiterate, each node develops in its own rhythm in the absence of a central pacemaker[95]. Therefore, some network nodes work more efficiently than others; some

---

[93] It is in this article that Irina V. Botvinko (1985) suggested the interdisciplinary concept of "the matrix" that is extensively used by me in this work.

[94] Cooperation of necessity prevailed over competition inside the network. Since the mainstream microbiological community provided little support to these "strange nerds", the conditions were so harsh that the whole network could only survive if its members constantly and actively supported one another. Most microbial biofilms face a similar challenge.

[95] Restrictions imposed on the differences between individual rhythms are due to the functioning of local and distant coordinating factors in a network (see 1.5). In the example of a network of microbiologists, these factors represent interactions between individuals or groups, such as exchanging messages, participating in joint research projects, attending conferences, and writing research articles together. Even weakly connected nodes may tend to become at least partly synchronized (entrained, Kuramoto, 1975, 1984; Arenas et al., 2008), and seemingly weak ties ultimately prove to be quite strong (Granovetter, 1973, 1985). Nevertheless, this synchronization is generally not as strong as that in a hierarchy, as exemplified by a bureaucracy working according to the schedule set by the boss.

of these highly efficient nodes may do their work quicker than a competing hierarchy, such as a research institute.

This was the case with some of the achievements made by the network, which were focused on the population-and-communication-centered paradigm in microbiology. For example, the concept of the microbial culture as a constantly communicating system that is capable of regulating its own development was advanced by Nikolai D. Yerusalimsky in the 1940s–1950s; similar concepts were suggested by James Shapiro, Eschel Ben-Jacob, and others in the 1980s and 1990s. The concept of autoregulatory substances put forward by Galina I. El'-Registan et al. in the 1960s–1970s (see, e.g., El'-Registan et al., 1979) was independently developed and reinterpreted in terms of the quorum sensing concept in the West in the 1980s and in the 1990s (Shapiro, 1988; Fuqua et al., 1994); this was the subject of a series of works on the history of microbiology (Kirovskaya, 2004; Kirovskaya & Oleskin, 2003, 2004; Oleskin & Kirovskaya, 2005).

Apart from the population-and-communication-centered paradigm discussed above, network structures spontaneously emerged in such diverse interdisciplinary fields as biosemiotics, bioethics (as exemplified by bioethics organizations such as the *International Centre for Ethics in the Sciences and Humanities* in Tübingen), biotechnology, and biopolitics. For instance, some of the biotechnology centers that have recently been established in a number of countries do not use bureaucratic organizational principles. They lack a central leader and tend to function as a decentralized interdisciplinary team that brings together specialists with different backgrounds.

This is exemplified by the networked interdisciplinary laboratory, *DNAX Research Institute of Molecular and Cellular Biology*, which was established in the early 1980s at Palo Alto, California, by Alejandro Zaffaroni (*Syntex*) and the Stanford Professors Charles Yanofsky, Paul Berg, and Arthur Cornberg. At the moment of its foundation, the decision was made to use unconventional organization principles. The laboratory combined fundamental research in molecular biology (cell growth control and DNA replication in bacteria) and immunology (lymphokines and T-lymphocytes) with biotechnology and medicine; marketable products were developed by *DNAX* in collaboration with *Schering-Plough Corporation*[96] as a "side effect" of scientific research, which was the main focus (Klausner, 1988). *DNAX* included a number of small groups of researchers that interacted according to network principles. During the course of the company's history, up to the present time, the main goal has been "to create medicines that serve the unmet needs of patients" (DNAX, 2014, http://venturebeatprofiles.com/ company/profile/dnax-research-institute).

From the beginning of its development, *DNAX* has encouraged its people to behave as one research team and it promoted cooperation rather than competition between individuals and groups. Cooperation was facilitated by encouraging free information exchange. Unlike most commercial labs and institutes, *DNAX* used to have no closed doors and sophisticated safety systems that would separate the research team into autonomous subgroups. The *DNAX* team included two partial leaders (cf. the *hirama* in 1.2.2), which were responsible for the molecular biology and the immunology area of research, respectively. *DNAX* also had an advisory board consisting of 15 prominent scientists, including 5 Nobel Prize winners. Compared to the *hirama* structure, the advisory board was a collective analog of both (1) the psychological leader, because advising usually involves psychotherapy; and (2) the external

[96]Currently called *Schering-Plough Biopharma*.

leader, because it represented the entire network structure in the scientific community. The 15 advisory board members established a non-hierarchical network (a subnetwork within the entire *DNAX* network). Therefore, *DNAX* can be regarded as containing a two-level network structure.

Networks dealing with interdisciplinary applied research may include researchers along with business people. Large biotechnology centers may also include politicians who aid with the coordination and financing of the business activities.

Modified *hirama* principles have actually been used by the network structure composed of software development teams that adopted the Manifesto for Agile Software Development on February 11–13, 2001, during a meeting at *The Lodge at Snowbird* ski resort in the Wasatch Mountains of Utah. According to their website (Beck et al., 2001, http://agilemanifesto.org), the conference involved representatives from Extreme Programming, SCRUM, DSDM, and a variety of other firms and organizations. It was emphasized that "a bigger gathering of organizational anarchists would be hard to find, so what emerged from this meeting was symbolic – a *Manifesto for Agile Software Development* signed by all participants". The new network was called *The Agile Alliance,* and it adopted a "set of values based on trust and respect for each other and promoting organizational models based on people, collaboration, and building the types of organizational communities in which we would want to work" (Ibid.). Software development was broken down into short stages (iterations) carried out by the network structure. Some of the 12 main principles set forth in the manifesto dealt with *The Agile Alliance*'s network organization. The interprofessional dialogue-promoting activities of *The Agile Alliance*'s are stressed in the principle that "business people and developers must work together daily throughout the project". In addition, the principle is adopted that "the most efficient and effective method of conveying information to and within a development team is face-to-face conversation". Therefore, such a development team looks like a primitive hunter–gatherer band engaging in collective hunting or waging war with their neighbors. Most *Agile* teams usually work in one room (the bullpen). The development team includes the client or his/her representative, as well as testers, designers, and managers (Cohn, 2010). The dynamic nature of the network organization, in contrast to the rigid character of bureaucracies, is stressed in the *Manifesto,* and it is assumed that "the best architectures, requirements, and designs emerge from self-organizing teams". Established over a decade ago, the network of software developers is currently increasing its influence, as evidenced by the fact that over 40 new signatories pledged their support to it during the February 15–28, 2014 period alone (http://agilemanifesto.org). Evidently, the *active teams of specialists plus the pool of mobilizable generalists* pattern (see the section on the eusocial systems of ants, subsection 2.4.7) appears to work with *The Agile Alliance.*

To reiterate, in the animal kingdom, this pattern is particularly characteristic of *eusocial* systems typified by social insect families. A creative network (a lab) built on similar principles resembles a service club with active members (partial project leaders) and larger support groups that can help the project leaders. A similar example is the *Club of Innovative Development (CID)* at the Institute of Philosophy of the Russian Academy of Sciences in Moscow. Its main goal is to develop the strategies and the methods through which to foster new Russian innovation-oriented elites, which will be made up of well-educated young people (Club of Innovative Development, 2014, http://www.reflexion.ru/club).

In addition, many networks dealing with interdisciplinary research are similar to *egalitarian structures* formed by primitive hunter–gatherers (analogous structures are characteristic of some apes; 2.6); the following features are to be emphasized:

1. *Respect for individual freedoms (particularly the freedom of choice) and rights.* The network of microbiologists discussed above respected the right of every individual or collective member to deal with their favorite area of research and to develop their own theories; this freedom was only limited by temporary obligations in terms of joint projects, publications, or conferences.
2. *Partial hierarchization of the structure associated with acknowledging the merits and degrees/titles of high-ranking network members* (analogs of silverback males in gorilla groups); however, no network member can become the central leader and play the dominant role across the entire structure.
3. *Loose links between network members;* in an analogy to fission–fusion groups formed by, e.g., chimpanzees, individuals or subgroups can choose to either join the network or quit it.

The fission–fusion organizational pattern can cause serious problems if a network structure pursues the goal of carrying out a long-term project that requires the stability of the network's composition. Obviously, this paradigm does not secure such network stabilization unless (1) either the network is completely hierarchized, or (2) it is stabilized on the basis of informal loyal interpersonal relationships between the few members it includes (resembling, to an extent, the relatively stable coalitions of some chimpanzee males).

Human "coalitions" that include prominent scientists, scholars, or experts can perform important political functions by providing guidelines for the political system either in the capacity of "think tanks" used by the government or, even more importantly, as independent creative network structures that form part of civil society. It is these scientists'/scholars' networks that promote the development of a new type of *meritocracy*, where its members become politically influential because they get the support of civil society (see 3.8.6). Actually, the networks of scientists/scholars can exert their influence on the hierarchy of the political system and on the policies pursued by the government, even though final political decisions rest with the political hierarchy.

In Russia, network structures formed by scientists and scholars such as the *Saint-Petersburg Scholars' Alliance* have considerable potential in political terms. Of potential political importance is the *Moscow Society of Natural Scientists (MSNS)*, which works in collaboration with Moscow State University. MSNS also has branch offices in several other towns in Russia. The main goal of MSNS is "to study nature in Russia, promote the development of science and education, familiarize the people at large with scientific knowledge, and bring together scientists and enthusiasts interested in nature" (MSNS, 2014, http://moipros.ru).

MSNS includes is the *Club of Biopolitics*, which is focused on environmental protection, health, and biotechnology, as well as on a number of other biopolitical projects (the structure of the Club of Biopolitics is depicted in Figure 3.4).

Throughout the world, most research institutes and, in many countries (for example, in the USA and in Russia), national Academies of Sciences are hierarchical structures. Eminent scientists exemplified by Full Academy Members are at the hierarchy's uppermost level.

In this situation, network structures can be set up to include the rank and file of the scientific community that are actively involved in laboratory experiments, field studies, and other important types of research. They include undergraduate and (post-)graduate students, technicians, assistant professors, and so on. The knowledge and expertise of some of these individuals may hold as much value as that of Full Academy Members. Horizontal network structures where academic titles and positions are relatively unimportant can defend the interests of the rank and file, and provide support for them in terms of their careers. Such networks could also give them a chance to publicly express their opinion on socially or politically important issues.

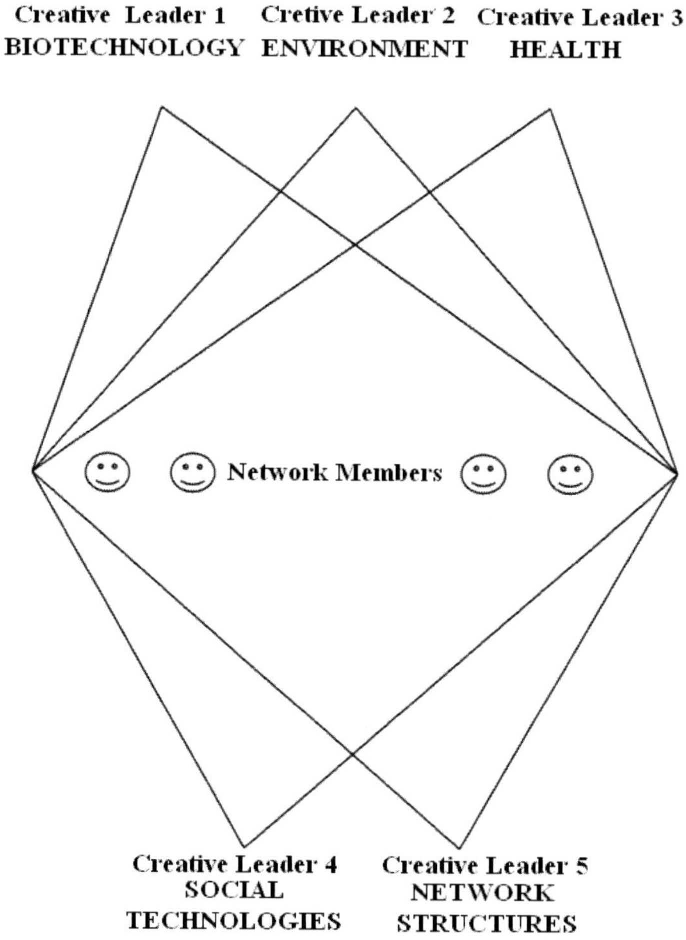

Figure 3.4. The network structure of the Club of Biopolitics (Club of Biopolitics, 2014; http://biopolitika.ru).

For instance, the aforementioned Club of Biopolitics—apart from its specific bioopolitical goals—also envisions providing "support to all those specializing in modern biology and its social implications such as students, lab assistants, professors, school teachers of biology, people working at other educational institutions, as well as social activists concerned with environmental or genetic engineering-related issues. The Club seeks the cooperation of all organizations and associations that defend the interests of the intellectuals as a whole and make efforts to improve the difficult life of researchers, lab technicians, postgraduate students, teachers, experts in the field of biology, and other groups that can face financial problems" (Club of Biopolitics, 2014, http://biopolitika.ru, click Club of Biopolitics).

Hence, spontaneously emerging or intentionally created network structures have considerable advantages over bureaucratic hierarchies with respect to interdisciplinary and innovative research (and development) projects. They encourage a creative and enthusiastic work style, broaden a scientist's horizons, make his/her work more important in terms of science and philosophy, and enable the scientist to develop innovative ideas that may be "ahead of time".

3.6. Modern Educational Technologies Based on Networks

3.6.1. Network Structures and Interactive Teaching

Network structures composed of students in the classroom[97], i.e., *decentralized project teams with split leadership*, have successfully been used in terms of *interactive teaching* by a number of innovative teachers including, e.g., the Russian educationalists Dmitriy Kavtaradze (1997, 1998), Alexander Kamnev (1997), and Ludmila Pivovarova (Oleskin et al., 2001; Pivovarova et al., 2002). Interactive teaching implies that there is interactive learning: apart from attending lectures and completing classroom tests, students are involved in the collective work on creative tasks (Angelo & Cross, 1993; Silberman, 1996; Kavtaradze, 1997, 1998; Morrison-Shetlar & Marwitz, 2001, Watkins, 2005).[98] For example, during geography classes, students are to compare two different parts of Northern China in terms of landscape, flora, and natural resources, with the help of the online materials they locate on the Internet. Several subgroups in the classroom report the results obtained by them. The results can be compared and evaluated. All students participate in making group-level decisions and they take responsibility for the results of the collective work; they should thus appreciate its meaning and be able to reflect on it.

"Instead of just giving the information to the students, teachers encourage them to come up with ideas on how it connects to their own world, thus constructing their own meaning of the material" (Pulsifer, 2013; http://www.ehow.com/about_5552709_interactive-teaching-learning.html). An important factor is a supportive and friendly climate that enables the students—apart from acquiring new knowledge—to improve their cognitive capacity and to

[97] Apart from their creative potential in the classroom, network structures made up of active students are also quite efficient in terms of *outdoor* classes.

[98] Actually, even lecturing can become at least partially interactive if the lecturer uses student feedback-promoting techniques, e.g., demonstrates a picture and lets the students discuss it, makes a statement and waits in silence for about 30 seconds, and encourages the audience to give a short answer to the lecturer's question. Such techniques enable some students to play the roles of temporary partial leaders and to mitigate the teacher–student hierarchy.

develop advanced forms of interindividual cooperation that currently involves using virtual communication facilities and online information.

Interactive teaching is facilitated by the formation of student teams whose goals are set by the teacher. This pedagogical approach is referred to as *project-based learning (PBL)*, and was suggested a century ago by Kilpatrick (1918). PBL is "a highly effective means of motivating students to learn independently" (Chang & Lee, 2010, p.961). Teams use a variety of interactive learning techniques ranging from brainstorming to role-playing games.

For example, one of the students might be asked to act the role of the CEO of an industrial firm, while another student is in charge of a commission that aims to expose the environmental pollution caused by such firms, and the rest of the group are assigned other important roles (Oleskin et al., 2001). The teacher explains the scenario of the role-playing game: "You live in a city with a high environmental pollution level. I'm showing you data on the concentrations of manganese, cadmium, and lead in the atmosphere. This map shows the location of environment-polluting factories. Your task is *to suggest ideas enabling us to prevent an environmental catastrophy"*. In a *hirama*-like fashion, the task is broken down into subproblems such as (1) *Making technological changes in the factories in order to improve the environmental situation;* (2) *Creating economic incentives for stimulating environment-friendly production scenarios;* and (3) *Restructuring the administrative system in order to facilitate its direct interaction with local industrial agents and environmental activists.* A partial creative leader can be assigned to each of the subproblems. There are a large number of alternative ways to subdivide the overall task, and the students can make their choice themselves, with the teacher playing the role of both a facilitator and a consultant.

A network (*sensu stricto*) can be a structure of choice with respect to student teams that are established during these interactive classes. However, such teams can also use the hierarchical (bureaucratic) organizational pattern; both patterns have their pros and cons.

The potential advantages and disadvantages of network structures in the classroom were revealed during a series of interactive classes that were conducted by me at a high school (school No.119, Moscow) on the subject of *City Environment and Neurochemistry.* The classes focused upon the effects of pollutants, e.g., heavy metals, on the human brain. They are known to disrupt the functioning of neurochemicals, i.e., substances involved in transmitting impulses from neuron to neuron in the brain's neuronal networks. For example, the metals lead, manganese, and cadmium decrease the activity of serotonin-dependent (serotonergic) networks, resulting in depression, anxiety, and increased aggressiveness, which may cause criminal behavior (see Masters, 1994, 2001).

The group of about 30 students (aged 15–16 years) was given a short lecture on the subject. It was followed by several creative tasks completed by temporary subgroups (teams), which included:

1. Assessing the environmental situation in several districts of Moscow using the following characteristics: (1) changes in the bark color of birch trees that absorb large amounts of dust and soot in industrial areas, and (2) the percentage of abnormal (writhen, dwarf, and so on) trees; designing scales upon which these characteristics could be evaluated (e.g., the bark can be light-colored, darkened, or dark; the trees can be tall, short, or dwarf); and drawing an ecological map of the Moscow districts involved;

2. Writing the draft of a legal document that would stipulate the rights of various animals, plants, and other life forms, drawing upon the articles of the Russian Constitution that deal with human rights[99]
3. Collectively deciding (and reporting the decision to the teacher) whether a woman who attempted to kill her husband in the state of serotonin deficiency-related depression was to be punished or subjected to compulsory medical treatment in order to improve her brain chemistry;

The whole student group was subdivided into two teams with approximately equal numbers of students. One of the teams implemented the more traditional *hierarchical (bureaucratic)* principle. It had a boss who was superior to the heads of the three subgroups. In accordance with the tasks given by the teacher (items 1–3 listed above), the three subgroups were called the *Eco-Assessment Subgroup*; the *Biolegislation Subgroup*; and the *Rehabilitation Subrgoup,* respectively. Each subgroup contained 5–6 students.

The other team was a *hirama-type network* (Figure 3.5). It had no boss and was *not* subdivided into subgroups, but it contained three partial leaders, each dealing with one of the tasks. These leaders facilitated and guided the activities of the whole team in terms of each respective task, and they subsequently recorded the results. The creative partial leaders in the networked group were responsible for the following subproblems whose solution involved all network members in the classroom: (i) Estimating the environmental pollution of the tested district of Moscow; (ii)Writing the text of the Ecological Constitution; and (iii) Resolving the dispute over the punishment for the attempted murder (see item #3 above). All team members could choose the partial leader to deal with at any given moment; the psychological leader of the network promoted the creative work on all the tasks, trying to make sure that no partial leader was left without supporters, and that all the tasks were efficiently completed. Finally, each team orally presented the results of its creative work. In compliance with the organizational principles of each team, it was the boss who reported the results of the bureaucratic team's work, and the "external affairs" leader who presented a talk concerning the progress made by the *hirama*-type team.

In the following, I compare the efficiency and creativity of the work of the two teams (based on my work: Oleskin et al., 2001):

- The bureaucratic team spent less time completing most parts of the assignment than did the network; the documents containing the results of the bureaucracy's work looked more official, accurate, and neat (particularly the legend to the ecological map of Moscow's districts) than did those produced by the network.
- However, the network was more creative, and its documents contained more interesting ideas when compared to those of the bureaucratic team (the same trend manifested itself during my classes on a different subject, *The Molecular Basis of Behavior,* Oleskin et al., 2001). The network made a more humane decision regarding the attempted murder case: the woman was to be released from prison and

[99] Similar lessons were conducted on other environmentally important topics. Dmitriy N. Kavtaradze (1997), who made a significant contribution to the development of interactive teaching techniques in relation to the environment, described a lesson with a game-based scenario aimed at developing a strategy to protect the gene pool of nut and fruit trees in the forests of Kyrgyzstan in Central Asia.

to undergo compulsory medical treatment for serotonin deficiency-related problems. In contrast, the "bureaucrats" condemned the criminal to 1 year of prison followed by compulsory treatment. The network produced a more biocentric document concerning the rights of the biosphere; they called it the *Constitution of the Biological Federation,* in an analogy to the Constitution of the Russian Federation. The bureaucracy created a more anthropocentric document prioritizing the interests of humans over those of other biological species.

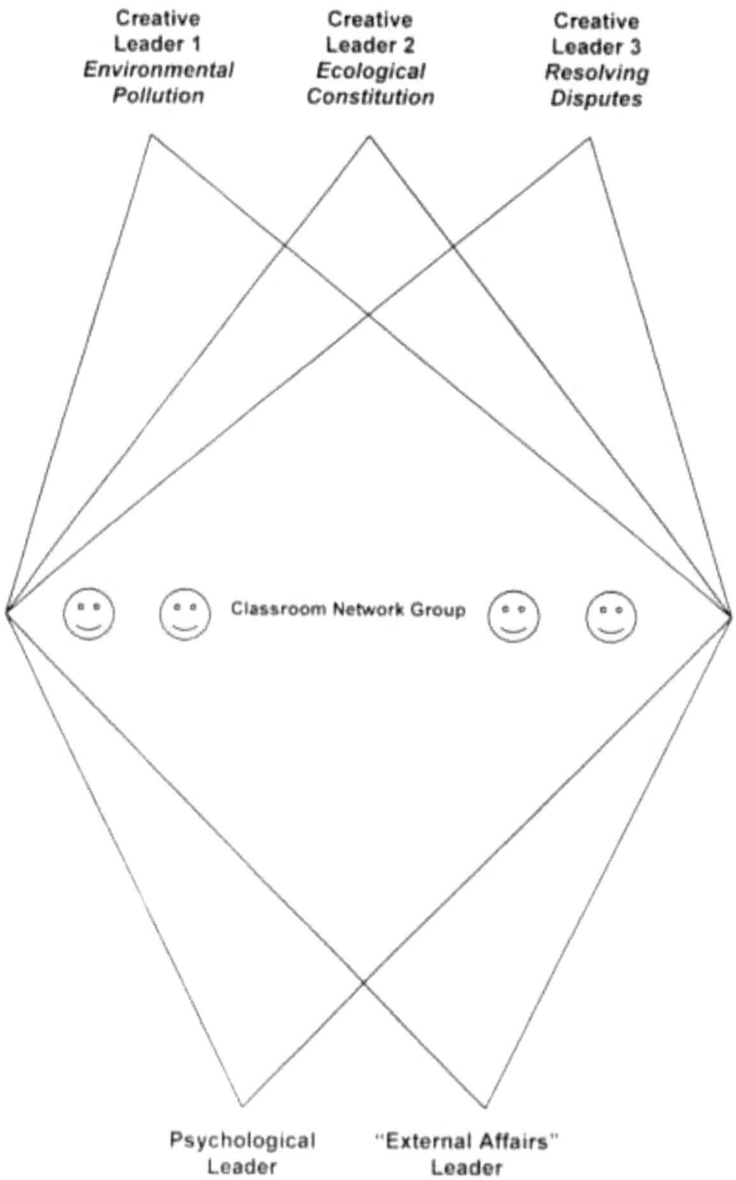

Figure 3.5. A network structure established in the classroom in order to carry out a creative task concerning the environment.

- The students that lacked sufficient experience with respect to the network-specific work style considered the presence of several creative leaders in the network to be confusing; the *hirama's* successful work was, to a large extent, dependent on the assistance of the teacher who performed the role of the organizational leader in the *hirama* (see 1.2.2), while the "bureaucrats" were more independent.
- The successful performance of both teams was also crucially dependent on the organizational skills of the people who played the key roles. The efficient work of the bureaucracy required the organizational skills of the boss; the psychological leader's efficient performance was crucial to the *hirama's* success.

It should be emphasized that these findings are consistent with the data available in the literature on networks and hierarchies such as bureaucracies. A bureaucracy operates in a more orderly fashion; it is more efficient in dealing with clearly formulated tasks and particularly with routine work. A network tends to outperform a bureaucracy when carrying out more challenging, fuzzy, innovative projects, especially if faced with a dynamic, turbulent environment where the development trend is difficult to predict (Mescon et al., 1986).

I also used the network scenario during discussions concerning *Genetic Engineering: the Pros and Cons* at two departments of Moscow State University:

- the State Management Department in 2005 where I conducted seminars on Organizational Principles of Biological Systems and their Implications for Human Society;
- the Global Studies Department in 2008-2014, where these discussions were on the Biopolitics curriculum.

During both discussions, the students set up a networked team with three partial creative leaders performing the following functions:

- Leaders 1 and 2 collected arguments presented by all students in the group for and against genetic manipulations, respectively;
- Leader 3 helped students critique both kinds of arguments and produce a well-balanced, unambiguous final document.

The students were encouraged to interact with any of the leaders, depending on whether they supported genetic engineers, protested against their activities, or preferred a middle-ground attitude.

In a similar fashion, network structures can be established to facilitate interactive teaching during classes in various subjects at a school or college. Teachers can vary both the organizational pattern and the subjects, which can include, for instance, environmental concerns, issues in biomedical ethics, or problems caused by the ineptness and corruption of the state apparatus.

Irrespective of the concrete subject dealt with by a network structure (team, workshop) in the classroom or a college seminar room, students' "creativity is closely related to opportunities for working collaboratively with their peers" (Davies et al., 2013). An additionally important positive factor is the development of a "supportive relationship with the teacher" (Ibid.) who actually plays the role of a *network-style partial leader*, a "hub".

By fostering creativity, students' networks produce a number of other beneficial effects: they make classroom work more diversified; give students more freedom in terms of planning their activities (Chan, 2013); increase students' confidence and resilience; enhance their motivation and engagement; help them develop proper social, emotional, and thinking skills; and improve school attendance (Davies et al., 2013). Importantly, the tendency towards setting up task-oriented network teams in the classroom often comes into conflict with the hierarchical environment characteristic of performativity culture that is widespread, for example, in the United Kingdom (McLellan & Nicholl, 2013). This culture is characterized by accountability mechanisms that are largely based upon standardized tests of students' performance and academic achievement.

A heavily bureaucratized version of performativity culture in Russia is typified by the so-called *Unified State Exam* (*ЕГЭ, Единый государственный экзамен*) that can potentially be abused by corrupt teachers and whole educational bureaucracies, unless it is independently controlled by educational activists' networks (see 3.6.2 below).

Recently, the development of computer technology has given a new impetus to the *"playful" or "game-based" approach* of interactive teachers (Davies et al., 2013). In its current computer-facilitated version, it is aimed at creatively using (in the classroom) the students' interest in playing computer games while "blurring the boundaries between learning and entertainment". Importantly, computer games do not only cause problems by distracting students from their studies; if properly used under a teacher's guidance in terms of interactive teaching/learning, they also promote the development of problem-solving behavior, motivation, and social networking.

Of particular interest are classroom scenarios in which students invent game stories of their own using game construction software such as *ScriptEase* (Carbonaro et al., 2008), rather than solely playing video games.They can form networked workshops that collectively construct such games.

In a similar manner, students work in pairs within larger teams in order to compose new pieces of music using the sample sequencer *eJay* (Gall & Breeze, 2008). Computer workstation-aided music classes are of particular importance in terms of informal network formation in the classroom because of the symbolic meaning of music: specific melodies can form part of the group identity-forming matrix, uniting emerging network structures in the classroom.

Apart from music classes, music can be creatively used to increase the efficiency of classes in a variety of other subjects. Music can promote social networking during interactive classes, which is in conformity with Vygotsky's (1997) views on the socially integrative role of culturally meaningful symbols, including pieces of music. Obviously, both music and visual arts can be used to promote students' creativity and, thus, the efficient functioning of their network structures. Actually, music is visualized using modern software such as *eJay*, because each piece of music corresponds to a visual image (a colored block). This can yield the combined functioning of several sensory modalities, enhancing the parallel information processing principle that is characteristic of neural networks and many other kinds of network structures.

Another creativity-promoting strategy is interactive teaching in an outdoor environment, as exemplified by forest schools. These schools promote network formation among students in a "wild setting" and their creativity at the individual and group level, due to the following factors: (1) *"Regular, frequent contact in the same setting over a significant period of time"*; (2)*"Providing freedom to explore using multiple senses and intelligences"*: (3) *"Time and space for individual learning styles to be recognised and nurtured"*; and (4)*"A low pupil:adult ratio"* (words in quotation marks are based on: Borradaile, 2006 and quoted according to: Davies et al., 2013, p.85).

In terms of interactive teaching/learning, the teacher plays the unusual role of a network hub whose functions include facilitating the network's operation by consulting the students, providing general guidelines in terms of the tasks involved, and supplying the necessary materials. The teacher tends to be 'less prescriptive' in lesson planning (Davies et al., 2013), allowing more room for student initiatives and self-organization, both in terms of group structure and the group's overall agenda.

The teacher's role is becoming even less hierarchical[100] if the *team-teaching* scenario is used. "Team teaching involves two or more teachers sharing teaching expertise in the classroom and engaging in reflective dialogue with each other" (Chang & Lee, 2010), as exemplified by the collaboration of a computer teacher and an English or a geography teacher in a school in Taiwan. Teachers form a subnetwork (cluster) within the fractally-structured, higher-order network established in the classroom in terms of a project-based interactive teaching scenario. The teachers give the students several different perspectives on the subject, but the problem is that a conflict between these perspectives can potentially arise; special measures are to be taken to cope with this type of conflict. A *hirama*-type strategy can be used to assign the role of the conflict-mitigating mediator (psychological partial leader) to an additional teacher.

As far as classroom network structures (creative student teams) are concerned, the most suitable biological analog is apparently the *neural* paradigm characterized by collective information processing, decision making, and learning/training. Emphasis on this paradigm would encourage parallel information handling by creative subgroups within the whole network composed of school/college students. Similar to a neural network, a network composed of students can begin to piece together the solution of a given problem (as proposed by the teacher) on the basis of fragments supplied by individual students and creative subgroups. In terms of the quasi-neural scenario of a student network's operation, special attention is to be given to the creative learning at the levels of the individuals, subgroups, and of the whole multilevel network as a "collective brain".

While dealing with a problem/task, students in the classroom can simulate the operation of an artificial neural network such as the perceptron (see 2.5.1). Like the ANN, students can form several distinct "layers" (i.e., subgroups). One subgroup can specialize in collecting task-related information, in an analogy to the perceptron's input layer. Another subgroup can process the information received from the "input layer", i.e., function as the "hidden layer". A third subgroup—the "output layer"—can generalize and verbalize the result obtained by the "hidden layer" subgroup and report it to the teacher. The scenario would obviously be still

---

[100] A contribution of a hierarchical organization to classroom work is mandatory because the teacher is responsible for organizing the whole interactive teaching session, although not in the prescriptive manner typical of more conventional lectures or seminars. The teacher provides general guidelines and goals, but the students have much freedom as far as the strategy of attaining these goals is concerned. The whole network including the teacher(s) is necessarily three-dimensional and, to an extent, similar to primitive hunter–gatherer *egalitarian* structures.

more interesting if the "output layer" could send messages back to the "input layer" and the "hidden layer", providing guidelines for their activity on the basis of the result already obtained. This would transform the "neural network" composed of students (as neuron analogs) into a Hopfield-type recurrent network structure.

However, the plurality of "biological" paradigms that are potentially applicable—and creatively utilizable—in human network structures has already been emphasized in the preceding sections. Apart from neural networks, students' networked teams/workshops share some features with *egalitarian* networks, whose versions exist in primitive hunter–gatherer bands and, to an extent, in some ape groups.

As already mentioned, classroom networks are moderately hierarchized. Not only are the teacher–student relationships necessarily hierarchical (although they form a mitigated hierarchy), students themselves have different social ranks, and some of them tend to become informal leaders. As a participant in a classroom group discussion noted, "as often happens in groups, one member decided to take over meeting times, delegation of responsibilities, and actions" (Neuman et al., 2009). In cross-age groups, older students "were seen positioning themselves as both teacher and researcher for younger children" (Dillon et al., 2007, quoted according to: Davies et al., 2013, p.86).

Similar to hunter–gatherer bands, hierarchical relationships in student groups (teams, workshops) in project-based interactive teaching are limited by egalitarian social norms introduced during such classes. Therefore, these relationships are prevented from evolving into monopolistic dominance. This concerns also teacher–student interactions because interactive teaching is facilitated by allowing students to "initiate their own activities or make their own choices within a loosely framed activity" (Davies et al., 2013), i.e., by respecting their individual freedom and rights, regardless of formal and informal rank and status.

If *hirama*-type network teams are set up during interactive classes, students feel free to join and quit each of the several temporary creative subnetworks (subgroups) centered around each partial leader during the collective work on a classroom project. This is similar to the *fission–fusion* pattern that is typical of chimpanzee groups, as well as of many hunter–gatherer bands with temporary leaders.

The creative work of a team of students is stimulated during an interactive teaching session if the students face sufficiently important challenges, and if teamwork and a sense of belonging are promoted. Both a challenging environment and teamwork are characteristic of the egalitarian network structures of primitive hunter–gatherers (cf. 3.1.1).

3.6.2. Network Structures of Educational Reformers and Defenders of Educators' Interests

Educators, including reformers of the education system, could form network structures at the local, national, and international levels. This would help them maintain business contacts; exchange ideas and educational developments; and, to an extent, coordinate their efforts.

Network structures (associations, organizations, etc.) can defend the professional interests of teachers and other people (ranging from lab assistants to theoreticians, who are conceptually developing innovative teaching techniques) who are involved in education. This is of paramount importance for countries where the funds allocated for education are currently being reduced, and teachers face unemployment or a decrease in their salaries.

Network structures can use *online* communication to help educators unite their efforts and effectively struggle against political hierarchies if these hierarchies disrespect educators' rights and ignore the importance of their professional activities.

Organizing educators' conferences and networked workshops seems to be an efficient strategy of collectively coping with problems faced by educators (including teachers' salaries and standards of living), as well as promoting the professional development of educators, including familiarizing them with modern interactive teaching techniques (see above, 3.6.1).

An example is provided by the recently established *Creative Science Teaching (CST) Labs III*, which deal with science teachers' continuing professional development (CPD). CSTIII were set up in Britain by a London-based organization, Performing Arts Labs (PAL), with the support of the NESTA (National Endowment for Science Technology and the Arts). Apart from science teachers, specialists from the arts, science, and technology were invited to improve the skills of the teachers (Chappel & Croft, 2009).

CSTIII were situated in the rural environment of Kent (Southern England). They held two meetings (labs) aimed at encouraging creative teaching and learning, and their activities were predominantly based on the network's organizational pattern. The *equality of status* principle was adopted: everybody was considered "a participant". "The definition of 'participants' within the Labs was unusual; it encompassed the teachers as well as the science and arts specialists present. As time went on even Lab Directors became participants" (Chappel & Croft, 2009, p.47). Roles were interchangeable, and there was no "upper hierarchy", in the words of one of the funders. The *co-leading* principle used by CSTIII was based on the partial leadership pattern that is typical of a *hirama*. "By being challenged ... to co-lead sessions with another teacher, where participants included science and arts specialists... teachers were provoked and given confidence to lead new activities in new ways for them" (Ibid.). The CSTIII climate was generally characterized by inclusivity (nobody was considered an outsider), immersion (promoting a primitive hunter–gatherer-like feeling of identification with the lab), "boundary busting" (crossing boundaries between fields of science and even between science and the arts), and spirituality.

International network organizations/movements that deal with educational issues are exemplified by the *Commission for Biology Education (CBE)*, which was established under the aegis of the *International Union of Biological Sciences (IUBS)* in the early 1970s. It develops innovative educational methods including interactive teaching/learning techniques, produces innovative syllabi concerning biology-related subjects, and defends the professional interests of biology teachers worldwide, with special emphasis on eradicating "bio-illiteracy" (a lack of basic biological knowledge). The CBE's main goal is "to formulate, initiate and facilitate effective methods of improving education in the biological sciences and allied fields, including the applications and implications of biological studies" [CBE (Commission for Biological Education), 2014, http://iubscbe.org]. The CBE holds international conferences and provides guidelines for improving education worldwide, particularly in the form of published books devoted to educational issues.

Networks of students can be formed in the classroom in terms of interactive teaching scenarios. Such creative student teams (analogs of neural networks) can implement various interactive teaching techniques ranging from brainstorming sessions to role-playing games in order to collectively piece together the results of the task given by the teacher(s). Educators, including education reformers, can establish network organizations/movements to promote the modernization of the educational systems and to defend the professional interests of all those involved in education.

3.7. Network Structures in Business

3.7.1. Defining Networks and Contrasting Them with Hierarchies and Markets

It is in the business sphere that the concept of networks *sensu stricto* was originally developed (Thorelli, 1986; Powell, 1990; van Alstyne, 1997, and other publications). Subsequently, the network concept was applied to politics (Börzel, 1998), public management (Meulemann, 2008), cultural organizations (Karpukhin, 1996), education (Kavtaradze, 1997, 1998; Oleskin et al., 2001), interdisciplinary scientific research (Oleskin & Kirovskaya, 2007), and other areas.

Network structures in business are often defined as "a coalition of autonomous but interdependent organizations that are willing to exchange information and coordinate some of their actions" (Trkman & Desouza, 2012, p.2); it is stressed that "relationships among network members are primarily non-hierarchical, and participants often have substantial operating autonomy. Connections may be informal and completely trust-based, or more formalized, as through a contract" (Ibid., p.3). However, this interpretation of the network concept in business omits consideration of the business network structures that are composed not of organizations, but of individuals or their small groups. Owing to the fractal properties of networks in general, such networks made up of "singletons" are to be regarded as reduced copies of larger networks consisting of whole organizations.

The work by Powell (1990) described a classic example of network structures composed of small enterprises (with less than 50 employees) in Northern Italy (Emilia-Romagna).

"These small firms are frequently grouped in specific zones according to their product, and give rise to industrial districts in which all firms have a very low degree of vertical integration (Brusco, 1982). Production is conducted through extensive, collaborative subcontracting agreements. Only a portion of the firms market final products, the others execute operations commissioned by the group of firms that initiate production. The owners of small firms typically prefer subcontracting to expansion or integration (Lazerson, 1988). The use of satellite firms allows them to remain small and preserve their legal and organizational structure as a small company. Often satellite firms outgrow the spawning firms. Though closely related and highly cooperative, the firms remain strictly independent entities. These industrial districts embrace a wide range of consumer goods and engineering components and machines: knitwear in Modena, clothes and ceramic tiles in Modena and Reggio, cycles, motorcycles, and shoes in Bologna, food processing machinery in Parma, and woodworking machine tools in Capri, to name just a few…" (Powell, 1990, p.310).

Networks are more likely to emerge in areas marked by high competition levels (because pooling the network members' resources enhances their competitiveness)[101]. If businesses

[101] Hence, market-type competitive interactivity is characteristic of the relationships between entire networks of firms rather than of those between individual firms within a network where cooperation prevails over competition

form a network, this reduces the uncertainty level that is typical of the market and helps the businesses access resources owned by other business agents. The formation of business networks is promoted by the globalization process and increasing global competition, as well as by rapid technological and structural changes in the business environment. Under the influence of all of these factors, "organizations are being increasingly integrated into dynamic networks connected by time-space compressing information and communication technologies" (Grinshaw et al., 2005, p.1).

In the business sphere, networks became prominent and received some attention from social scientists in the 1980s. Before this point, the business world was mainly regarded as only including hierarchical structures (firms) and markets.

In some respects, networks possess intermediate properties that fall between hierarchies and markets. For instance, networks are obviously less vertical than business hierarchies typified by traditional capitalist firms. Nevertheless, most networks represent three-dimensional structures with partial leaders; therefore, a majority of business networks are not as horizontal as classical markets that only include autonomous agents.

However, although networks and markets both are non-hierarchical decentralized systems, only networks promote "institutionalized cooperation which reflects the collective solution of complex problem structures and processes, an institution to yield coordination through cooperation" (Elsner et al., p.14).

Networks in business are not comparable with either hierarchies or markets in terms of the degree of trust, loyalty, and affiliation[102] that are essential for the networks' survival and successful operation.

"In markets, the standard strategy is to drive the hardest possible bargain in the immediate exchange... Within hierarchies, communication and exchange is shaped by concerns with career mobility – in this sense, exchange is bound up with considerations of personal advancement... *In networks, the preferred option is often one of creating indebtedness and reliance over the long haul*" (Powell, 1990, p.302; the order of the sentences quoted is changed; emphasis is added – O.A.). This quotation emphasizes reliable long-term interactivity, which is contrasted with "the temporarily limited exchange relations of markets" (Kahler, 2009a). At the same time, interaction among network members is not centrally controlled by a hierarchical leader.

Trust, loyalty, and *reciprocity* are prerequisites for the accumulation of *social capital.* Although a number of interpretations of this term has been presented in the literature, most scholars agree that this term deals with "the value of connections" (Borgatti & Foster, 2003), and social capital is often regarded as the ensemble of "ties to resource-filled others" that is used by an individual (a network member) or a whole network. Burt (1992) related the term to structural holes (see 1.1.3.3) in a network, i.e., the network's parts that are located between dense clusters and characterized by sparse links between nodes so that "the neighbors of a node are not connected to one another" (Newman, 2012, p.202). Social capital, in these terms,

[102] Indisputably, guile and dishonest behavior can be characteristic of some network members (of free riders exploiting other, cooperating, network members). However, such free riding obviously results in violating the principles upon which a network structure is based, and it endangers its very existence (therefore, calling for immediate sanctions) while, in competitive market-type systems, such behavior, unless considered illegal, is often regarded as "part of the game". As emphasized by Parker (2008, p.629), "non-reciprocation by network members ('free riders') can also cause networks to disintegrate, but this can be countered by designing networks to punish free riders by excluding them from further participation."

is based on the few valuable internode connections that span such "structural holes". Generally, social capital "refers to values, norms, and social networks that influence the capacity of people to cooperate" (Badescu & Uslaner, 2003, p.3).

These features of networks enable efficient cooperation among network members that prevails over the competition among them. Cooperation in networks is facilitated by cospecialized assets under joint control of network members and by collective goals shared by all of them (van Alstyne, 1997).

Establishing business networks is often associated with the following organizational changes:

Figure 3.6. Business networks are similar to a Möbius strip, where the outer and inner side cannot be delimited (permission is granted to copy, distribute and/or modify this document under the terms of the GNU Free Documentation License, Version 1.2. Copyright © 2005 David Bebennick).

- *Creating "organizations without boundaries"*, i.e., establishing interactive expanding networks enlisting external agents such as suppliers, marketers, customers, and even competitors that carry out long-term contracts. As a result of such *outsourcing*, external and/or internal borders (borders between firms or between firm departments) become permeable. For instance, suppliers are allowed to obtain direct information about the use of their goods in the production process, which prevents stockout or supplying the goods in excess of demand (Lee, 2002). The blurring of firm boundaries is in conformity with the typical behaviour of an interactive network (see 1.7.2.2 above). Such networks are similar to a Möbius strip where the outer and inner side cannot be delimited (Figure 3.6). As for internal boundaries, they can be almost completely eliminated if a business entity represents a single creative team or, alternatively, consists of several teams with open boundaries that constantly interact and deal with overlapping tasks, as is the case with the *Semco Group* in Brazil (to be described below). In strategic alliances formed by several firms, each partner behaves like this type of team. In addition, boundaries between specialties and professional roles are also blurred, so that employees in a network business structure can combine jobs and positions in different organizations or subdivisions ("temps", "part-time jobs"). Importantly, blurring many formal boundaries does not result in eliminating the boundary that separates the network members and nonmembers; moreover, networks make efforts to preserve their identity. For example, the employees of many networked entreprises wear uniforms, organize social events, and may even perform special rituals, in an analogy to a hunter–gatherer band. Naturally, the main criterion used to distinguish members and nonmembers is whether they accept the goals, behavior norms, and other aspects of a network's matrix.

- *Managing without controlling.* This feature of a networked business is due to the predominance of horizontal interactions both at the intra- and interorganizational level. Efficient business links should be established with those people and organizations that cannot be controlled; however, a firm inside a business network can *influence* other "nodes" and, importantly, it can *be influenced* by them. The fractal properties of networks manifest themselves in the tendency to develop horizontal interactivity not only at the interorganizational, but also at the intraorganizational level—and to overcome bureacratic hurdles. Managers on all rungs of intraorganizational stepladders increasingly play the roles of moderators, mediators, experts, and conflict managers vis-á-vis their employees. This is in conformity with a new business paradigm that emphasizes satisfying the needs of the people involved in a business network and promoting their personal development. In contrast, people were often considered only as means to carry out the production process in terms of the more traditional bureaucratic paradigm (particularly if the assembly line system was used, as typified by the Ford Motor Company in the early 20th century).

Of relevance to the "managing without controlling" principle are the following somewhat paradoxical features of networked enterprises (Håkansson & Ford, 2002):

First, "companies within a network are not free to act according to their own aims", because other network nodes exert their influence on them. Moreover, the whole network behaves as a single agent whose collective will significantly influences each of the nodes involved.

Second, the strategy of the companies that form part of a network develops under the influence of the whole network. "The Swedish telecom company, Ericsson, and the largest Swedish telephone operator, Telia… have had a close relationship for 100 years and this has had profound effects on both of them. They developed their first automatic exchanges together in the 1920s. The later AXE exchange, developed by Ericsson in co-operation with Telia had a major effect on its international success. Later, mobile phones were developed within this relationship. From this perspective, the relationship has formed both Ericsson and Telia" (Ibid., p.136).

Third, although each network node is interested in attaining its own goals and tries to control the activities of other nodes and the whole network, exercising too much control over the network is likely to decrease its efficiency. Hierarchization caused by such controlling behavior deprives the network of its specific advantages: decentralized organizational patterns enable entrepreneurs in networks to more efficiently respond to demand changes, as well as to minimize risks caused by supply problems because goods can move along several parallel routes in a networked business organization.

Networks can be described in terms of conventional *transaction cost economics (TCE)*: by establishing network structures, business people aim to minimize their costs and maximize their profits. Networks help them attain these goals because establishing them involves long-term interactions with partners that typically cause fewer expenses than market price-based contracts; additional financial advantages are associated with information exchange between formally independent entities that represent network nodes.

However, we can account for the development of networks in terms of noneconomic factors associated with the social environment in which business enterprises are *embedded*, as emphasized in the classical work by Mark Granovetter (1985). People are motivated to create networks because of their personal relationships, which are based upon trust and reciprocity, and that constitute their social capital. Such informal relationships do not usually result in the formation of rigid hierarchies, nor do they boil down to market-style exchange interactions. I reemphasize that Granovetter (1973) distinguished between *strong ties*, such as the relationships between bosses and their subordinates, and *weak ties* that, for example, can result from interaction between people that belong to different business firms.

Some of the spontaneously emerging, originally weak, ties between human individuals and groups can secondarily become strong enough to provide the basis of a business network *sensu stricto*. It is of relevance that network structures play a more important role in countries lacking developed business institutions and regulations, i.e., in emergent rather than advanced economies (Danis et al., 2011). In the work cited, a relationship was revealed between associational activity and the development of new businesses.

Network structures in business are characteristic of postsocialist countries such as Russia and China (Jansson et al., 2007). The transition from state socialism to capitalism promotes the development of spontaneous network structures. In Russia and China "already weak institutions have become even weaker, which means that firms cannot rely on formal institutions to solve conflicts, but must instead solve them by relying on informal institutions" (Ibid., p.959).

Network structures in business, like other kinds of network structures in human society (see the introduction to Chapter three) can be classified into formal and informal networks. *Formal* networks in business are typified by "organizations that bring entrepreneurs together in order to share business information and experience /as well as other resources – O.A./ for mutual advantage" (Parker, 2008, p.628). Formal business networks are exemplified by chambers of commerce where groups of business people in a town or area meet to discuss measures to ameliorate business. Significantly, such "formal networks not only enhance the performance of entrepreneurs who join them, but also promote efficiency and social welfare in the wider economy" (Ibid., 627).

Informal network structures in business are characterized by horizontal organization and "lateral coordination", and they are project-oriented and self-organized; they are similar in this respect to primitive hunter–gatherer bands (Bernhard & Glantz, 1992).

3.7.2. Structural Levels of Business Networks

In the business world, the network organizational principle applies to at least three different structural levels: (1) *Networks composed of whole enterprises (firms, business organizations)*; each of their nodes may be internally organized as a hierarchy, a network, or even as a market-type structure involving competing entities; (2) *Network structures formed within the boundaries of a single business entity* (internal networks, Snow et al., 1992); their nodes represent enterprise subdivisions/departments; and (3) *Small-sized networked enterprises* whose nodes are individuals or small groups.

3.7.2.1. Interorganizational Business Networks

A widely used business strategy that results in the establishment of interorganizational networks is *outsourcing*, i.e., the contracting out of a business activity to an independent business entity; this may implicate the transfer of employees and assets to another enterprise. Thanks to outsourcing, a firm can concentrate on its core activities, delegate functions requiring special skills or expertise to a network partner, and reduce expenses associated with hiring and training specialists. A related strategy is to use interorganizational networks to access products, distribution channels, project funding, capital equipment, and intellectual property owned by other network members. In terms of the *synergy* concept (see 1.2) considered by Corning (1983, 2003a, b, 2005, 2007), interorganizational networks cause cooperative effects (synergies) that can provide benefit for their members.

The formation of an interorganizational network often involves the establishment of "ties among organizations through a member of one organization sitting on the board of another" (Borgatti & Foster, 2003, p.996), i.e., *board interlocks*. Such board interlocks are characteristic of network-dominated businesses in emerging economies, e.g., in China where interlocking firm directorates provide the foundations for small business groups that tend to be regionally fragmented (Ren et al., 2009).

In the literature, there are a plethora of terms applied to networks consisting of whole enterprises as nodes, including alliance networks, business nets, collaborative or cooperative arrangements, co-opetition, external knowledge sourcing, innovation outsourcing, interorganizational knowledge networking, strategic alliances, and supply networks (reviewed, Trkman & Desouza, 2012). While discussing all of these terms in detail is beyond the scope of this work, emphasis should be placed on the widely used term "strategic alliance" that applies to a wide variety of interorganizational network structures and can be defined as a *business coalition based on stable connections between independent partners* (Radaev, 2002b, p.169).

Strategic alliances are widely spread among all kinds of companies, but they are particularly characteristic of giant firms that aim to gain competitive advantages on the global scale. Several examples were given in the classical work by Powell (1990). *Boeing* and *Rolls Royce* formed an alliance to create *Boeing 757*, and most production processes were carried out in terms of joint projects involving partners in Japan and Italy. Such networks become partly hierarchized if the alliances enlist one or more large firms along with smaller partners. This is exemplified by the alliance that included *General Motors* and the comparatively smaller company, *Teknowledge*, which specialized in manufacturing artificial intelligence systems (Powell, 1990).

The development of strategic alliances among firms results in the formation of an interorganizationally distributed intelligence system that stores information and know-how, which does not belong to any single network member. Using this collectively stored knowledge, a strategic alliance can learn/train as a whole, adapting to the often dynamic environment. An analogy can be drawn between the organizational pattern of such interorganizational business networks and the neural paradigm (2.5.2), which implies learning/training at the level of the whole neural network or its analog.

Interorganizational business networks include *dynamic* and *stable* networks (Snow et al., 1992), and most long-term strategic alliances belong to the second subtype.

Dynamic network structures are analogous, in organizational terms, to fission–fusion groups considered above with respect to the egalitarian societies of primitive hunter–gatherers (3.1.1) and groups of some primates (2.6.4). In the business world, they are typified by temporary interfirm alliances that are formed for achieving a specific goal; thereupon, firms terminate their relationship in order to establish another temporary union (Millner, 2006). Dynamic networks are widely used, for instance, in the fashion industry (van Alstyne, 1997). In this setting, network nodes establish no stable long-term relationships. Many dynamic networks are partly hierarchical, particularly if their hubs, i.e., broker or lead firms, acquire degree centrality and/or betweenness centrality, in terms of the network analysis. Temporary alliances are established among suppliers, lead/broker firms, and marketers; if they also involve representatives of the political system and law-enforcing bodies, they may become hierarchized because these bodies tend to dominate the entire structure.

Stable networks "consist of firms engaged in long-term relationships with external suppliers who bring expertise into the parent company". Participants may be "organized around a single large firm as with Japanese auto manufacturing" (van Alstyne, 1997). Alternatively, the structure may be less hierarchical if composed of firms of similar size. Stable networks in business are typified by alliances of spatially distributed firms, including Japanese *keiretsu*, that are stable unions of firms of various sizes. They specialize, e.g., in different stages of producing cottonwool textiles. Stable networks are also exemplified by Scandinavian interorganizational alliances composed of large industrial companies such as *Volvo, Ericsson, Saab-Scandis,* and *Fairchild.* Analogous stable interfirm alliances are actively formed in present-day Russia; their degree of hierarchization is highly variable and often situation-dependent.

Some characteristic features of stable networks actually evoke the *modular paradigm*, discussed in this work in the example of cnidarians (2.2). The long-term interactivity typical of stable interorganizational networks suggests the development of a strong *matrix* composed of a set of routinized behavior rules, business norms, and formal and informal regulations. The matrix is responsible for the integrity and viability of the whole network facing the challenges of the turbulent business world (exemplified by industrial network structures in Emilia-Romagna, which were considered at the beginning of this section).

The matrix of stable business networks is functionally analogous to the *coenosarc*, which connects individual polyps or medusae (*zooids*) in the modular organism (*cormus*) of cnidarians. During a difficult period, some of the small enterprises can go bankrupt, but the whole structure will survive; in a similar fashion, cnidarian cormuses sacrifice their zooids but retain the viability of the whole structure during a starvation period (Marfenin, 1993). Different enterprises within a stable network can be functionally differentiated, similar to gastrozoids, gonozoids, and other functional types of polyps/medusae; alternatively, they may represent multifunctional uniform modules. Such uniform modules can compete with one another, but the whole business network structure mitigates this competition, and cooperative interaction prevails in the network.

In the section on modular networks (2.2), it was emphasized that the behavior of each zooid only weakly influences the activities of the whole system. However, its effect is potentiated if its rhythm coincides with that of the majority of other zooids. An analogous phenomenon is *network-facilitated leverage.* Each member who contributes some resources to the network may obtain access to a much larger pool of resources belonging to the whole network (Chuchkevich, 1999, p.26-27). Network-facilitated leverage can deal with material goods, information, social status, or communication facilities, depending on the types of resources that are collectively used by the network. Unfortunately, the drawbacks of individual network nodes can also combine, creating a negative general image of the network (this is "negative leverage").

Both dynamic and stable networks can be subdivided into at least four groups that were singled out by Lee (2002) with respect to alliances of suppliers, manufacturers, marketers, and consumers. This reasoning was later applied to all kinds of interorganizational networks, and the following classification was suggested (Trkman & Desouza, 2012):

- *Efficient (functional)* networks aimed at maximizing cost efficiency in the supply chain (Lee, 2002; Trkman & Desouza, 2012). A prerequisite for increased cost efficiency is information exchange among network members. For instance, if the supplier is familiar with the situation at the manufacturer firm level (and vice versa), this will prevent efficiency-decreasing problems such as stockout, as well as supplying goods in excess of the real consumer demand for them. Manufacturers can create supplier hubs[103] that enable them to easily monitor the situation at the supplier level. "For example, at their former manufacturing site in Fountain, Colorado, Apple Computer created a supplier hub that was operated by a third-party logistics company, Fritz Companies... The use of the hub has allowed the suppliers to have much better information about Apple's needs and consumption patterns of the parts as well as about the inventory in transit" (Lee, 2002, pp.112-113).
- *Risk-hedging* networks. They pool their members' resources to share risks (Trkman & Desouza, 2012). In this respect, not only alliances between firms dealing with different production stages (they may be referred to as "vertical[104] alliances" in the literature), but also networks bringing together several firms concerned with the same stage ("horizontal alliances"), are of paramount importance. In a neural network-like fashion, the parallel functioning of several entities performing the same function, i.e., dealing with the same production stage, increases the reliability and resilience of the whole network structure. "A single entity in a supply chain can be vulnerable to supply disruption, but if there is more than one supply source or if alternative supply resources are available, then the risk of disruption would be reduced" (Lee, 2002, p.114).
- *Agile* networks aimed, according to Trkman & Desouza (2012), at "improving the ability of an organization in responding rapidly to changes in demand".[105] This is exemplified by Original Equipment Manufacturers (OEMs), companies specializing

---

[103] The specific meaning of the term "hub" used by business people in this context is actually in conformity with its more general meaning with regard to scale-free networks where hubs are nodes having numerous links to other nodes (see 1.1.3.2)

[104] Not to be confused with the other meanings of the words "vertical" and "horizontal" that are widely used in this work (vertical as hierarchical, and horizontal as non-hierarchical, respectively).

[105] Lee (2002) denotes such networks as "responsive": in his opinion, an agile network should combine responsiveness to consumers' needs with risk hedging by sharing resources.

in producing parts, e.g., hoods, for the automotive industry in Europe. In terms of the more traditional hierarchical scenario—"in a centralised supply chain"—assembly tasks can be performed only by the OEM, "who delivers the product to the dealers so that it can be sold to a customer" (Mourtzis et al., 2012, p.294; see Figure 3.7). Forming a strategic alliance with suppliers and dealers makes the whole network more responsive to customers' needs, particularly with respect to customizing production according to their individual preferences. For example, the customization of the hood (including an ornament) can be outsourced to suppliers

- *Innovative* networks "are problem sharing networks and comprise parties that exchange problems and solutions. An example of such a network is Procter and Gamble (P&G)'s Connect + Develop network. Instead of using formal alliances to find the best research and innovations, P&G now circulates problem stories throughout a network. The sources of innovation in the network are technology entrepreneurs around the world, suppliers and open networks (e.g., NineSigma, YourEncore, and Yet2.com). The problem stories are presented to these groups and anyone with an answer can respond" (Trkman & Desouza, 2012, p.8). Networks can make good use of network-level distributed intelligence; active work within the framework of a "collective neural network" is obviously promoted by collectivist attitudes and a sense of belonging.

The different kinds of networks can actually be used in combination.

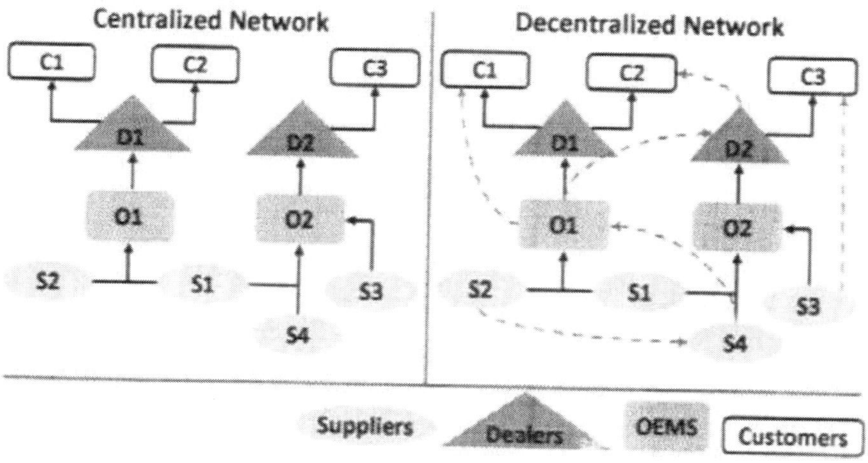

Figure 3.7. Hierarchical (left) and networked (right) production chains, exemplified by car equipment manufacturers in modern Europe. S1–S4, suppliers; O1–O2, OEMs, i.e., original equipment manufacturers; D1–D2, dealers; and C1–C3, customers. Reprinted from *Procedia CIRP, 3,* Mourtzis, D., Doukas, M, & Psarommatis, F. Design and planning of decentralised production networks under high product variety demand, 293–298, Copyright © 2012, with permission from Elsevier.

3.7.2.2. Intraorganizational Networks

Intraorganizational networks, or *internal* networks (Snow et al., 1992), "are loose associations of assets and business units contained within a single company and which subject themselves to market forces", typified by large-scale oil companies (van Alstyne, 1997).

Bureaucratic companies can be dehierarchized by creating internal networks, particularly if they are characterized by *divisional* internal structures. Hierarchical structures of this kind consist of semi-autonomous subdivisions focusing on different types of products, different areas, or different kinds of consumers. Internal networks are established if the subdivisions become legally independent to the point of entering into contracts with one another.

Nevertheless, such contracts between former subdivisions of one firm do not necessarily imply that unmitigated market relationships will be established between them. They usually retain their partly informal ties; the internal market is "domesticated".

While many modules connected by one stalk perform the same function in colonial cnidarians (e.g., many contracting polyps are involved in pumping the liquid through the whole coenosarc cavity), a large number of suppliers, marketers, or product designers can form part of the same network. The whole process is carried out by them in parallel and not only consecutively.

The fractal properties of networks manifest themselves in analogous phenomena occurring at the levels of both inter- and intraorganizational networks. The four network kinds existing at the interorganizational level (see 3.7.2.1), have their analogs at the intraorganizational level. For example, many intraorganizational networks enable increasing cost-efficiency, i.e., represent analogs of efficient (functional) networks composed of whole organizations; this effect is partly due to improved information exchange between legally autonomous firm subdivisions.

If a hierarchical scenario with a large number of rungs in the stepladder is adopted, "successive layers of supervisor-subordinate relationships in organizations often distort information, increase monitoring costs, and lead to a cumulative loss of control" (Jindal, 2011, p.549). In contrast, an intraorganizational network lets subdivisions (that typically communicate only via the central office in a hierarchy) directly exchange information and, moreover, make independent decisions locally, while otherwise enjoying the advantages afforded to them by their autonomous status.

This has been one of the prerequisites for the success of an internal network created by a number of Russian companies, including *Sportmaster* that has 10 regional centers providing services for *Sportmaster*'s local offices and for two subnetworks called *O'stin* and *Funday*, which are connected by horizontal ties to it (Pertseva, 2013; http://www.kommersant.ru/doc/2235941).

Agile networks at the intraorganizational level are exemplified by the *Zara* company, which has granted independent status to its local subdivisions. The company deals with clothes; it directly contacts customers and efficiently meets their needs. The company shortened the period between designing and marketing new articles of clothing to 10–15 days. Instead of letting one designer do all the work, the company has established an internal designer subnetwork. The subnetwork includes several local workshops. Taken together, they contain up to 200 employees that develop new trendy fashions. A modernized business strategy using goods inventory-monitoring software helps connect individual outlets and protects against financial risks associated with producing and storing large quantities of goods that may be difficult to market. Therefore, this company also functions in the capacity of a risk-hedging network, an analog of those existing at the interorganizational level (Zara, 2013, http://www.zara.com/ru/en; and Alexander V. Bubnov, businessman, personal communication).

Alternatively, the goals attained by creating internal networks can be also achieved by inviting external partners, i.e., by establishing an inter- instead of intraorganizational network. Problems caused by the multilevel hierarchical stepladders of big firms can be mitigated using the subfranchising strategy: independent firms play the role of master franchisees that are franchised by the "big firm"; they assume control over a certain area or group of outlets and subfranchise retailers dealing with each outlet separately. Instead of hiring supervisors centrally controlling the operation of each outlet, the main firm lets the master franchiser exercise control at the local level (Jindal, 2011).

Generally, the fractal nature of networks is emphasized by the "interchangeability" of the intra- and interorganizational level in terms of increasing the network's efficiency and risk hedging.

3.7.2.3. Small-Sized Networks in Business

Large-scale business network structures consist of structural units that represent whole firms or their subdivisions. In contrast, small-sized networks are directly composed of individuals or their small groups that establish non-hierarchical relationships. Such small-sized networks can behave as "microcapitalist" actors in the business realm. Small-scale networks in business can make good use of network-specific features, such as spontaneous creativity and adaptability. An example was provided by *The Vision Web,* a Dutch consultancy enterprise set up in the late 1990s that lacked an internal hierarchy; its employees were empowered to make decisions concerning their own salaries; *The Vision* Web's business activities were primarily based on trust (Derix, 2000; Meulemann, 2008).

Small-scale networks can also form second-order networks, resulting in the formation of sufficiently large competitive business entities (this process was described as the formation of hiramiads from hiramas in 1.2.2 and 1.8.5). Apart from small-sized networks, such second-order networked businesses can incorporate originally hierarchical firms; the network organization of the whole business is likely to promote the internal "networkization" of such originally hierarchical inclusions.

In organizational terms, small-sized networks can be structured as egalitarian organizations with partial leaders and a pool of members. They can follow the organizational pattern of a hirama with its set of partial leader roles. Apart from the commercial leader, who is of paramount importance under these circumstances, the organizational leader can also be expected to be extremely useful. This individual will be responsible for all the legal procedures involved, from promoting the official establishment of the networked enterprise to filing commercial lawsuits—a very frequent practice in an "uncivilized" market environment in emergent economies. In such fields as computer software, the pharma industry, biotechnology, and other high-tech areas, networked enterprises have the potential to outcompete larger, bureaucratized firms.

An important subtype of business networks are *cooperatives* (*co-ops*) that are owned and managed by the people who work there or who use their services (these are *worker*[106] and *consumer* cooperatives, respectively). A large number of co-ops are characterized by an internal egalitarian structure. These "ethical businesses" dating back to the consumer co-op *Rochdale Pioneers* (established in 1844 in Britain) espouse the principles of voluntary and open membership, democratic member control, autonomy, and independence (Birchall, 2004), which encourages people to apply analogous principles elsewhere.

[106]*Agricultural* co-ops whose members are farmers are often considered a special type of cooperative. Other important types include *housing* co-ops that collectively build houses and *health* co-ops that provide insurance to enable people to afford health care or actually provide the care (see Birchall, 2004).

According to the "Member's Guide" of the *Consumer Cooperative Society* in Hanover, New Hampshire, that operates the Co-op Food Stores, Co-op Community Food Market, and Co-op Service Center: "Members of a cooperative support it with their patronage, participate in decision-making, and share in the profits generated by the organization's activities". This co-op is affiliated with a higher-order *Cooperative Grocer Network* that operates in compliance with the principles laid down by the International Co-operative Alliance and represents "… an autonomous association of persons united voluntarily to meet their common economic, social, and cultural needs and aspirations through a jointly owned and democratically controlled enterprise" (International Co-operative Alliance Statement on the Co-operative Identity; see Cooperative Grocer Network, 2014, http://www.cooperativegrocer.coop).

Outside the USA, similar cooperative structures have recently been developed in the Basque-dominated North-Western Spain (the *Mondragón* model). These co-ops form a horizontal network of self-governing enterprises owned by the workers.

The Mondragón model envisages overcoming the barriers between different specialties and combining a variety of social roles, and "its business philosophy is contained in its Corporate Values: Co-operation; Participation; Social Responsibility; Innovation" (Mondragón Corporation, 2014; http://www.mondragon-corporation.com/eng/about-us). Apart from working at industrial enterprises, the Mondragón co-op members dealt with health care, housing, and food stores (Whyte & Whyte, 1988). Currently, it is the top Basque business group and the seventh biggest in Spain; it incorporates four main overlapping subnetworks dealing with finance, industry, retail, and knowledge, respectively, according to the official website. Mondragón businesses have flattened hierarchies, and the difference between the incomes of managers and rank-and-file workers is minimized. Mondragón is characterized by direct democracy: important decisions are made by worker assemblies, similar to what is found in Israeli *kibbutzim*. Importantly, Mondragón is currently establishing new links outside Spain. For instance, in the USA, "the United Steelworkers, working with the Mondragón Corporation, has proposed a nationwide effort to create unionized worker-owned co-ops that is beginning to bear fruit in Cincinnati, Pittsburgh and elsewhere" (Alperovitz & Bhatt, 2013, http://truth-out.org/opinion/item/18908-what-then-can-i-do-ten-steps-toward-transforming-the-system).

As mentioned at the beginning of this subsection, small business networks typified by co-operatives can form second- (and third-, fourth-, etc.) order network structures. "The established principle of 'co-operation amongst co-operatives' provides the means for these interests to 'scale up' by building more extensive networks" (Simmons & Birchall, 2008, p.2134).

This "networking of networks" tendency is exemplified by co-ops that establish unions such as the *Co-op Grocer Network* in the USA that brings together members of 300 food co-ops across North America, holds conferences, sets up discussion groups, and includes an active job board (Cooperative Grocer Network, 2014, http://www.cooperativegrocer.coop). Co-ops can be envisioned as important "alternative economic institutions" within the general framework of a project that is aimed at building up a new system—"one centered around building egalitarian wealth, nurturing democracy and community life, avoiding climate catastrophe and fostering liberty through greater economic security and free time", i.e., implementing *socialist principles* in conjunction with an environmental agenda (Alperovitz & Bhatt, 2013, http://truth-out.org/opinion/item/18908-,what-then-can-i-do-ten-steps-toward-transforming-the-system). Although partly concentrating on political issues, the online article cited addresses networked business entities. Among the 10 steps to be made for changing both the economy and the regime, the authors placed emphasis on "helping build a worker co-op or encourage interested businesses to transition to employee ownership" because "worker-owned co-ops bring democracy and democratic ownership into the economy and into community life" (Ibid.).

In turn, co-op alliances can form higher-order networks that ultimately—at the top level—converge on global organizations such as the International Co-operative Alliance, the World Council of Credit Unions, and the International Mutual Aid Cooperation Insurance Federation.

Generally, networked enterprises can promote mutual aid, cooperation, group-level responsibility, a sense of belonging, collectivism, and communalism, as well as an integral vision of each individual's role in society. The network's structure-specific principle of split (partial) leadership fosters the internal democratization of networks—to the point of adopting the model of a collective farm, a kibbutz, or a Kropotkin-style anarchist social system. From a business viewpoint, small-sized networks either represent collectively-owned enterprises (without shareholders) or are legally considered a subtype of open joint-stock companies.

To reemphasize, the hirama model often appears to be a suitable organizational pattern, although real-life networked enterprises modify this model in various ways. Instead of one commercial leader (see 1.2.2), they can include several commercial leaders with specialized functions. In addition, network organization can be used in combination with a flattened hierarchy where superiors are tolerant and supportive of subordinates. The resulting mixed structure is organizationally similar to a partly hierarchized egalitarian hunter–gatherer band. An example is provided by the *Semco Group* in Brazil.

The Semco Group was established in the 1950s and originally specialized in manufacturing centrifuges for industrial plant oil production. Subsequently, the company made investments in other business areas, especially in the service sector, including in environmental consultancy, real estate business, and equipment repair. Unlike conventional capitalist enterprises, the Semco Group had a mitigated hierarchy and made efforts to make the relationships inside the company less formal and partly horizontal. The employees enjoyed much freedom and were treated with respect and tolerance (SEMCO, 2014, http://www.semco.com.br/en; see also Bernhard & Glantz, 1992). From 1989, all Semco businesses have been developed by semiautonomous business units. This "modular" pattern provides greater freedom and also more responsibility for managers and employees. Rico Semler, the company's founder, believed in the following principles:

- Democracy, implying the employees' involvement in managing the company and their access to the information concerning the company's business;
- Splitting the company into business teams;
- Enhancing horizontal ties and mitigating hierarchical relationships: only three levels separated an employee from a top-level manager, and the difference between their wages/salaries was insignificant; moreover, a part of the company's profit was distributed among the employees.

Recently, the Semco Group has established the *Semco Partners* organization that includes six directors who had worked at the company for many years. Its goal is to establish interorganizational networks by inviting those interested in doing business in Brazil to form alliances with Semco Partners (SEMCO, 2014, http://www.semco.com.br/en/). Hence, the network organizational pattern is implemented at both the intra- and interorganizational level.

3.7.3. Network Socialism?

It should be reemphasized that many co-ops implement the principle of collective ownership. Therefore, they actually promote some basically *socialist* ideas and values, despite the capitalist environment in which they are embedded. Importantly, a large number of other networks, despite the pragmatic goals they set themselves, comply with socialist rather than capitalist principles to the extent that, in economic terms, their development entails partial collectivization of the property of their members, i.e., joint control over some of their assets, whereas capitalism per se—at least in its classical form—is based upon the principle of predominantly private property. Establishing networked "strategic alliances" among capitalist enterprises often implies that some of their resources become accessible for all members of the alliance. The development of trust, loyal relationships, and social capital foster collectivism and the sense of belonging ("communalism") that are distinct from both hierarchical and market-style interaction and are compatible with the principles of socialism.

Self-governed socialism is to be predominately considered in this context. It is based upon (1) the operation of autonomous self-regulated economic actors, e.g., cooperatives and similar self-governed businesses and (2) a decentralized mechanism where economic and political decisions are made. Despite the wide variety of subtypes of self-governed socialism, which range from worker-owned enterprises in Yugoslavia under Tito to Israeli kibbutzim (see 3.2.4), they are all characterized by the principle of collective ownership with respect to production means at the enterprise level. Such enterprises are often small in size (or are composed of smaller-sized modules) and are devoid of a rigid centralized bureaucratic hierarchy. To an extent, they are organizationally analogous to primitive hunter–gatherer bands.

3.7.4. Potential Disadvantages of a Network Structure in Business

Generally, a business network structure possesses a number of advantages including the fast distribution of information and the implementation of innovative technologies within the network, decreased production costs (for instance, if suppliers and marketers join the same network), flexibility in response to demand dynamics, adaptability in a changing environment and, in many interorganizational networks, increased transparency to which the externality of the relations among firms can contribute, although it "does not guarantee it" (Ekbia & Kling, 2005, p.167).

However, networks may cause serious problems that relate to their systemic properties (considered above, in 1.8.2, in general terms). In the business world, these problems may discourage entrepreneurs from creating network structures because "an extremely large fraction of network forms of organization do not perform the function for which they were designed" (Podolny & Page, 1998, p.71).

Networks tend to disregard organizational boundaries and build bridges across them (see 1.8.2, item #1); they are predisposed to share confidential information with outsiders, including competitors. Confidential information is likely to leak if several different networks share some of their nodes; i.e., there are organizations joining more than one alliance. In this situation, information easily flows through the whole "meganetwork" composed of several overlapping networks. A related problem is caused by *risks associated with knowledge sharing*. Admittedly, knowledge/information sharing can help a network increase its effectiveness and efficiency by updating its "nodes" (business people, their groups, or entire enterprises) on current designs, client lists, prices, customer profiles, and sales forecasts (Trkman & Desouza, 2012).

However, information sharing, particularly within an interorganizational network, poses serious risks for the network nodes involved. This primarily concerns innovative networks whose members share problems and difficulties with one another. "The exposition of knowledge problems reveals an organization's knowledge deficiencies, which could prove to be costly (e.g., if a competitor can then identify how to take advantage of the knowledge gaps)… In addition, as the focal organization may not be in a position to evaluate a solution to a knowledge problem, the provider has an opportunity to act with guile (e.g., by installing a Trojan horse on a software program) or even to charge abnormal fees for the knowledge" (Trkman & Desouza, 2012, p.8).

A lack of centralized control over a network's finances poses the threat of budget overrun and even money embezzlement inside the network; it is often difficult to pinpoint the network node(s) causing these problems. In literature that concerns various kinds of business networks, this potential problem is considered in conjunction with the currently important issue of *corruption* in business. Establishing interorganizational horizontal ties (instead of carrying out business operations intraorganizationally) has a dual effect: it tends to make business transactions more transparent, but it also promotes informal relationships among interorganizational network members. Informal relationships may encourage clandestine interaction among these organizations. Attempts to fight corruption in business networks may actually encourage corrupt network members to skillfully use accounting to conceal illegal interactivity. "In the case of money laundering, large-scale networks are often assembled to circulate, hide, and repatriate financial flows to their original owners" (Neu et al., 2013).

The dark side of business networks is illustrated by the scandalous story of the *Enron* firm that was established by merging Houston Natural Gas and Internorth, which dealt with the exploration, production, and transfer of oil and gas. Enron included a network of a large number of special purpose entities (SPEs). A relatively long period of the network's successful development was followed by Enron's bankruptcy in December 2001. An investigation revealed that the firm had been repeatedly involved in false accounting. "In partnerships, for example… Enron shifted failed businesses onto the partnership's accounts, and moved some assets, such as a water plant or a broadband unit, into the partnership… It borrowed loans, funded by investors such as J. P. Morgan, and booked them as earnings on assets" (Ekbia & Kling, 2005, p.167).

A closely related problem is *free riding*. If an interorganizational network contains several nodes performing overlapping functions, e.g., several franchisees use the same brand name and are responsible for outlets in the same area under the supervision of the master firm (the franchiser), the efforts made by some of the franchisees have the positive *spillover effect* on other franchisees (Jindal, 2011). If one of them successfully promotes the brand name, the other franchisees will benefit from these efforts and can exploit the promoter without making their own contributions.

However, these potential problems caused by business networks can be mitigated in several ways:

- Efforts should be made to *build up trust and accumulate social capital*, which is essential for the operation of network structures in human society;
- *Chaperones*[107] are regulatory molecules that help other biological molecules assume a correct functional conformation. Of paramount importance are the *human analogs of biological chaperones* that help other network members perform their functions, promote the accumulation of social capital, discourage network members from destructive or illegal activities, and resolve conflicts inside a network; such human "chaperones" are typified by the psychological leaders of hiramas or by the moderators of networked clubs. It is conceivable that special networked organizations (business network-supervising "chaperones") will monitor, as "third parties", the interaction between business networks or subnetworks (factories, outlets, etc.) inside them; unlike law-enforcing bodies, they should pay attention to the moral rather than to the purely legal norms to be obeyed during this interactivity; free riders, embezzlers, or other corrupt business people are to be exposed by them. Such "chaperone" networks can form part of a country's civil society (see 3.8.5 below).
- In addition to the efforts of such "chaperones" (if they become a reality), *legal norms and official regulations concerning business networks* are to be developed in order to provide guidelines enabling the successful functioning of business networks.

Decentralized management may result in the desynchronization ("arrhythmia") of a network nodes' behavior and a lack of coordination of their activities. However, there are situations in which this type of "arrhythmia", as well as other systemic properties of networks, can be beneficial rather than detrimental. The desynchronization of the nodes of a business network may enable the implementation of a flexible subcontracting scenario during production.

The complexity level of business networks may be high enough to confuse their members, in contrast to the relatively simple organization of hierarchies. Network organization calls for the involvement of skilled moderators (psychological leaders). However, the complexity of the nexus of links inside a network can provide an impetus for the creativity, invention, and initiative of its members. A network creates favorable conditions for using techniques that are aimed at promoting the generation of new ideas; these techniques are collectively referred to as *maieutics* (originally suggested by Socrates in Ancient Greece).

Because of a lack of vertical (dominance–subordination) relationships between network members, an efficient and productive operation of a network requires that their motivation

[107] The word "chaperone" originally meant "an older woman… who went out with a young unmarried woman on social occasions and was responsible for her behavior" (Longman Dictionary, 2003, p.246).

level be sufficiently high. The situation is similar to that in an egalitarian hunter–gatherer band (or in a chimpanzee group) where most individuals realize how important their collective activities, e.g., hunting or fishing, are and, therefore, do not shirk—even without coercion. A network exists as long as the critical number of its members is motivated to engage in collective activities (Chuchkevich, 1999, p.21).

Generally, problems characteristic of networks demonstrate that, in human society and particularly in the sphere of business, networks are situated in a stress field generated by pairs of opposing tendencies including *Trust vs Deception, Empowerment vs. Coercion,* and *Cooperation vs Competition* (Ekbia & Kling, 2005, modified). Obviously, an efficient and socially useful network structure should be predominantly characterized by the former rather than the latter feature (in each pair of them), but real-life situations are more complex, and networks face the threats of hierarchization (if coercion replaces voluntarism) and marketization (if competition is stronger than cooperation; Table 4). Strengthening network-specific tendencies should be one of the goals of moderators/psychological leaders supported by an adequate framework of legal norms.

Network structures in business are characterized by non-hierarchical cooperative interactions among their nodes (firms, their subdivisions, or individual business people) that are often based upon trust, reciprocity, and social capital. Networks can be subdivided into interorganizational (dynamic or stable), intraorganizational (internal), and small-sized networks; the latter subtype is typified by worker cooperatives. Many networks are characterized by permeable internal and external boundaries and the "management without control" principle; they creatively use the informal relationships that exist among their members. Regardless of the motivations of the developers of business networks, many networks promote some of the basic principles of self-governed socialism, *and this is particularly evident in* co-ops. *Network structures do not only offer their advantages, but they also have their disadvantages, which are associated with general systemic properties of* sensu stricto *networks.*

Table 4. Opposing tendencies and organization of business networks (based on Table 4 in: Ekbia & Kling, 2005; modified and supplemented). Cf. Table 3 in 2.7 (concerning networks in biological systems)

<table>
<tr><td>Pairs of tendencies</td><td>Implications for network organization</td></tr>
<tr><td>Empowerment ↔ Coercion</td><td rowspan="2">Network ↔ Hierarchy Interconversion</td></tr>
<tr><td>Decentralization ↔ Power Concentration</td></tr>
<tr><td>Cooperation ↔ Competition</td><td rowspan="3">Network ↔ (Quasi-)Market Interconversion</td></tr>
<tr><td>Team Spirit ↔ Individualism</td></tr>
<tr><td>Trust ↔ Deception</td></tr>
<tr><td>Transparency ↔ Secrecy</td><td rowspan="2">Interactive ↔ Isolationist Network Interconversion</td></tr>
<tr><td>Information ↔ Strategic Misrepresentation</td></tr>
</table>

3.8. Network Structures in the Political Sphere

To the author's knowledge, the term "network structure" was used in political science for the first time in the first issue of *Social Networks* (de Sola Pool & Kochen, 1978–1979), although Rice (1927) had actually explored political networks 50 years earlier. He described "blocs" (network clusters in modern terms) within political organizations.

Political scientists' interest in networks increased at the turn of the 21st century. This was partly due to the publication of works on the role of decentralized cooperative network structures in terms of public policy (Börzel, 1998; Meulemann, 2008) and social capital (Putnam, 2000; Putnam et al., 1993). Research on network structures from the political–science perspective has recently gained in importance and won official recognition. This is emphasized in the program materials of the 2013 meeting of the "American Political Science Association. A perusal of the program would have yielded few if any explicitly network-related papers in any given year in the 1990s. In contrast, in the 2013 program there are 23 network-themed *panels,* and over 100 papers explicitly evoking network concepts and data" (McClurg & Lazer, 2014, p.1).

However, it is the broad sense (*sensu lato*) in which the term "networks" has predominantly been used recently, particularly in the Anglo–Saxon political science[108] community. It has often been considered, e.g., in the field of public policy, as "a generic term which applies *to all kinds* of relations between public and private actors" (Börzel, 1998, p.255). Indisputably, the *sensu lato* network concept has considerable potential in the political realm because "power—the central construct of political science—is intrinsically relational",... "power exists between actors and among actors in a complex, differentiated fashion" (McClurg & Lazer, 2014). For instance, network analysis is a potentially fruitful approach in the field of international relations "ranging from broad reaching analyses of the international systems to the flow of goods between nations to the nature of intrastate violence" (Ibid.).

Nevertheless, the term "network" has been used in its more specific sense in a number of political–science works (see, e.g., Kenis & Schneider, 1991; Kooiman, 1993; Rhodes, 1997; Börzel, 1998; Olson, 2005; Meulemann, 2008). Networks are construed as structures "based on non-hierarchical co-ordination, opposed to hierarchy and market" (Börzel, 1998, p.255). Non-hierarchical organization is characteristic of many informal network structures that are receiving increasing attention in terms of political science.

Currently, informal political networks are promoted by new information and communication technologies (ICTs) that also provide researchers with new techniques of analyzing such networks. For instance, "the exchange of e-mails between individuals in organizations reveals how people interact and therefore provides a map of the real network structure behind the formal chart" (Guimerà et al., 2006, p.653). Monitoring online communication using email messages, blogs, homepages, or *Twitter* enables the disentangling of informal links between political party members (Hsu & Park, 2012) or university staff members (this is a nonpolitical, but illustrative example; Guimerà et al., 2006).

---

[108] In contrast, using the *sensu stricto* network concept is more characteristic of political scientists (e.g., specializing in public policy issues) from Germany (Börzel, 1998), the Netherlands (Meuleman, 2008), and other European countries.

There is a wide variety of political network structures (*sensu stricto)* ranging from alliances formed by states to grassroots organizations of political activists. Like networks in general (see 1.1 and 1.2), political networks can be flat or three-dimensional, dense or sparse, homo- or heterogeneous; they can have a manifest community structure or they may lack it. Similar to other networks in human society, political networks can arise spontaneously or they might be set up deliberately. "To take some extremes as examples: there are networks that form because people tend to go to the same place for lunch and have the same eating schedule, or because a group of people, located in different countries and sharing similar values and resentments, want to coordinate with one another in using terrorist techniques to challenge what they see as the global domination of particular countries and their ideological and cultural agendas" (Bunce & Csanádi, 2007, p.136).

Political networks can be *inactive (dormant)* or *active*. For instance, state alliances or global political organizations significantly differ in their degree of activity that often drastically changes during the stages of their life cycle. Dark networks exemplified by terrorist organizations, as considered in 3.2.8, undergo alternate periods of "dormancy" (making them more difficult to detect) and extremely high activity, in an analogy to microbial colonies/biofilms (Oleskin et al., 2011) and ant families (Zakharov, 1991) that display similar collective dormancy–activity rhythms.

An additional classification criterion that only concerns political networks is whether they are "committed to subverting the existing order or to affirmation of the status quo" (Bunce & Csanádi, 2007, p.136).

Like microbial colonies and other biological networks (see Chapter two), human networks (*sensu stricto*) differ in terms of their size. Small network structures are similar in size to primitive hunter–gatherer bands. Their nodes are human individuals.

A relevant political example is an incipient political party that has not yet established a hierarchical structure, e.g., the *Russian Workers' Social Democratic Party*, during its First Congress in Minsk in March 1898 when only 9 people attended the Congress.

In contrast, large networks are composed of smaller networks, although they can also include hierarchical structures. To cite a nonpolitical but illustrative example, the International Centre for Ethics in the Sciences and Humanities established under the aegis of Tübingen University (Germany) included several hierarchical research institutes as well as a subnetwork of independent scientists. The ethical and educational aspects of science were the primary focus of its attention (IZEW, 2014, http://www.uni-tuebingen.de/en/facilities/zentrale-einrichtungen/international-centre-for-ethics-in-the-sciences-and-humanities.html).

3.8.1. Multilevel Networks and Fractal Properties

Political networks can be subdivided into: (1) *supra- and interstate* networks, including global political bodies, state alliances, and international/transnational governmental and nongovernmental organizations; (2) *state-level* networks involved in the functioning of the state apparatus; they are exemplified by various networked consultancy organizations such as kitchen cabinets, think tanks, and more official Centers of Public Policy; (3) *substate-level*[109]

[109] The term "substate-" in the sense "related to network structures formed inside political parties and other officially hierarchical political bodies" is the author's neologism.

networks including spontaneous or deliberately established network structures (clubs, informal cliques, etc.) inside political parties and other officially hierarchical political bodies; and (4) *intrastate* political networks exemplified by political system-independent network structures forming part of civil society.

Many political network structures possess manifest fractal (self-similar) properties. Similar phenomena and structures occur at different organizational levels: the subnetworks/clusters inside them represent reduced copies of the whole network. As stated in 1.8.6, topological self-similarity manifests itself in the formation of analogous branching patterns at various structural levels of a network graphically represented as a tree.

Quantitatively, this may imply that branches at all levels ramify into the same number of lower-order branches (i.e., they have the same bifurcation ratio, *Б*). This was demonstrated in a study concerning the internal structure of University Rovira i Virgili (Tarragona, Spain). "The fact that the community structure is topologically self-similar means that the organization is similar at different levels. In other words, it means that individuals form teams in a way that resembles very much the way in which teams join to form departments ... the way in which departments organize to form colleges, and ... the way in which the different colleges join to form the whole university" (Guimerà et al., 2006, p.665).

Of direct political relevance are data on the structural and functional similarity between networks belonging to different levels, e.g., between interstate and state-level networks.

For instance, both state-level and interstate/suprastate-level political networks face the same organizational dilemma: these networks may consist either of hierarchies, like eusocial systems such as ant societies (1.8.6 and 2.4.4), or of smaller network organizations.

At the state level, the former option corresponds to the *consociational* model: the political system is split into several groupings representing hierarchical structures. The leaders of all the groupings establish horizontal relationships, allowing them to make political decisions on the basis of negotiations and compromises. The consociational model was characteristic of Holland in the mid-20th century, where the society included four major blocs (pillars) that united Calvinists (which were further subdivided into two subgroups because Calvinists belonged to two different political parties), Catholics, Liberals, and Socialists (Lijphart, 1968, 1977).

The alternative *pluralist* model assumes the coexistence of a large number of networked political movements and organizations involved in making state-level political decisions.

At the inter-/suprastate levels, states whose apparatuses represent hierarchical structures can establish multicentral alliances exemplified by the League of Nations in the 1920s and 1930s and, subsequently, by the United Nations. This suprastate-level analog of the consociational model "is risky, as disputes and misunderstandings /between the states that form the alliance – O.A./ may easily give rise to all-out conflict" (van der Wusten, 2004, p.679).

These risks could be mitigated if the alternative pluralist (*networks consisting of networks*) scenario were adopted: diverse transnational networks including nongovernmental organizations would form part of inter-/suprastate-level political bodies.

Multilevel network organizations underlie the *polycentric system* concept suggested by Vincent Ostrom (1962; V. Ostrom et al., 1961) and his wife, Elinor Ostrom (2005, 2010; E. Ostrom et al., 1978). Political decisions—for instance, measures aimed at reducing greenhouse gas emissions (E. Ostrom, 2010) and protecting coral reefs in Palau (Gruby & Basurto, 2014)—are made as a result of negotiations involving political bodies at different levels, ranging from local organizations to national and supranational authorities. The multilevel fractal organization of political networks can actually be *"horizontalized"* in such a nested political structure: agents belonging to different levels (e.g., representatives of a local organization, of a state apparatus, and a supranational body) can participate, on an equitable basis, in political decision making. Moreover, they are encouraged to jointly carry out political projects, e.g., those concerned with the environment.

"In theory, a nested polycentric system is advantageous because, through the involvement of resource users, local knowledge can inform the design of diverse, context-specific rules, while larger organizations (including but not limited to governments) can enhance local capacity to deal with non-contributors or local tyrants, share and invest in information, and coordinate cross-boundary problems" (Gruby & Basurto, 2014). "A nested system where decision-making is centralized is less likely to succeed in supporting sustainability and resilience of the resources it governs" (Ibid.).

3.8.2. Interstate and Suprastate (Supranational) Network Structures

These networks include *interstate structures* and *transnational organizations* as well as higher-order (regional, global) networks that are composed of them.

3.8.2.1. Interstate Networks

Interstate networks represent "collaborative pursuits between parts of different state apparatuses. They may be shaped on the basis of *intergovernmentalism* or *suprastatism* depending on the level of final autonomy of all partners" (van der Wusten, 2004, p. 678, emphasis added – O.A.), i.e., whether the main power rests with the individual nodes (states) or with the whole network. An important subtype of the interstate network is a state alliance that is a political analog of strategic alliances in the business world (see 3.7). Such global or (multi)regional networks consist of states as structural units. They are exemplified by the Asia-Pacific Economic Cooperation (APEC) or the European Union. Most interstate networks as well as transnational organizations (see below) are neither completely flat network structures, nor are they rigid centralized hierarchies. A majority of them represent three-dimensional networks with partial leaders. Such hubs represent the most influential partners in state alliances, e.g., France and Germany in the European Union

To cite a historical example, the Delian League was established in Ancient Greece in the early 5th century BC to repel the Persian aggression. The officially egalitarian alliance of independent city–states was actually hierarchized due to the dominance of Athens that forcefully imposed its political model on other member states. In contrast, the Boeotian League that was established in the late 5th century represented a more egalitarian network structure. It included eleven groups of sovereign cities and associated townships as its nodes; its hierarchization occurred in the early 4th century with the rise of Thebes.

3.8.2.2. Transnational Organizations

Transnational organizations can be governmental and nongovernmental. They can aim to subvert the existing order or they can affirm the status quo (Bunce & Csanádi, 2007). *Status quo-maintaining networks* can form higher-order networks dealing with politics at the suprastate level, together with the state apparatuses of the countries involved, in conformity with Ostrom's polycentric principle.

Many transnational organizations have influential hubs in capital cities, important cultural centers such as Geneva or Barcelona in Europe, or traditional university towns (van der Wusten, 2004). The role of university towns demonstrates the linkage between the academic world and the development of network structures in the modern-day world.

A large number of global organizations use the network organizational principle. An illustrative example is Amnesty International. "We are introducing a new, global way of working – with a distributed centre and Regional Hubs of research, campaigns and communications – because we owe it to the people we work for to be the most effective force for freedom and justice that we can, globally" (Amnesty International, 2014, http://www.amnesty.org/en/who-we-are).

Transnational networks do not necessarily represent global or regional organizations. They can minimally contain only two nodes, as exemplified by municipal international cooperation, or city-to-city cooperation (C2C). An efficient link has been established between the towns of León (Nicaragua) and Utrecht (Holland), which is focused on "activities that enhance general skills and specific urban planning, management and implementation capacities in the municipal apparatus of León" (van Lindert, 2009, p.173). Interestingly, this two-node system already has the main advantages of a full-scale network structure, including mutual trust and reciprocity that generate enthusiasm and "the relatively high level of involvement of the general public on both sides" (Ibid.).

Subversive networks are exemplified by anti-globalist and alter-globalist organizations. For instance, in Central America, some of them are linked with the *Zapatista movement (Ejército Zapatista de Liberación Nacional, EZNL).* From the very beginning of its existence in 1994, EZNL has aimed to granting autonomy to Chiapas, the southernmost state of Mexico, and they have hoped to establish a true democracy across the entire country. One of the first protest actions was an armed campaign against the North American Free Trade Agreement (NAFTA) signed by Mexico on January 1, 1994. Signing the Agreement meant that Mexican farmers would lose their subsidies and face competition with American exporters.

With the exception of the initial period of uprising, the Zapatistas predominantly relied on nonviolent methods, making good use of modern information technologies to network with people around the globe who held similar political views. "We are the network," declared the Zapatistas, "all of us who resist" (Networks, 2003). The Sixth Declaration of the Lacandon Jungle adopted by the EZNL on June 28, 2005, stresses that the Zapatistas support the global alter-globalization movement and protest organizations in Cuba, Bolivia, Ecuador, and other countries.

Although the Zapatistas, as well as analogous anti-globalist organizations/movements, officially adopt the principle of decentralization, the EZNL includes important leaders such as the legendary Subcommandante Marcos. However, due to the predominance of the network organizational pattern that enables collective decision making and participatory democracy, they are perceived as partial hirama-type creative leaders (cf. 1.2.2). Subcommandante Marcos stressed that his leadership was based on obeying those who collectively function as the "real commander". To a certain extent, Marcos is similar to a hirama's spiritual leader (the "guru") whose inspiring image is more important than his/her current activities.

Indisputably, Emiliano Zapata, the Mexican hero of the early 20th century, was another important "guru" whose deeds and ideology still inspire the people.

The Zapatista and some other alter-globalization movements use biopolitical ideas. They compare political network structures to neural networks and, to a greater extent, to the structures formed by insects, such as ants, as was mentioned at the beginning of Chapter three. Actually, the anti-globalists use not only the eusocial paradigm implemented by ants, but also some elements of other biological paradigms:

1. In an analogy to biological networks that should contain a sufficient number of nodes (individuals, groups) to produce cooperative effects, a politically active network in human society should enroll a sufficient number of participants. "Increase their numbers and interconnect them and you'll have something which behaves quite differently – you'll get systematic change – a movement that can cause an entire summit to be cancelled, or the entire corporate accounting system to come crumbling down. Many interacting smaller pieces create the exponential magic of emergence: swarm logic" (Ibid.).
2. Like many types of biological network structures, networks in human society are characterized by a clustering tendency; they tend to separate into smaller subnetworks ("teams"). If a network contains too many individuals, "communication tends to break down and hierarchies develop. We must learn to divide like cells before this happens; big is unwieldy, small and connected is what we should aim for" (Ibid.).
3. Special emphasis is placed upon spontaneous chaotic interactions among network members. Individual behavior is stochastic in an anthill, a fish shoal, a cnidarian colony, and a bacterial biofilm. For example, the movement direction and the rhythm of some individuals may differ from those of the majority. With respect to human network structures, it is emphasized that "haphazard encounters are key to network-building – they are where creativity lies".
4. Network members should pay attention to the behavior of their neighbors. Fish schools can form complex structures if they follow simple behavioral rules based on "short-range order" (see 2.3.8). These rules can be described in terms of algorithms.

However, it should be stressed that networks in biological systems, as well as in human society, are also characterized by "long-range order" dependent on the signals spreading within the whole system, e.g., the queen's pheromones in an ant family, as well as on the modern means of communication used to mobilize network members in human society, e.g., during a flashmob campaign.

It should be also reemphasized that, unlike animals, humans can creatively recombine different network organization scenarios, modify them in accordance with their goals, and supplement them with uniquely human elements.

The idea that complex behavior patterns can be approximated using sets of relatively simple algorithms (item #4 in the list) applies not only to the individual but also to the whole network level. Network structures, including those that pursue subversive goals, can actually follow relatively simple rules. For example, an anti-globalist network can behave in compliance with the rule: *If any single superpower attempts to dominate the whole planet, so that the ostensibly multipolar world becomes unipolar, we must immediately perform protest actions!* If there is a large number of subversive networks that operate in the international arena, their combined behavior may be quite complex. If they succeed in coordinating their efforts (like ant teams in one colony—a "superorganism")—to the point of "filling in for another worker /i.e., another protester or entire protest movement – O.A./ that has halted" (Hölldobler & Wilson, 2009, p.120), their political performance is likely to be very effective.

Networks that concentrate on biopolitical issues—for example, on protecting the environment (see 3.9.1 below)—can follow, e.g., the rule whereby networks can *Protest whenever any country fails to implement the Kyoto protocol concerning greenhouse gas emissions.* Apart from interstate networks, the idea that network behavior can be algorithmized also applies to intrastate networks that form part of civil society (see 3.8.5).

Modern ICTs promote both conservative and subversive transnational networked organizations. Of considerable importance for the modern-day world is the involvement of online networks in preparing "*color revolutions*". A free decentralized network consisting of small teams is organized; centralization is intentionally avoided because a centralized organization can easily be disrupted. Emphasis is placed upon nonviolent methods, including *swarming* (i.e., "the method of deploying mass mobs of digitally-linked youth in hit-and-run protest formations moving like swarms of bees"; Engdahl, 2011, http://www.engdahl.oilgeopolitics.net).

3.8.3. State-level Networks

State-level networks can be affiliated with the state apparatus. Alternatively, they can be organizationally independent and they can assist the apparatus in carrying out state-level political projects. Network structures can develop long-term strategies that concern various interdisciplinary issues ranging from assessing the ecological situation in some parts of the planet to developing optimal strategies for Brazil in the Middle East. They can involve experts with different professional backgrounds that represent various scientific, business, and cultural insitutions. The RAND Corporation and the Hudson Institute in the USA represents efficient networks involved in political planning.

Networks can provide guidelines for social, political, economic, and cultural progress. They can function as generators and promoters of new values and politically important ideas in society. Such idea-generating networks are already at work at the national and global levels in various parts of the world.

Network organization is characteristic of many *think tanks.* They are nonprofit political organizations that can evaluate political projects and develop political guidelines (Rich, 1999). Think tanks consist of expert teams carrying out educational, evaluative, creative, communicative, and promotional projects (Rimsky & Sungurov, 2002; Gorny, 2006). The networked *Strategy Center* (Saint-Petersburg, Russia) is an example of this type of think tank. Its main goal is "promoting the development of civil society, the rule of law, and public policy in Russia by implementing projects and programs aimed at facilitating public participation and social partnership, enhancing the government's responsibility, and developing Centers of Public Policy" (Sungurov, 2006). Despite their professed autonomy, many think tanks cannot consider themselves politically independent. Some think tanks, e.g., the Brooklyn Institute for Social Research associated with the Democratic Party, are closely linked with the political system; some other think tank networks are associated with business firms.

Think tanks are similar to *Centers of Public Policy (CPPs)*. They include pools of experts that analyze the current situation, make predictions, and develop new strategies and doctrines. CPPs also function as mediation structures between the hierarchical state apparatus and networked public organizations (1.8.3; to be revisited in 3.8.6). Interaction between public organizations and the business sector may also be facilitated by CPPs.

The state apparatus can be partly restructured in compliance with the *network organizational principle* in terms of Ostrom's polycentric system, which was briefly discussed at the beginning of this section. To reiterate, "'polycentric' connotes many centers of decision making that are formally independent of each other... To the extent that they take each other into account in competitive relationships, enter into various contractual and cooperative undertakings or have recourse to central mechanisms to resolve conflicts, the various political jurisdictions in a metropolitan area may function in a coherent manner with consistent and predictable patterns of interacting behavior. To the extent that this is so, they may be said to function as a 'system'". (V. Ostrom et al., 1961, pp. 831–832). Hence, apart from the "central mechanisms" of a hierarchy, this system is based on the horizontal and cooperative interaction between autonomous units (network nodes) as well as, to a certain extent, (quasi-)market "competitive relationships."

This type of system implies that political issues are resolved, on an equitable basis, by the state apparatus and lower-level political bodies such as local administrative apparatuses. As mentioned above, the system is applicable to environmental issues, e.g., the protection of marine resources in the western Pacific island of Palau. From this perspective, small marine protected areas (MPAs) should be managed by autonomous political decision-making bodies that form a decentralized network (Gruby & Basurto, 2014). The MPAs' network directly interacts with the national government, as well as with conservationist nongovernmental organizations (NGOs). Using the example of the greenhouse gas emission issue, Elinor Ostrom (2010) stressed that "instead of focusing only on global efforts (which are indeed a necessary part of the long-term solution), it is better to encourage polycentric efforts... Polycentric approaches facilitate achieving benefits at multiple scales as well as experimentation and learning from experience with diverse policies."

Nonetheless, despite its potential advantages in certain situations, the polycentric system might also cause problems relating to potentially negative features of network structures in general (cf. 1.8.2) that include, according to Ostrom (2010), "inconsistent policies" of the different-level bodies involved (related to what I called the "arrhythmia" of network structures above), "gaming the system" (e.g., embezzling its money) and, in particular, free riding.

However, the problem of free riding seems to be mitigated in the polycentric system thanks to the cooperation-promoting and social capital-strengthening properties of horizontal cooperative decentralized relationships between "network nodes." "A combination of structural features leads many of those affected to trust one another and to be willing to take an agreed-upon action that adds to their own short-term costs because they do see a long-term benefit for themselves and others and they believe that most others are complying" (Ostrom, 2010, p.551).

3.8.4. Substate-Level Networks

The network structures I call *substate-level networks* relate to political parties and other similar political bodies that form "pillars" (to use Lijphart's term) supporting the political system under a democratic regime; they compete for influence and political domination. Successful competitors are involved in the functioning of the state apparatus.

Despite the hierarchical organization of most political parties[110], networks spontaneously arise within them, or they are created deliberately, e.g., in the form of clubs. The data obtained by researchers often demonstrate a largely non-hierarchical organizational pattern of interpersonal links that are based on cooperation, common interests, and affiliation. Typically, "the network of interactions within an organization is considerably more complex than implied in the formal chart" (Guimerà et al., 2006, p.653). Internally, a political party often exhibits a community structure complete with clusters (cliques) and influential hubs (Guimerà et al., 2006; Hsu & Park, 2012).

New information and communication technologies that enable efficient online communication apparently play two different roles with respect to the formation of network structures within political parties (and similar organizations):

1. They increase the visibility of small- and middle-sized parties and encourage the participation of "interested citizens who might not have had the time or capacity to become involved in the more conventional forms of party membership" (Römmele, 2003, p.10), thereby creating a more decentralized and pluralist political landscape. For instance, *Twitter* "was used extensively for social networking during the 2009 national parliament election in Germany and reflected the election results" (Tumasjan et al., 2010; quoted according to: Hsu & Park, 2012, p.170).
2. They make strong political parties even stronger: the "coordination of different party branches becomes easier" (Römmele, 2003); they also promote the formation of closed cliques that protect their information using passwords.

In accordance with Putnam's (2000) ideas, roles #1 and #2 correspond, respectively, to the *bridging* and the *bonding* functions that are performed by the new links that are established between network nodes (see 1.7.2). The bridging function is more important for an *interactive* network, whereas an *isolationist* network, e.g., a closed clique inside a party, should prioritize the bonding function.

ICT-mediated communication is more hierarchical (top–down) in *vote-maximizing* parties that are attempting to win elections by increasing the number of voters supporting them: "broadcasting via new ICTs is the dominant communication strategy" (Römmele, 2003, p.12); similar hierarchization is typical of parties aimed at *office maximization*, "distinguished by their focus on holding positions in a coalition government" (Ibid., p.13) if a multiparty system is adopted.

---

[110] Notable exceptions to this rule are some non-hierarchical "virtual" parties that are based on online networks. They are exemplified by "the German party 'Die Digitalen', which ran in the 1999 local government election in Berlin. This party operated on an Internet basis only, with an 'open policy' programme whereby policy formulation was subject to any interested person online. So-called party members were solely in charge of organizing and coordinating the process of policy development" (Römmele, 2003, p.10).

In contrast, *policy-seeking parties,* i.e., parties prioritizing policy goals (for instance, environmental protection), are characterized by bottom–up ICT-mediated communication. If attaining policy goals is considered more important than advertizing for specific candidates for political offices, then ICTs "serve as a useful channel for informing a broad audience, as well as targeted groups, about what the party stands for. The message, therefore, will be more policy-oriented and less concentrated on political candidate(s)" (Ibid., pp.14-15).

In a study conducted in South Korea, it was revealed that ICT-promoted communication predominantly proceeded within political party boundaries; therefore, the ICT links that were provided performed a bonding function rather than a bridging function. Members of the same political party chiefly used blogs and homepages to exchange information. *Twitter*-based communication was somewhat less selective, so that members of different parties tweeted with one another relatively frequently. The representatives of the Korean political party elite preferred to communicate with one another, paying considerably less attention to their constituents. For example, "National Assembly members tended to communicate more with other politicians than with citizens" (Hsu & Park, 2012, p.179). These findings suggest that, currently, "citizens are increasingly less likely to be involved in institutional politics and that new media are encouraging such trends" (Ibid., p.180), although additional studies need to be conducted in order to confirm this statement.

Generally, it seems that the development of ICTs just highlights trends that were formerly based upon traditional "offline" interactions.

3.8.5. Intrastate Networks and Civil Society

This subsection deals with political system-independent network structures in the form of nongovernmental associations and alliances based upon the principles of self-organization, self-government and, as a rule, financial self-sustainability (Mezhuev, 2008). Such network structures are established on a voluntary basis; they comprise a variety of academic, religious and cultural associations, initiative groups, and so forth (Habermas, 1990). Taken together, these network structures lie at the core of *civil society as* "the aggregate of non-governmental organizations and institutions that manifest interests and will of citizens" (Dictionary.com, 2014, http://dictionary.reference.com). They exert a democratic influence on the state apparatus and its regulators (Motroshilova, 2009, p.16) and filter the citizens' demands vis-à-vis the political system (Easton, 1965).

Interstate networks forming part of civil society may be concerned with a wide spectrum of pressing problems (human rights, constitutional law enforcement, environmental protection, health care organization, etc.) and they may develop strategies to solve them. Apart from their specific goals, they carry out an important political mission by *building social capital and trust and, therefore, promoting the development of civil society.* This point was emphasized by Robert Putnam (Putnam et al., 1993). If the mission is successful at the micro-level (within the boundaries of a single network), this encourages an analogous mission at the macro-level, i.e., dealing with the political system of a whole state or even with the global political arena.

Interstate networks also include networked businesses (see 3.7.2.3) and especially *co-ops.* Their political mission involves promoting participatory democracy and collectivism ("communalism"). In order to become politically influential, co-ops form alliances (higher-order networks that were mentioned in 3.7). By providing jobs and material support to their members, they can help a large number of formerly poor people become sufficiently well-to-do and, therefore, politically active. Their increased political participation is bound to enhance the political influence of civil society and it can provide further support to the principle of political power decentralization.

Networks within the framework of civil society (including co-ops) can promote the achievement of the *Millennium Development Goals* (*MDGs*), which were formulated at the turn of the 21st century, in the *United Nations Millennium Declaration*. Apart from poverty eradication, these goals include other politically important objectives such as achieving democracy and good governance, promoting respect for nature, observing international human rights law and international humanitarian law, and developing a global partnership for development (United Nations Millennium Declaration, 2000), to which co-ops and other network structures can contribute (Birchall, 2004).

Ultimately, all network structures, irrespective of their specific goals, can converge on one focal point: they can predict the direction of the future development of global civilization—the single "body of humankind" (Vlavianos-Arvanitis, 1985, 2003)—and positively influence this process. Cooperatives and related network structures promote "the development of public self-government and the broadest possible involvement of popular masses in administering industry and production, and society as a whole, without which the interests of the individual and society cannot coexist in harmony" (Bestuzhev-Lada, 1990, p.64).

In particular, civil-society networks can help the state apparatus cope with a wide variety of social and political problems including various humanitarian issues, e.g., providing aid to the homeless and victims of catastrophes, as well as to orphanages and nursing homes. Nevertheless, apart from *helping* the state machinery, the network structures of civil society should also be able to *struggle* with it, if the state apparatus makes inadequate or anti-democratic decisions.

In developed civil society, a large number of networks with different agendas can interact in a complex fashion. Some of them, despite the network-specific tendency towards an interdisciplinary integrated approach to any issue, can confine themselves to more specialized tasks and missions, particularly if they involve specialists in relevant fields.

The response of some civil society-level networks to decisions made by the political system may be, in some cases, based on binary choice. Therefore, the behavior of these networks can be described in terms of relatively simple sets of algorithms. For instance, a network structure that operates inside a single country and deals with human rights can follow this simple rule: *Immediately perform protest actions once the number of political convicts in the country exceeds a certain limit!* The coexistence of a sufficiently large number of networks with different attitudes and agendas can result in an extremely complex pattern of political activities.

3.8.6. Network Meritocracy?

To reiterate, civil society in a democratic country can exert a sufficiently strong influence on the political system. Its networks and their representatives acquire a significant amount of political power without anyone electing or appointing them (Zaleski, 2006). The author cited doubts that the power of these network structures is compatible with the principles of democracy. Nevertheless, despite a lack of a formal election or appointment procedure, networks and their influential members ("hubs") are politically active because they receive sufficient support from the people at large thanks to their good reputation, i.e., to the social capital accumulated by these individuals (Putnam, 2000). Apart from their honesty, sense of responsibility, organizational skills, and other socially important features, this reputation is also based upon the *professional competences* of a network's members with respect to socially/politically relevant areas. These competences are at the forefront, if politically influential networks are composed of professionals (scientists, scholars, experts) whose decisions on political issues are based on their knowledge and expertise.

Generally, network structures can increase the influence of meritorious professionals (scientists and scholars) in society, strengthen the social and political impact of their ideas and, therefore, promote the development of *meritocracy* in modern-day society. In contrast to the more traditional concept of meritocracy that was considered, for instance, by Daniel Bell (1973), and which was implemented in Singapore after attaining independence in 1965 (Bell & Chanyang, 2013, http://www.singapolitics.sg/views/compassionate-meritocracy), it is not a government-appointed commission that assigns the high social status to meritorious professionals. It is the networks and civil society as a whole that make this decision; it is their trust and support that enable socially recognized experts to speak on the network's behalf with regard to political, economic, social, cultural, humanitarian, or environmental problems and issues[111].

The activities of network structures within the framework of Russian civil society have considerably increased the political weight of a number of socially recognized scientists/scholars including Alexander Yu. Sungurov, a biologist (biophysicist) and a political scientist. Starting in 1994, he has been in charge of the *Strategy Center* in Saint-Petersburg, which is focused on humanities- and political science-related issues. The support of civil society networks has been a prerequisite for the political careers of some opposition leaders in Russia and the Ukraine.

Importantly, not only intrastate but also interstate and transnational network structures tend to involve qualified specialists. Of relevance is the fact that their hubs are often located, apart from political centers, in major university cities (van der Wusten, 2004), so that such network structures have an academic image.

The animal organism contains two regulatory systems: the nervous system and the relatively slow hormone (endocrine) system. In contrast to the nervous system, the hormone system does not "give orders" that are to be carried out immediately. The hormone system adjusts the organism's activity level, influences its long-term strategy, and helps it overcome stress. The political system with the state apparatus is analogous to the nervous system and the brain; local political bodies are analogs of peripheral ganglia. As for the hormone system, its functions are comparable to those performed by civil society.

In an analogy to hormone-releasing endocrine glands, the network structures of civil society spread important social, political, and cultural ideas and values in society. Indoctrinating the people at large with such ideas and values produces an effect on the long-term strategy of the

[111]Currently, the level of trust and the degree of social recognition of professional merits by the members of relevant networks composed of competent professionals can be easily estimated using online questionnaires and voting procedures.

whole society, leading the citizens to accept or reject the government's decisions and policies. The networks that form the basis of civil society—the "hormone system" of human society—develop an *ideology* in the broad sense of this term. They help people answer such questions as *What is the meaning of human life? How can we optimize the political regime? Can we believe in a bright tomorrow and what should it be like? What principles should underlie relations between human individuals and between whole nations? What should be our attitude to the environment?*

The functioning of civil society networks, as well as of political system-aiding think tanks (see above), involves the ongoing learning/training of these networks, the acquisition and processing of enormous amounts of information by their "collective brain" and, at each given moment, the singling out of the issues to be prioritized. These features of human political networks evoke the *neural paradigm* of the network's organization that exhibits similar features.

Human network structures also seem to apply modified elements of other paradigms considered in Chapter two. Politically influential networks consisting of a small number of important people (academicians, business people, etc.) that develop innovative political projects (e.g., concerning public policy) usually emphasize the individuality of each of their members who are free to join and quit such a network structure. Therefore, these networks are, to an extent, structurally similar to fission–fusion groups of, e.g., chimpanzees, and they conform to the *egalitarian* paradigm.

3.8.7. Interaction between Networks and Hierarchies: The Role of Mediators

Networks and hierarchies interact in various political situations; for example, networked expert organizations evaluate the political system's decisions or networks within civil society make their political demands or publicly critique the government's policies. Obviously, the hierarchy–network interaction may not be harmonious. For example, there may be serious conflicts if the hierarchical political system fails to meet the demands or expectations of civil society. One way to mitigate or prevent such conflicts is to make good use of mediators, or *meta-governors* (Meulemann, 2008; see 1.8.3). Mediators/meta-governors guarantee that each of the two partners (a hierarchy and a network) has legally acknowledged rights, that they interact on an equitable basis, and that they aim to jointly make consensus-based decisions.

As emphasized in 1.8.3, a mediator structure facilitating network–hierarchy interactions can consist of three components: a subnetwork of the network, a special department of the hierarchical structure, and an intermediate component sandwiched between them (1.8.3). All the three components can form part of one political body that is exemplified by the NGOs–Government Interaction–Coordinating Committee (Saint-Petersburg, Russia). Although established by the city government's initiative, the Committee, which consists of 24 members, includes only 4 government representatives. The other Committee members represent NGOs. The intermediate third component is a mixed dynamic network structure that forms whenever a decision is to be made. Decision making is based on reaching a consensus, even though the Vice-Governor of Saint-Petersburg is considered the formal leader of the structure (Rimsky & Sungurov, 2002).

The potential role of mediators (meta-governors) is not confined to hierarchy–network interactions. Mediators can also facilitate and regulate *network–network interactivity*. For instance, the networks of civil society can contact various transnational and global network structures. Civil society networks dealing with human rights can interact with similar international structures, e.g., Amnesty International. Promoting creative interactivity between networks, including the establishment of combined networks such as transnational think tanks, could be another important mission of mediator structures.

Internetwork mediator structures could protect the rights of each of the networks involved and they could prevent one of the networks from suppressing, exploiting, or engulfing other networks. This protective function assumes particular importance if a network–network conflict reflects a more general political conflict. By suppressing a country's own networks and replacing them with foreign structures, the foundation can be set to put the country under a foreign aggressor's control.

3.8.8. Potential Role of Networks in Preventing, Mitigating, or Overcoming Political Conflicts

Not only networks, but also political hierarchies, as exemplified by the governments of two different countries, can come into conflict with one another. Establishing a higher-order network involving both conflicting parties might mitigate their diagreement by promoting cooperative rather than agonistic interactions, irrespective of the network's agenda. For example, protecting the planet's biodiversity (bios) could be a suitable agenda for a network, as it can bring together representatives of potentially hostile nations or religious movements. A global network structure can collectively represent an independent "third party", a global-level judge that deals with a political conflict. The obvious problem with some of the presently existing global conflict-mitigating networks is caused by their insufficient political influence. To an extent, this problem can be allayed by strengthening the global networks' ties to local (intrastate) networks in order to be able to influence dangerous political regimes both from above and from below, according to the *glocalism*[112]principle that is actually a variation on the *polycentric governance* theme discussed above.

3.8.9. Network Society?

The term "network society" has been used in the literature starting from the 1990s; it gained currency under the influence of Manuel Castell's works (see, e.g., Castells, 1996, 2000, 2004, 2009). It is assumed that the dominance of network structures over more traditional bureaucracies is due to a drastic increase in the amount of information spreading across the world, particularly because of the development of the Internet. As a result, political and business hierarchies are gradually losing their power; they are being replaced by decentralized structures that have no boss at their top level, and they use modern communication facilities provided by the Internet.

To an extent, the author of this work shares the enthusiasm of the prophets in predicting the advent of a network society. However, at least in the nearest future, centralized hierarchies will

[112] In political terms, *glocalism* is a current trend that weakens state- (nation-) level political structures by activating two mutually complementary processes: (1) the strengthening of local-level social and political structures; (2) the establishment of international structures including politically influential global organizations.

remain important parts of human society, including its political system. In particular, administrative bodies, as well as defense and law enforcement agencies, will probably prefer the hierarchical structure. Nonetheless, the whole social system will probably use a *mixed organizational pattern* in which hierarchies will coexist and interact with networks and (quasi-)market structures. To reiterate, it will be a mediator's/meta-governer's job to optimize the interactivity of different organizational structures.

Centralized hierarchies, such as the state government, perform the important function of condsolidating their citizens and promoting national self-identification in the face of potentially dangerous "others." Presumably, despite the development of network structures and civil society, the central political hierarchy will still play a consolidating role that is associated with promoting the state's ideology (although developed with the help of network structures such as think tanks), representing the state on the international arena, and organizing its defense. It is pertinent that the political systems of a number of countries, such as Great Britain and Japan, are constitutional monarchies, and in this context, the monarch represents a symbol of national unity.

Importantly, network structures have both their advantages and disadvantages. As I emphasized in the preceding sections, some of their peculiarities, such as the desynchronized operation of network nodes ("arrhythmias") and their limited controllability, present serious difficulties for social and political organizations that should engage in quick, well-orchestrated activities. They are exemplified by defense and law enforcement agencies where network structures are likely to be useless and even harmful, except for counter-terrorist organizations. The successful functioning of counter-terrorist organizations should be facilitated if they are structurally similar (isomorphic) to that of their enemy, who typically prefers the network organizational pattern (dark networks; 3.2.8).

3.8.10. Network Warfare?

In the future, conventional wars between two countries or two coalitions will probably still be waged by hierarchically organized armies. Nevertheless, despite the importance of military hierarchies, the network organizational principle will also be used in conducting military operations. This is in conformity with the *Network-Centric Warfare (NCW)* doctrine developed by the United States Department of Defense in the 1990s. The NCW was defined as the "concept of operations that generates increased combat power by networking sensors, decision makers, and shooters to achieve shared awareness, increased speed of command, higher tempo of operations, greater lethality, increased survivability, and a degree of self-synchronization" (Alberts et al., 2000, p.2). Attaining these goals is promoted by granting considerable freedom of action to combat troops, decentralizing the command-and-control system, and flattening the army hierarchy. Data concerning the enemy as well as the friendly troops should be accessed in a more decentralized way, enabling those directly involved in military operations to efficiently respond to them without waiting for centralized agencies to relay the information to them (Kovalev et al., 2013).

3.8.11. Legal Conditions Promoting the Development of Political Networks and Civil Society

In the author's opinion, the following social and legal conditions can enable network structures within civil society to successfully carry out their political mission:

- Network structures should enjoy a *legally acknowledged independent status*; representatives of the state apparatus can form part of mediator structures facilitating network–hierarchy interactions, but *they should be forbidden from becoming members of these networks* per se in order to avoid their dominance in civil society networks.
- *Special attention should be given to* civil society and its networks *in the media*; these networks should be granted basic political rights including, e.g., rights to organize public demonstrations.
- Civil society should create *its own independent think tan*ks that involve highly qualified specialists and deal with politically important issues; they should not be directly affiliated with any business or governmental organization.
- Effective *mechanisms of consensus-based decision making* at the intra- and internetwork levels are to be developed; networks should come up with innovative ideas and independent political doctrines during their dialogues and debates with the central political hierarchy.

Under these conditions, network structures and those civil societies based upon them are likely to effectively perform their functions in society, including (1) independently evaluating the performance of the political system as well as its decisions; (2) making scientifically grounded recommendations and demands addressed to the political regime; and (3) preventing corruption, particularly at the state apparatus level.

The mission of network structures also includes a psychological aspect. They can mitigate the feelings of loneliness, alienation, and helplessness that are unfortunately widely spread among people in present-day society. In a network, its members develop a sense of belonging. Network structures provide social protection, promote loyal relationships based on trust, stimulate social and political activity, and foster grassroots initiatives. Network structures also have considerable potential with respect to humanitarian and charitable projects.

An example is the networked *Good Old Age (GOA) Foundation* that deals with the aged and disabled in Russia. According to the GOA's official website, their "main goal is to prevent people staying in rest homes and old people's homes from feeling lonely and abandoned, to convince them that somebody really needs them, ... to improve their living conditions and—above all—to convince the people at large that the issue concerning the aged and the disabled is of paramount importance and, therefore, should receive sufficient attention" (GOA, 2014, http://starikam.org).

Network structures develop at various levels of human society and its political system. They can be subdivided into: (1) supra- and interstate networks, including global political bodies, state alliances, and international/transnational governmental and nongovernmental organizations; (2) state-level networks involved in the functioning of the state apparatus, which are exemplified by networked consultancy organizations such as kitchen cabinets, think tanks, and more official Centers of Public Policy; (3) substate-level networks including spontaneous or deliberately established network structures (clubs, informal cliques, etc.) inside political parties and other officially hierarchical political bodies; and (4) intrastate political networks, exemplified by political system-independent network structures that form the basis of civil society. Network structures can help the state apparatus cope with a wide variety of social and political problems. Nevertheless, they should also be able to struggle with it if the state apparatus makes inadequate or anti-democratic decisions. Politically active network members are neither elected nor appointed by political bodies; nevertheless, their political status is based on the social recognition of their professional qualifications and personal merits. Interaction between networks and hierarchies, particularly between network-structured civil society and the hierarchical state machinery, can be facilitated and regulated by mediator structures (meta-governors). In order to enable network structures to successfully carry out their political mission, adequate legal and social conditions are to be created.

3.9. Network Structures and Biopolitics

In this section, the mission of network structures is illustrated in the example of their applications in biopolitics, which is *the whole totality of interactions between biology and politics, including the impact of modern life science data and concepts on politics as well as the influence of politics on human biology and the entire environment* (see Introduction)[113]. A broad spectrum of biopolitical issues are on the agenda for a number of local, national, regional, or global organizations. Some of the networked organizations dealing with biopolitics are listed below (the list is certainly not complete):

- The *Association for Politics and Life Sciences* (APLS, 2014; http://www.aplsnet.org) established in 1980 as an "organized section" of the American Political Science Association. It presently represents "an international and interdisciplinary association of scholars, scientists, and policymakers concerned with evolutionary, genetic, and ecological knowledge and its bearing on political behavior, public policy and ethics". Some Association representatives are also members of other biopolitical organizations. The Association publishes the journal entitled *Politics and the Life Sciences.*
- The *Gruter Institute for Law and Behavioral Research.* Founded in 1981 by Margaret Gruter and Roger Masters. This is a networked "research community that fosters collaboration across disciplines in order to advance our understanding of the interplay between law, institutions and human behavior" (Gruter Institute, 2014; http://www.gruterinstitute.org). The headquarters are situated in the USA (Portola Valley, CA), and a branch office is in Germany. The organization chiefly deals with the legal and criminal aspects of biopolitics. "To that end, the Institute brings

[113]*Note*: This section presents, in an abridged and modified version, part of the materials discussed in more detail in my work, which was previously published by *Nova Science Publishers, Inc.* (Oleskin, 2012).

together academics and practitioners from diverse fields ranging from evolutionary biology, economics, anthropology, finance and law to exchange knowledge from the diverse disciplines, collaborate on core questions, and develop tangible outputs, including publications, policy recommendations and the architecture of innovative institutional and legal regimes" (Ibid.).

- The *Institute for the Study of Complex Systems (ISCS)* founded in 1994 in the USA (Palo-Alto, CA). The director of the ISCS is Peter A. Corning. In terms of systems theory, attempts have been made to conceptualize human beings and human society as the products of the "teleonomic evolution" process, which is common to all life forms. The Institute develops evolutionary/functional approaches to complexity. "Other work at the Institute includes a new approach to the relationship between thermodynamics and biology called 'thermoeconomics', a new, cybernetic approach to information theory called 'control information', and research on basic needs under the 'Survival Indicators' Program" (ISCS, 2014; http://complexsystems.org).
- The *Biopolitics International Organization* (B.I.O.) whose center is located in Athens (Greece). It currently includes an "extensive network of supporters and affiliates (leading representatives, academics, diplomats, civic and business leaders, and institutions) in 165 countries around the world" (B.I.O., 2014, http://www.biopolitics.gr). It was founded in 1985 by Agni Vlavianos-Arvanitis. Under its auspices, the *International University for the Bio-Environment (I.U.B.E.)* has been established. The organization focuses on the practical dimensions of biopolitics (biopolicies) dealing with efforts to protect the manifold of life on the planet, genetic technologies, bioethics, bioesthetics, and other biopolitical aspects.

Some biopolitical organizations represent almost flat network structures. For instance, the APLS, apart from ordinary members, includes only the Council, which currently (in 2014) consists of 13 people, including the Editor-in-Chief of its journal and the APLS webmaster (see APLS, 2014; http://www.aplsnet.org); others, e.g., the B.I.O., include a hierarchical core with limited power and a sufficiently strong peripheral network.

The network's organizational pattern is explicitly implemented by the local and, prospectively, transnational *Club of Biopolitics* established in 2010 at Moscow State University (MSU) under the auspices of the Moscow Society of Natural Scientists. According to its website (Club of Biopolitics, 2014, http://biopolitika.ru, click Club of Biopolitics), the Club aims "to promote biopolitical ideas in society, including the diverse aspects of the impact of modern biology in the social, cultural, and political realms." Importantly, the Club of Biopolitics "represents a networked organization. All major issues are resolved during a meeting of all Club members or of those whom they concern. Decisions can be also made online, during an Internet conference." The Club's structure includes *project teams* with *creative leaders* dealing with specific tasks or areas of research; their tasks overlap, and Club members can join several project teams, like *hirama* members (see 1.2.2); the creative leaders are considered partial leaders within the framework of the higher-order network of the whole Club (in a similar fashion, clans of worker ants form part of a decentralized flat structure called a column; see 2.4.5).

In the following, network structures concerned with some concrete aspects of biopolitics will be considered. Since network structures can work in parallel on several problems or issues, they can actually deal with the entire spectrum of biopolitics, irrespective of a network structure's concrete goal.

3.9.1. Environmental Biopolitics

Environmental issues have been the focus of attention of biopoliticians, beginning from the seminal work of Lynn Caldwell (1964). Currently, many international networks pay much attention to environmental problems including water, atmosphere, and soil pollution, as well as deforestation and urbanization-related issues.

An important example is the transnational network, *Living Lakes*, which deals with the conservation of the ecosystems of 28 bodies of water and wetlands. The network is supported by the *Global Nature Fund* (Internationale Stiftung für Umwelt und Natur, Radolfzell, Germany). "The main purpose of this international initiative is the conservation and protection of natural resources, chiefly the drinking water reservoirs of the earth" (*Living Lakes*, 2014; http://www.globalnature.org). This global network consists of local second-order networks dealing with each lake/wetland involved. In the Baikal Region of Russia, it is organizationally linked to the local network called *Club Firn* that concentrates on protecting Lake Baikal with its ecosystems. However, like many other networks, *Club Firn* is a versatile organization that combines several different activities. Apart from protecting Lake Baikal, it also sets itself other goals such as promoting local small-scale enterprises; in this capacity, it cooperates with local business-stimulating organizations of the Baikal Region (*White Book*, 2004).

The *hierarchical core + network periphery* scenario is used by a large number of environmental organizations. The network periphery exerts a sufficiently strong influence on the core, which limits the bureaucratization of the whole structure. For instance, the *Boreal Forest Conservation Network* in Siberia includes the core, which is "connected to a resource providing center", and the periphery; overall, this is a dispersed, spatially distributed network (Davydova et al., 2008, p.119). Network structures can efficiently perform the following environment-related functions:

- *Developing and promoting philosophical concepts that provide the foundations for effective environmental activities.* In particular, environment-centered network structures can foster biocentrism, where the worldview gives top priority to the Earth's bios (life), so that humankind is considered part of the bios as a coherent global entity (the "body of bios", Vlavianos-Arvanitis, 1985, 2003).
- *Coping with pressing environmental problems.* Network structures can successfully deal with a wide variety of concrete tasks, ranging from assessing ocean pollution caused by surfactants, e.g., laundry detergents and shampoos, to improving the environmental situation in many third-world cities by building a network of public toilets (an additional way of implementing network principles). It should be stressed that a large number of environment-related issues are interdisciplinary and present difficulties for traditional bureaucratic organizations that prefer a more narrow focus.
- *Developing an educational system concentrating on the environment and, in more general terms, on biopolitics,* including the whole spectrum of socially and

politically important implications of the life sciences. Of interest in this context are interactive teaching methods, including those that envision the formation of network structures (teams) in the classroom (3.6.1).

The *hirama* pattern, as described in 1.2.2, can be a structure of choice for an environmental network organization. This *hirama* can, for example, pursue the goal of optimizing the living conditions of some biological species, e.g., the panda in China or the koala in Australia. In terms of the *hirama* scenario, the network structure should include several partial creative leaders that deal with subproblems. As for a panda-protecting network, these subproblems could be as follows (Oleskin, 2012): (1) stopping poachers from hunting pandas; (2) setting up feeding sites to be visited by pandas; (3) monitoring their population dynamics (does their number tend to increase or decrease?); and (4) enlightening local people as to pandas' lifestyle and habits and encouraging them to join the network. In conformity with the *hirama* scenario, all network members are expected to work in parallel on several of the subproblems. The psychological leader's role is to stimulate the members' creative activity, and the "external affairs" leader should get in touch with other similar organizations.

3.9.2. Genetic Engineering-Related Biopolitical Issues

In recent decades, genetic engineering has been applied to various living beings including microorganisms, fungi, plants, and animals. Modern biotechnology and medicine benefit from recent developments in the field of genetic engineering. Genetic engineers have enabled the cells of bacteria (e.g., *E. coli*), yeast, insects such as the silkworms, and a variety of other species to produce human-specific proteins. They include insulin, interferons, antithrombogenic factor VIII, growth hormone, plasminogen, and other substances used as drugs. Such developments are collectively referred to as the "DNA industry" in the field of pharmacology. Recent developments in genetic engineering have caused a large number of ethical, legal, and political problems; these multifaceted, interdisciplinary issues call for the involvement of networks such as *hirama*-type structures that include partial leaders concentrating on different specific aspects of these problems. One of the main concerns is the threat of accidentally and—still worse—intentionally producing genetic monsters. This threat was already discussed during a conference on genetic technologies at Asilomar (USA) in 1975. However, the threat still exists in the present-day world.

In EU countries, the legal framework is set by the still valid *Directive 2001/18/EC of the European Parliament and of the Council of 12 March 2001 on the deliberate release into the environment of genetically modified organisms and repealing Council Directive 90/220/EC* or, in short, the *GMO Directive* (2001), emphasizing that "the protection of human health and the environment requires that due attention be given to controlling risks from the deliberate release into the environment of genetically modified organisms (GMOs)." If a new project in the field of genetic engineering is in progress, the main responsibility for its biosafety is taken by the leader of the research team that carries it out. The team leader is assisted by biosafety experts affiliated with a "competent authority" that "organises inspections and other control measures as appropriate, to ensure compliance with this Directive" (GMO Directive, 2001). The biosafety principle-implementing structure becomes more decentralized if representatives of public movements and organizations step in.

Genetic engineering-related issues have recently become much more socially/politically important because (1) the large-scale production of transgenic organisms has started; (2) animals located close to humans on the evolutionary stepladder (monkeys) have been used as research subjects (for example, in 2009, a squirrel monkey was cured of color-blindness with the help of a viral vector); and (3) research on human DNA has made genetic diagnostics and, to an extent, therapy technologically feasible (Blank & Merrick, 1995; Ryabchuk, 2005). *Genetic therapy*, i.e., curing hereditary diseases by changing the DNA, "raises many policy issues regarding what role the government ought to play in encouraging or discouraging such research and application" (Blank & Hines, 2001, p.90). Genetic therapy applied to practically healthy people (*genetic enhancement*), is particularly problematic.

The functions that could be performed by biopolitical network organizations/movements could include (a) sensitizing public opinion to (bio)political issues concerning genetic technologies and their applications and (b) encouraging competent specialists in various relevant areas (genetics, health care, environment, public policy, etc.) to participate in developing the appropriate policies. Albert Somit (1972, p.235) asserted in his seminal paper "Biopolitics" that biopoliticians could develop precise and reliable methods to monitor public opinion that concern important biological problems with political aspects, including genetic innovations and their social and political implications. Biopoliticians can help us dispel groundless fears that are inspired by sensation-seeking journalists and sometimes poorly informed politicians. Informing the people at large of the actual situation and of real—not invented—threats associated with genetic engineering and biotech developments, biopoliticians should ideally be concerned with the interests of all social classes and all regions of the world that are affected by genetic technologies—or by their unavailability.

An example is provided by the activities of the Danish Board of Technology (Teknologirådet). One of the goals of this organization is to help the Danish government assess the whole spectrum of issues caused by technological innovations. The Board holds conferences and organizes workshops dealing with a specific set of issues. For example, they "manage a project on genetically modified plants and food together with other technology assessment institutions in Europe. The project is carried out under the EPTA network (European Parliamentary Technology Assessment). The aim is to provide information on regulatory challenges, points of public debate in the future and approaches to technology assessment fitted to handle these issues" (Teknologirådet, 2007, http://www.tekno.dk). In organizational terms, the Danish Board is a network structure. Technology assessment involves contacting and consulting scientists, scholars, business people, and politicians.

In a similar fashion, networks can contribute to the assessment of various ethical and political issues related to *biomedical technologies*, such as organ transplantation, extracorporeal fertilization, surrogate motherhood, and others. Working on several issues in parallel (in a neural network-like fashion), even a single network structure with partial creative task-specialized leaders has the potential to deal with a large number of different aspects of biomedical developments and their ethical and political implications.

3.9.3. Researching Social Behavior in Biopolitical Terms

One of the main research directions in the field of biopolitics is focused upon elucidating "connections between our species' genetic constitution and our species' political behavior" (Carmen, 1997, p.173). Our evolutionary legacy still predisposes us to engage in aggressive or cooperative behavior, form hierarchies or dismantle them in favor of networks, and so on, and biopoliticians aim to unravel the mechanisms involved in actualizing these genetically fixed predispositions. Such biopolitical research is based on interdisciplinary studies that are to be conducted by a mixed collective of specialists including geneticists, neurophysiologists, ethologists, psychologists, political scientists, and sociologists. These issues are considered in detail in my recent book (Oleskin, 2012).

The Club of Biopolitics mentioned above is conducting research on a number of behavioral projects. One of the projects involving the *Network of Independent Scholars, Educators, and Experts in Biopolitics (NISEEB)*, which is affiliated with the Club, envisages "social technologies that involve human evolutionarily concerned behavioral trends and aim at mitigating human aggression, resolving political conflicts (particularly, ethnic clashes and civil unrest), and stimulating altruistic, loyal, cooperative behaviors on all levels of human society" (http://biopolitika.ru, click NISEEB). Beginning in 2012, the Club of Biopolitics, including the NISEEB, has been carrying out a research project concerning network structures, which is the subject of this book. This project is formulated as follows: *Hierarchical, network, and quasi-market structures. Their ratio in biological, social, technological, and ecological systems: optimizing the ratio for the purpose of solving practical problems*. Along with research activities, the Club of Biopolitics is planning to make practical use of the results of its research.

Network organizational principles can be implemented by organizations dealing with a wide spectrum of issues related to the interaction between the life sciences and politics (biopolitics). For example, network structures can deal with environmental problems; ethical and political concerns related to genetic technologies and biomedical developments; and various issues concerned with the relationship between evolution-molded behavioral predispositions, human genetic characteristics, and political behavior. All these biopolitical issues are on the agenda for a number of global and local biopolitical networked organizations.

Conclusion

To reiterate, the term "network structure" is used in the literature according to at least two different meanings. The broader meaning (denoted by this author as a network *sensu lato*) refers to any system composed of nodes (vertices) connected by links (edges). In contrast, a network structure, per its narrow meaning, is a decentralized, non-hierarchical, cooperative system (a network *sensu stricto*). Only in its broader meaning has this term been used with respect to objects unrelated to human society, including various biological species. Research on these species in terms of network analysis has indisputably yielded valuable results related to various kinds of systems. The analytical tools that deal with centrality measures, clustering- and community structure-related criteria, small world behavior, and other network characteristics have provided important insights into the organization and functioning of various objects including biological systems (see, e.g., Barabási, 2002; Barabási & Oltvai, 2004; Almaas et al., 2007; Croft et al., 2008; Newman, 2003, 2012; Newman et al., 2006; Wey et al., 2008).

In contrast, the concept of a network structure as a non-hierarchical cooperative system has been relatively widely used—to the author's knowledge—only with respect to human social structures.

A novel feature of this work is that the "*sensu stricto*" network concept is applied, apart from human society, to biological systems. Network structures both in human society and in the biological realm are considered in conjunction with other types of structures, i.e., hierarchical and quasi-market (competitive) structures, which constantly interact with networks and often form combined or mixed systems; in addition, the three different types of structures can interconvert.

This work is also aimed at demonstrating how our knowledge concerning biological network structures can contribute to our understanding of the networks that are spontaneously or deliberately formed in human society.

This type of comparative approach is related to modern *biopolitics*, which was defined at the beginning of this work as "the totality of all kinds of interactions between the life sciences and politics" (Oleskin, 2012). Biopolitics includes, as an important subfield, research concerning the political potential of biology, i.e., "the approach of those political scientists who use biological concepts, with neo-Darwinian evolutionary theory at the center, and biological research techniques to study, explain, predict, and sometimes even to prescribe political behavior" (Somit & Peterson, 2011, p.3).

On the basis of the biopolitical comparative approach to network structures, this work demonstrates that non-hierarchical, cooperative networks do not represent a principally new social phenomenon that manifested itself in human society towards the end of the 20th century. It is more than just "the third alternative" to hierarchies such as corporate bureaucracies and market structures in contemporary Western society. Despite the obvious importance of online social networks, it is evident that network structures exist not only in the virtual world.[114].

In a number of biopolitical works, it is stressed that human nature combines biological and nonbiological (social, cultural, spiritual, etc.) components. Significantly, network structures reflect not only the cultural influences of a particular historical period, but also humankind's prehistory. The structural features of many modern-day networks are still similar to those characteristic of primitive hunter–gatherer bands. To the extent that the Internet is a *sensu stricto* network, it appears to revive some features of such archaic egalitarian[115] structures with partial leaders specializing in hunting, fishing, dancing, faith-healing, warfare, etc., which were primitive analogs of present-day moderators in virtual networks. However, they did not invent networks either because their establishment—in conformity with various paradigms—is also within the ability of chimpanzees, bonobos, white-faced capuchins, and some other primates, as well as schooling fish, social insects, colonial cnidarians, and even biofilm-forming unicellular organisms.

Apart from networks, various biological systems form other types of structures, i.e., hierarchies and competition-dominated quasi-market structures. Different structure types interact in a complex manner and can interconvert. Some primitive human societies apparently included not only egalitarian structures but also rigid hierarchies. The argument concerning the relative importance of egalitarian and more hierarchical structures in primitive societies, as well as in other biological and social systems, is ongoing in the literature including in biopolitical works (see, e.g., Boehm, 1993, 1997, 1999, 2004; Somit & Peterson, 1997, 2001b; Corning, 2003a, b; Summers, 2005; Ames, 2010).

Data on the existence of multiple organizational patterns both in biological systems and (primitive) human society highlight a biopolitically important point: evolution has endowed us with *mixed (ambivalent)* propensities. Importantly, we are predisposed to both agonistic (hostile, aggressive) and loyal behavior and to the establishment of hierarchical and horizontal, cooperative and predominantly competitive (market-type) structures. The idea that many (bio)social systems are polystructural, i.e., combine different organizational principles (see Zakharov, 1987, 1991) also applies to human society insofar as human society combines hierarchies, networks, and (quasi-)markets. Each structural type can be subdivided into several subtypes. Networks can be flat or three-dimensional and dense or loose (see 1.2); moreover, hierarchies include agonistic and hedonic structures (Chance, 1967 and see 1.3.2). In addition, markets can be regulated or not. Both biologists and scholars in the social sciences can be involved in studying the mechanisms of interconversion associated with the three structure types of interest and their subtypes.

---

114 As noted above, the World Wide Web does not only include *sensu stricto* networks. It is characterized by a tendency of hierarchization with the emergence of a clandestine power structure (see Bard & Söderqvist, 2002) and also by the development of quasi-market structures marked by competitive interactions between users or whole online networks.

115 There are networks in modern human society that are related to other biological paradigms than the egalitarian paradigm.

The extrapolation of data obtained with human social structures to biological systems seems to be of considerable importance to biologists. Generally, a large number of concepts and terms originally used in the humanities and social sciences have recently been applied to biological systems. This is exemplified by the introduction of such originally "human" terms as "altruism" or "quorum" into biology. The "human" terms "caste," "queen," or "retinue" have already been in use for a long time with regard to social insect systems. Ethologists traditionally used concepts such as "dominance" and "leadership" in works on various animal species, including invertebrates.

Comparative research on networks in human society and in the biological realm actually contributes to this process of interdisciplinary concept/term transfer. As relevant (sub)sections of the work demonstrate, various paradigms of network structures in biological species, ranging from bacteria to apes, have their analogs in human society. This enables a biologist to view these biological systems from a novel perspective, taking into account relevant knowledge concerning human social structures.

As far as the main subject of this work is concerned, not only is the *applicability* of network-related concepts to biological systems of relevance, per se; of interest is also the fact that there are multiple subtypes (paradigms) of decentralized cooperative structures in the biological realm. Different variants of networks are denoted in this work by different terms that are routinely used in the literature. Accordingly, there are cellular, modular, equipotential, eusocial, neural, and egalitarian networks (Table 2 in 2.7); the author admits that other paradigms may exist that are not considered in this work. There is no strict correspondence between network paradigms and taxonomic groups. Some of these paradigms apply to a relatively wide variety of species. In addition, the social structures of some species can be described in terms of more than one network organization paradigm.

The diversity of network structure types (paradigms) in biological systems might provide some food for thought not only to biologists, but also to scholars in the social sciences and the humanities. The implication is that "network structures," a popular cliché in present day society, can be subdivided into a number of substantially different organizational variants. Before promoting network structures in various areas ranging from the worldwide Web to networked businesses, we should decide what kind of networks are more useful in terms of our specific goals.

For instance, a neural paradigm-like scenario has significant advantages in terms of interactive teaching in the classroom (3.6.1), and the human analog of the member individuality-enhancing egalitarian ("ape") paradigm could be preferred if our goal is to set up a network composed of a small number of important individuals, such as prominent scientists (3.5) or business managers (3.7).

Therefore, relevant biological data provide important knowledge that is applicable not only to biology, but also to the humanities and the social sciences. The information obtained by exploring network structures in living systems can be creatively used while developing network structures in human society. Such network structures can also be used to pursue different goals, ranging from increasing the efficiency of the educational system to helping HIV-infected patients and to reforming political structures.

Despite the diversity of potential applications of network structures in human society, they all are predisposed—owing to their organizational pattern—to approach any issue from an interdisciplinary and systemic perspective, taking into account many different aspects of the issue at once (in an analogy to a neural network). Irrespective of their specific focus, network structures in human society can produce important social effects caused by their structural properties, which can lead to positive economic and psychological consequences (building up trust and social capital, developing a sense of belonging and collectivism, promoting mutual support and informal personal relationships among the members, and so on), as noted in several sections of this work.

In political terms, decentralized, cooperative networks tend to promote direct participatory democracy, regardless of whether they protect the environment, implement interactive teaching methods, or help the government improve the political regime (by playing the role of a think tank). In each of the domains of modern-day society, including the political and the business sphere, the network structures and civil society predominantly formed by them can develop and implement scientifically based approaches and expert assessment methods, provided that these structures include a core composed of highly qualified professionals and intelligentsia in general.

Politically active members of such networks form part of a nation's *meritocracy.* The idea that socially recognized professionals should enjoy high social/political status dates back to the teachings of Confucius and Plato. However, unlike the system envisaged by these visionaries, and which has been revitalized and implemented in various parts of the modern world (e.g., in Singapore), it is not a group of high priests or a government-appointed commission that assigns this status to them. Rather, it is the network structure-based civil society that empowers socially recognized experts to contact the central government on its behalf with regard to political, economic, social, cultural, or environmental problems. Thanks to the support of network structures within civil society, a number of scientists and scholars have made successful political careers.

In business, the potential impact of network structures actually amounts to the promotion of some basic *socialist* ideas and values, despite their capitalist environment. This primarily concerns the subtype of small-scale business networks called *cooperatives* (*co-ops*; see, 3.7.2.3). A large number of these businesses explicitly support the socialist principle of collective ownership. However, many other networked businesses, regardless of the goals of their developers, also comply with socialist rather than capitalist principles to the extent that:

- In economic terms, their development entails partial collectivization of the property of their member firms—i.e., joint control over some of their assets—whereas capitalism per se, at least in its classical form, is based upon the principle of predominantly private property.
- In social terms, relationships among business network members are mostly based neither on hierarchical principles (typical of the capitalist firm described, e.g., in the classical work of Ronald Coase, 1937), nor on market principles that involve self-interest and, in some situations, even guile and strategic misrepresentation rather than *trust, long-term loyalty, and social capital* that are essential for the survival and successful functioning of most network structures.

Some of the modern-day business networks evoke the principles of *self-governed socialism* that is based upon the operation of autonomous self-regulated economic actors—a decentralized mechanism of making economic and political decisions—and upon collective ownership with respect to the means of production used at the enterprise level.

A further important aspect of networks' mission in the present-day world should be reemphasized. Since network structures promote cooperative bonds among people, they mitigate the feelings of loneliness, alienation, and helplessness that are unfortunately widely spread among people in present-day society. These networks are expected to counterbalance the undoubtedly important hierarchical structures that are characteristic of political systems, as well as the omnipresent, competition-promoting (quasi-)market structures of the capitalist world. Each of these nonnetwork structures has its own advantages and performs its own functions in human society. However, it is the non-hierarchical cooperative networks that this work is mainly focused upon.

As for network structures that are concerned with various aspects of biopolitics (as exemplified by the *Club of Biopolitics* that forms part of the Moscow Society of Natural Scientists), they actually tend to promote the idea that humans are "citizens of the biosphere," or representatives of the Earth's biodiversity. However, unlike other species, humans carry out a special global mission with respect to both the whole biosphere and their own society (Oleskin, 2012). In particular, many biopoliticians support the democratic pluralism principle in human society that emphasizes and legitimizes the diversity of humankind in terms of social ideas, ethical values, and political views —a human analog of the biodiversity principle used by nature on the planet. Biological facts also lend support to the idea that evolution has predisposed us to cooperate with one another, which conforms to the organizational principles of many structures formed by various biological species.

In sum, this work demonstrates the applicability of the term "network structure" in its narrow sense (a decentralized, non-hierarchical structure with predominantly cooperative interactions among its elements/nodes) to biological systems. Importantly, such systems differ in their organizational principles and they can, therefore, be subdivided into at least six different variants (paradigms). The diversity of network paradigms in biological systems provides food for thought not only to biologists, but also to scholars in the social sciences and the humanities. Before promoting network structures in various domains of society, we should decide what kinds of networks should be more useful in terms of our specific goals. Of paramount importance are the potential positive social and political effects of network structures in society, including the development of civil society and direct participatory democracy, as well as the enhancement of the social prestige and political weight of the intellectuals.

References

Adang, O. M. J. (1999). *De machtigste chimpanzee van Nederland: leven en dood in mansapengemeenschap.* Amsterdam: Nieuwezijds.

Adler, N. (2001). Competition in a deregulated air transportation market. *European Journal of Operational Research, 129(2),* 108–116.

Aguilar, C., Vlamakis, H., Losick, R., & Kolter, R. (2007). Thinking about *Bacillus subtilis* as a multicellular organism. *Current Opinions in Microbiology, 10,* 638-643.

Akaizin, E. O., Voskun, S. E., Panova, L. A., & Smirnov, S. G. (1990). The heterogeneity of an *Escherichia coli* population during induced autolysis. *Microbiology (Moscow), 59(2),* 283-288.

Alberts, D. S., Garstka, J. J., & Stein, F. P. (2000). *Network Centric Warfare: Developing and Leveraging Information Superiority.* CCRP Publication Series.

Alexander, R.D. (1979). *Darwinism and Human Affairs.* Seattle: University of Washington Press.

Alexeenkova, E. A. (2005). The formation and evolution of hierarchized terrorist networks in the example of Italian Red Brigades. *Political Studies (Polis), No. 6,* 53-65.

Almaas, E., Vázquez, A., & Barabási, A.-L. (2007). Scale-free networks in biology. In: F. Képés (Ed.), *Biological Networks. Complex Systems and Interdisciplinary Science. Vol. 3* (pp.1-21). Singapore, Hackensack (NJ), & London: World Scientific Publishing Co. Pte. Ltd.

Alperovitz, G. & Bhatt, K. (2013). *Ten Ways to Democratic economy.* http://truth-out.org/opinion/item/18908-what-then-can-i-do-ten-steps-toward-transforming-the-system

Ames, K. M. (2010). On the evolution of the human capacity for inequality and/or egalitarianism. In: T. D. Price & G. M. Feinman (Eds.), *Pathways to Power* (pp. 15-44). Berlin & Heidelberg: Springer Science+Business Media.

Amnesty International. (2014). Official Website. http://www.amnesty.org/en.

Anderson, C. & Brown, C. E. (2010). The functions and dysfunctions of hierarchy. In: A. P. Brief & B. M. Staw (Eds.), *Research in Organizational Behavior. Vol.30* (pp. 55-89). Greenwich, CT: JAI Press.

Anderson, M., Haar, J. & Gibb, J. (2010). Personality trait inferences about organizations and organizational attraction: an organizational-level analysis based on a multi-cultural sample. *Journal of Management and Organization, 16,* 140.

Angelo, T. A. & Cross, K. P. (1993). *Classroom Assessment Techniques.* San Fransisco: Jossy-Bass.

Angelucci, A. & Sainsbury, K. (2006). Contribution of feedforward thalamic afferents and corticogeniculate feedback to the spatial summation area of macaque V1 and LGN. *Journal of Comparative Neurology, 498 (3),* 330–351.

AntEra. (2013). *AntEra Association. Institute of Clinical Medicine and Social Work n.a. Maxim P. Konchalovsky.* http://celenie.ru/index_eng.html.

Anuchin, A. M., Chuvelev, D. I., Kirovskaya, T. A., & Oleskin, A.V. (2008). Effects of monoamine neuromediators on the growth-related variables of *Escherichia coli* K-12. *Microbiology (Moscow), 77,* 674-680.

Aoki, S. K., Webb, J. S., Braaten, B. A., & Low, D. A. (2009). Contact-dependent growth inhibition causes reversible metabolic downregulation in *Escherichia coli. Journal of Bacteriology, 191(6),* 1777-1786.

APLS. (2014). *Association for Politics and the Life Sciences. Official Website.* http://www.aplsnet.org.

Appleby, M. C. (1983). The probability of linearity in hierarchies. *Animal Behaviour, 31,* 600–608.

Arcadi, A. C., Robert, D., & Boesch, C. (1998). Buttress drumming by wild chimpanzees: Temporal patterning. Phrase integration into loud calls and preliminary evidence for individual distinctiveness. *Primates, 39,* 505–518.

Arenas, A. Díaz-Guilera, A., Kurths, J., Moreno, Y., & Zhou, C. (2008). Synchronization in complex networks. *Physics Reports, 469,* 93-153.

Arshinov, V. I. (2003). Synergetics as a networked thinking paradigm. In: *The Network and Biopolitics as Metaphors of Interdisciplinary Philosophy* (pp.21-39). Moscow: Academy of Innovative Management.

Artemieva, O. Yu. (2002). Initial phases of politogenesis. In: D. M. Bondarenko & A. P. Koroteev (Eds.), *Civilizational Models of Politogenesis* (pp. 58-74). Moscow: Center for Civilizational and Regional Studies of the RAS, Institute for African Studies of the RAS.

Attolini, C. S.-O. & Stadler, P. F. (2006). Evolving towards the hypercycle: a spatial model for molecular evolution. *Physica D, 217,* 134-141.

Axelrod, R. (1984). *The Evolution of Cooperation.* New York: Basic Books Inc. Note: R. Axelrod is both the editor and the only author of the book, except for the chapter coauthored by W. D. Hamilton (see reference below).

Axelrod, R. & Hamilton, W. D. (1984). The evolution of cooperation in biological systems. In: R. Axelrod (Ed.), *The Evolution of Cooperation* (pp.88-105). New York: Basic Books Inc.

B.I.O. (2014). *Biopolitics International Organisation. Official Website* http://www.biopolitics.gr.

Badescu, G. & Uslaner, E. (2003). Introduction. In: G. Badescu & E. Uslaner (Eds.), *Social Capital and the Transition to Democracy* (pp.1-9). London & New York: Routledge.

Bai, A. P. & Ouyang Q. (2006). Probiotics and inflammatory bowel diseases. *Postgraduate Medical Journal,* 82, 376-382.

Baker, W. E. (1994). *Networking Smart. How to Build Relationships for Personal and Organizational Success.* New York & San Fransisco: Mc. Graw-Hill Inc.

Balakrishman, S. N. & Weil, R. D. (1996). Neurocontrol: a literature survey. *Mathematical & Computer Modelling, 23(1/2),* 101-117.

Barabási, A.-L. (2002). *Linked: The New Science of Networks.* New York: Perseus.

Barabási, A.-L. & Albert, R. (1999). Emergence of scaling in random networks. *Science, 286(5439),* 509–512.

Barabási, A.-L. & Oltvai, Z. N. (2004). Network biology: understanding the cell's functional organization. *Nature Reviews Genetics, 5,* 101-113.

Bard, A., & Söderqvist, A. (2002). *Netocracy: The New Power Elite and Life after Capitalism.* London, New York, & Toronto: Pearson Education. Ltd.

Barkow, J. H. (1982). Return to nepotism. The collapse of a Nigerian gerontocracy. *International Political Science Review, 3(1),*33–49.

Barlow, J. W. (1989). Has sociobiology killed ethology or revitalized it? In: *Whither Ethology? Perspectives in Ethology* (pp.1-45). New York & London: Plenum Press.

Barner-Barry, C. (1977). An observational study of authority in a preschool peer group.*Political Methodology, No. 4,* 415–449.

Barner-Barry, C. (1983). Zum Verhältnis zwischen Ethologie und Politik: Macht, Dominanz, Autorität und Aufmerksamkeitsstruktur In: H. Flohr & W. Tönnesmann (Eds.), *Politik und Biologie. Beiträge zur Life-Science-Orientierung der Sozialwissenschaften* (S.101-111). Berlin & Hamburg: Verlag Paul Parey.

Barnes, J. (1954). Class and committees in a Norwegian island parish. *Human Relations, 7,* 39–58.

Barthélemy, M. (2011). Spatial networks. *Physics Reports, 499,* 1–101.

Barton, D. G. (1989). What is a cooperative? In: *Cooperatives in Agriculture* (pp.1-19). New Jersey: Englewood Cliffs.

Bechara, A. (2002). Editorial. The neurology of social cognition. *Brain, 125(8),* 1673–1675.

Beck, K., Grenning, J., Martin, R. C. & et al. (2001). *Manifesto for Agile Software Development.* http://agilemanifesto.org.

Beklemishev, V. N. (1950). On the issue of the individual in biology. The colony concept in animals with bilateral symmetry. *Uspekhy Sovremennoi Biologii (Advances in Modern Biology), 29(1),* 91-120.

Bell, D. (1973). *The Coming of Post-Industrial Society. A Venture in Social Forecasting.* New York: Basic Books.

Bell, D. & Chanyang, L. (2013). Compassionate meritocracy: the Singapore model. *Singapolitics.* http://www.singapolitics.sg/views/compassionate-meritocracy.

Ben-Jacob, E. (1998). Bacterial wisdom. *Physica A., 249,* 553-577.

Ben-Jacob, E. (2003). Bacterial self-organization: co-enhancement of complexification and adaptability in a dynamic environment. *Philosophical Transactions of the Royal Society. A,. 361,* 1283–1312.

Ben-Jacob, E., Becker, I., Shapira, Y., & Levine, H. (2004). Bacterial linguistic communication and social intelligence. *Trends in Microbiology, 12(8),* 366–372.

Ben-Jacob, E., Coffey, D. S., & Levine, H. (2012). Bacterial survival strategies suggest rethinking cancer cooperativity. *Trends in Microbiology, 20(9),* 403-410.

Ben-Rafael, E. (1997). *Crisis and Transformation: The Kibbutz at Century's End.* Albany, NY: State University of New York Press.

Berent, M. (2002). Greece (11-4th century B.C.). In: *Civilizational Models of Politogenesis* (pp.223-247). Moscow: Center for Civilizational and Regional Studies of the RAS, Institute for African Studies of the RAS.

Berkman, A. (1929). *Now and After: The ABC of Communist Anarchism.* New York: Vanguard Press.

Bernhard, J. G., & Glantz, K. (1992). *Staying Human in the Organization. Our Biological Heritage and the Workplace*. Westport, CT & London: Praeger.

Bestuzhev-Lada, I. (1990). Whither socialism in the 21st century? *Futures, January/February,* 57-68.

Beveridge, T. J. (1999). Structures of gram-negative cell walls and their derived membrane vesicles. *Journal of Bacteriology, 181,* 4725-4733.

Bezrukova, O. (2004). Network interactions among agents involved in business transactions: basic principles and implementation strategies. In: *Online Conference on Network Forms of Interfirm Integration: Strategic Challenges & Competitive Advantages of New Organizations in the 21st Century.* March 10–April 29, 2004. http://www.ecsocman.edu.ru.

The Bible. King James Version. (2011). Cambridge: Cambridge University Press.

Bingham, P. (1999). Human uniqueness: a general theory. *Quarterly Review of Biology, 74,* 133–169.

Birchall, J. (2004). *Cooperatives and the Millennium Development Goals.* Geneva: International Labour Office. Comnission for the promotion and Advancement of Cooperatives.

Blango, M. G. & Mulvey, M. A. (2009). Bacterial landlines: contact-dependent signaling in bacterial populations. *Current Opinion in Microbiology, 12,* 177–181.

Blank, R. H., & Hines, S. M. (2001). *Biology and Political Science.* London: Routledge.

Blank R. H, & Merrick J. C. (1995). *Human Reproduction: Emerging Technologies and Conflicting Rights.* Washington (DC): CQ Press.

Boehm, C. (1993). Egalitarian society and reverse dominance hierarchy. Current Anthropology, 34, 227–254.

Boehm, C. (1997). Egalitarian behavior and the evolution of political intelligence. In: R. Byrne & A. Whiten (Eds.), *Machiavellian Intelligence II: Extensions and Evaluations* (pp 341–364). Cambridge: Cambridge University Press.

Boehm, C. (1999). *Hierarchy in the Forest.* Cambridge, MA: Harvard University Press.

Boehm, C. (2004). What makes humans economically distinctive? A three-species evolutionary comparison and historical analysis. *Journal of Bioeconomics, 6,* 109-135.

Boerlijst, M. C. & Hogeweg, P. (1991). Spiral-wave structure in pre-biotic evolution: hypercycles stable against parasites. *Physica D, 48,* 17-28.

Bogdanov, A. A. (1980 [1921]). *Essays in Tektology: The General Science of Organization.* Seaside, CA: Intersystems Publications.

Bohlhalter, S., Fretz, C., & Weeter, B. (2002). Hierarchical versus parallel processing in tactile object recognition: a behavioural-neuroanatomical study of perceptive tactile agnosia. *Brain, 125(11),* 2537–2548.

Bonabeau, E., Dorigo, M., & Theraulaz, G. (2000). Inspiration for optimization from social insect behavior. *Nature, 406,* 39-42.

Bondarenko, D.M. (2004). Alternativity in cultural history: heterarchy and homarchy as evolutionary trajectories. In: *Hierarchy and Power in the History of Civilization* (p. 5). Moscow: Center for Civilizational and Regional Studies of the RAS, Institute for African Studies of the RAS.

Borgatti, S. P. & Foster, P. C. (2003). The network paradigm in organizational research: a review and typology. *Journal of Managemant, 29(6),* 991-1013.

Borgatti, S. P., Everett, M. G., & Johnson, J. C. (2013). *Analyzing Social Networks.* London, Thousand Oaks (CA), New Delhi, & Singapore: SAGE Publications, Ltd.

Borradaile, L. (2006). *Forest school Scotland: An evaluation. Report to Forestry Commission Scotland and Forest Education Initiative Scotland.* http://www.forestry.gov.uk/pdf./forestschoolfinalreport.pdf/$file/forestschoolfinalreport.pdf.

Börzel, T. (1998). Organizing Babylon — on the different conceptions of policy networks. *Public Administration, 76,* 253–273.

Botvinko, I. V. (1985). Bacterial exopolysaccharides. *Uspekhi Mikrobiologii (Advances in Microbiology), 20,* 79-122.

Brandner, J. P. & Kroos, L. (1998). Identification of the Ω 4400 regulatory region, a developmental promoter of *Myxococcus xanthus. Journal of Bacteriology, 180,* 1995-2002.

Brehm, A. (2001). *Reptiles.* Moscow: AST.

Bresch, C, Niesert, U, & Harnasch, D. (1980). Hypercycles, parasites and packages. *Journal of Theoretical Biology, 85(3),* 399–405.

Bridges, E. L. (1948). *Uttermost Part of the Earth.* London: Hodder & Stoughton.

Briggs, J. L. (1970). *Never in Anger.* Cambridge (MA): Harvard University Press.

Brusco, S. (1982). The Emilian model: Productive decentralization and social integration. *Cambrian Journal of Economics, 6,* 167-184.

Budrene, E. O. & Berg, H. C. (1991). Complex patterns formed by motile cells of *Escherichia coli. Nature, 349,* 630-633.

Budrene, E. O. & Berg, H. C. (2002). Dynamics of formation of symmetrical patterns by chemotactic bacteria. *Nature, 376,* 49-53.

Buhl, J., Gautrais, J., Deneubourg, J. L., & Theraulaz, G. (2004). Nest excavation in ants: group size effects on the size and structure of tunnelling networks. *Naturwissenschaften, 91,* 602–606.

Buhl, J., Deneubourg, J. L, Grimal, A., & Theraulaz, G. (2005). Self-organized digging activity in ant colonies. *Behavioral Ecology & Sociobiology, 58,* 9–17.

Bunce, V. & Csanádi, M. (2007). Social networks: ideas for future research from the vantage point of political science. *Physica A, 378,* 135-137.

Burt, R. S. (1992). *Structural Holes: The Social Structure of Competition.* Cambridge, MA: Harvard University Press.

Butovskaya, M. L. (2000). The evolution of human behaviour: the relationship between the biological and the social. *Anthropologie, 38(2),* 169–180.

Butovskaya, M. L. (2002). Biosocial premises of social and political alternativity. In: *Civilizational Models of Political System Formation* (pp.35-57). Moscow: Center for Civilizational and Regional Studies of the RAS, Institute for African Studies of the RAS.

Butovskaya, M. L. (2013). Aggression and reconciliation as manifestations of sociality in primates and humans. http://staff.ttu.ee/~bgordon/oppetegevus/valjendus/ref_rus/orig_article_6.pdf.

Butovskaya, M. L., & Fainberg, L. A. (1993). *At the Roots of Human Society.* Moscow: Nauka.

Byrnes, R. W. & Bates, L. A. (2007). Sociality, evolution, and cognition. *Current Biology, 17,* R714–R723.

Byrnes, R. W. & Bates, L. A. (2010). Primate social cognition: uniquely primate, uniquely social, or just unique? *Neuron, 65,* 815-830.

Caldwell, L. K. (1964). Biopolitics: science, ethics and public policy. *The Yale Review, 54,* 1-16.

Camazine, S., Deneubourg, J. L., Franks, N. R., Sneyd, L., Theraulaz, G., & Bonabeau, E. (2001). *Self-organization in Biological Systems. Princeton Studies in Complexity.* Princeton, NJ: Princeton University Press.

Carbonaro, M., Cutumisu, M., Duff, H., Gillis, S., Onuczko, C., Siegel, J., Schaeffer, J., Schumacher, A., Szafron, D., & Waugh, K. (2008). Interactive story authoring: A viable form of creative expression for the classroom. *Computers & Education, 51,* 687-707.

Carmen, I. H. (1997). Biopolitics: the newest synthesis? *Genetica, 99,* 173-184.

Carmona-Fontaine, C.., Theveneau, E., Tzekou, A., Tada, M., Woods, M., Page, K. M., Parsons, M., Lambris, J. D., & Mayor, R. (2011). Complement fragment C3a controls mutual cell attraction during collective cell migration. *Developmental Cell, 21,* 1026–1037.

Carneiro, R. L. (1970). A theory of the origin of the state. *Science, 169,* 733-738.

Caruso, R. & Schneider, F. (2013). Brutality of Jihadist terrorism. A contest theory perspective and empirical evidence in the period 2002–2010. *Journal of Policy Modeling, 35(5),* 685-696.

Castells, M. (1996). *The Rise of the Network Society, The Information Age: Economy, Society and Culture. Vol. I.* Cambridge, MA & Oxford, UK: Blackwell.

Castells, M. (2000). Materials for an explanatory theory of the network society. *British Journal of Sociology, 51(5),* 5-24.

Castells, M. (2001). *The Internet Galaxy: Reflections on the Internet, Business, and Society.* Oxford (UK): Oxford University Press.

Castells, M. (2004). Informationalism, networks, and the network society: a theoretical blueprint. In: M. Castells (Ed.), *The Network Society: a Cross-Cultural Perspective* (pp.3-45). Northampton, MA: Edward Elgar.

Castells, M. (2009). *Communication Power.* Oxford, New York, Auckland et al.: Oxford University Press.

Caton, H. (1998). Reinvent yourself: labile psychological identity and the lifestyle marketplace. In: I. Eibl-Eibesfeldt & F. K. Salter (Eds.), *Indoctrinability, Ideology and Warfare* (pp.325-343). New York & Oxford: Berghahn Books.

CBE (Commission for Biological Education) (2014). *IUBS Commission for Biological Education.* Official Website. http://iubscbe.org.

CBE International (2014). *Christians for Biblical Equality.* Official Website. http://www.cbeinternational.org.

Chan, Z. C. Y. (2013). A systematic review of creative thinking/creativity in nursing education. *Nurse Education Today, 33(11),* 1382-1387.

Chance, M. (1967). Attention structure as the basis of primate rank orders. *Man, 2(2),* 503-518.

Chang, L.-C. & Lee, G. C. (2010). A team-teaching model for practicing project-based learning in high school: Collaboration between computer and subject teachers. *Computers & Education, 55,* 961-969.

Chappel, K. & Croft, A. (2009). Creative science teaching labs: New dimensions in CPD. *Thinking Skills & Creativity, 4,* 44-59.

Charney, I. & Palgi, M. (2013). Sorting procedures in enclosed rural communities: admitting 'people like us' into renewing kibbutzim in Northern Israel. *Journal of Rural Studies, 31,* 47-54.

Chauvin, R. (1963). *Les societiés animals de l'abeille au gorille.* Paris: Plon.

Chen, H. & Fink, G. R. (2006). Feedback control of morphogenesis in fungi by aromatic alcohols. *Genes & Development, 20,* 1150—1161.

Chernyshova, M. P. (2009). Using the modular-network organizational principle in terms of a course of lectures in Physiology. In: E. S. Lobakova et al. (Eds.), *The Life Sciences and Education. Fundamental Integration Problems* (pp.343-347). Moscow: MAX Press.

Chrisholm, D. (1989). *Coordination without Hierarchy. Informal Structures in Multiorganizational Systems.* Berkeley, Los Angeles, & London: University of California Press.

Chuchkevich, M. M. (1999). *Fundamentals of Network Organization Management.* Moscow: Institute of Sociology of RAS.

Churakov, A. I. (2001). Analysis of social networks. *Sociological Studies (Socis), No.1,* 109–121.

Club of Biopolitics (CB). (2014). Biopolitics Website. http://biopolitika.ru, click Club of Biopolitics.

Club of Innovative Development (CID). (2014). *CID Official Website.* Moscow: Institute of Philosophy, Russian Academy of Sciences,http://www.reflexion.ru/club/.

Coase, R. (1937). The nature of the firm. *Economica, 4,* 386-405.

Cohn, M. (2010). *Succeeding with Agile: Software Development Using Scrum.* Upper Saddle River, NJ, Boston etc.: Addison-Wesley.

Collins English Dictionary. (2009). http://dictionary.reference.com /browse/hierarchy.

Connor, R. C., Smoker, R. A., & Bejder, L. (2006). Synchrony, social behaviour and alliance affiliation in Indian Ocean bottlenose dolphins (*Tursiops aduncus*). *Animal Behaviour, 72,*1371–1378.

Cook, G. (2004). Heterarchy and homoarchy in Maya village politics. In: I. L. Alexeev, D. D. Beliaev, & D. M. Bondarenko (Eds.), *Hierarchy and Power in the History of Civilization* (pp.9-10). Moscow: Center for Civilizational and Regional Studies of the RAS, Institute for African Studies of the RAS.

Cooperative Grocer Network. (2014). http://www.cooperativegrocer.coop.

Corning, P. A. (1983). *The Synergism Hypothesis. A Theory of Progressive Evolution.* N.Y., St. Louis, San Fransisco & Auckland: McGraw-Hill.

Corning, P. A. (2000). The synergism hypothesis or the concept of synergy and its role in the evolution of complex systems. *Journal of Social and Evolutionary Systems, 24(2),* 133-172.

Corning, P. A. (2003a). *Nature's Magic. Synergy in Evolution and the Fate of Humankind.* Cambridge (Mass.): Cambridge Univ. Press.

Corning, P. A. (2003b). *The Evolution of Politics.* Presentation at the annual meeting of the Association for Politics and the Life Sciences. Palo Alto (CA).

Corning, P. A. (2004). Synergy and the evolution of "superorganisms," past, present, and future. In: N. K. Udumyan (Ed.), *Challenge to Cognition. Strategies of Science Development in the Modern World* (pp.184-221). Moscow: Nauka.

Corning, P. A. (2005). *Holistic Darwinism: Synergy, Cybernetics and the Bioeconomics of Evolution*. Chicago, IL: University of Chicago Press.

Corning, P. A. (2007). Synergy goes to war: a bioeconomic theory of collective violence. *Journal of Bioeconomics, 9,* 109–144.

Cournot, A. (1838). *Recherches sur les principles mathematiques de la théorie de richésses.* Paris: Chez L' Hachette, Libraire de'l Université Royale de France.

Couzin, I. D., Krause, J., James, R., Ruxton, G. D., & Franks, N. R. (2002). Collective memory and spatial sorting in animal groups. *Journal of Theoretical Biology, 218,* 1–11.

Croft, D. P., Krause, J., & James, R. (2004). Social networks in the guppy (*Poecilia reticulata*). *Proceedings of the Royal Society London B (Biology Letters), 271,* S516–S519.

Croft, D. P., James, R., Ward, A. J. W., Botham, M. S., Mawdsley, D., & Krause, J. (2005). Assortative interaction and social networks in fish. *Oecologia, 143(2),* 211-219.

Croft, D. P., James, R., & Krause, J. (2008). *Exploring Animal Social Networks. Princeton University Press.* Princeton, NJ: Princeton University Press.

Cronhjort, M. B. & Blomberg, C. (1994). Hypercycles vs parasites in a two-dimensional partial differential equations model. *Journal of Theoretical Biology, 169,* 31–49.

Cronhjort, M. B. & Blomberg, C. (1997). Cluster compartmentalization may provide resistance to parasites for catalytic networks. *Physica D, 101,* 289–298.

Crossley, N., Prell, C., & Scott, J. (2009). Social network analysis: introduction to special edition. *Methodological Innovations Online, 4,* 1-7.

Crumley, C. L. (1995). Heterarchy and the analysis of complex societies. Im: R. M. Ehrenreich, C. L. Crumley, & J. E. Levy (Eds.), *Heterarchy and the Analysis of Complex Societies* (pp.1-6). Washington (D.C.): American Anthropological Association.

Cummins, D. D. (2001). The impact of the social environment on the evolution of mind. In: H. R. Holland III, (Ed.), *Conceptual Challenges in Evolutionary Psychology. Innovative Research Strategies* (pp.85–118). Dodrecht, Boston, & London: Kluwer Academic Publishers.

Danis, W. M., De Clerk, D., & Petricevic, O. (2011). Are social networks more important for new business activity in emerging than developed economies? An empirical extension. *Business Review, 20,* 394-408.

Davidson, K. (1970). The development of a cocoon star. *Astrophysics & Space Science, 6(3),* 422-435.

Davies, D., Jindal-Snape, D., Collier, C., Digby, R., Hay, P., & Howe, A. (2013). Creative learning environments in education – a systematic literature review. *Thinking Skills & Creativity, 8,* 80-91.

Davis, A., Gardner, B. B., & Gardner, M. R. (1941). *Deep South.* Chicago: University of Chicago Press.

Davydova, S. I., Mardar', I. B., & Usacheva, O. A. (2008). Leaders and networks in modern-day environmental movements. *Istoriya I sovremennost' (History & Modernity), No.2,* 117-127.

Dawkins, R. (1976). *The Selfish Gene.* New York: Oxford University Press.

de Sola Pool, I. & Kochen, M. (1978–1979). Contacts and influence. *Social Networks, 1(1),* 5–51.

de Vries, H., Stevens, J. M. G., & Verdaecke, H. (2006). Measuring and testing the steepness of dominance hierarchies. *Animal Behaviour, 71,* 585–592.

de Waal, F. B. M. (1982). *Chimpanzee Politics: Power and Sex among Males.* London: Jonathan Cape.

de Waal, F. B. M. (1996). *Good-Natured. The Origins of Right and Wrong in Humans and Other Animals.* Cambridge (Mass.), L.: Harvard University Press.

de Waal, F. B. M. (1997). *Bonobo: The Forgotten Ape.* Berkeley (CA): University of California Press.

de Waal, F. B. M. (2001). *The Ape and the Sushi Master: Cultural Reflections by a Primatologist.* N.Y.: Basic Books.

de Waal, F. B. M. (2006). *Primates and Philosophers; How Morality Evolved.* Princeton, NJ: Princeton University Books.

de Waal, F. B. M. (2008). Putting altruism back into altruism: the evolution of empathy. *Annual Review of Psychology, 59,* 279-300.

de Waal, F. B. M. (2009). *The Age of Empathy. Nature's Lessons for a Kinder Society.* New York: Harmony Books. Random House.

de Waal, F. B. M., & van Roosmalen, A. (1979). Reconciliation and consolation among chimpanzees. *Behavioral Ecology & Sociobiology,* 5, 55–66.

Deneubourg, J. L & Goss, S. (1989). Collective patterns and decision-making. *Ethology, Ecology & Evolution, 1,* 295–311.

Derix, G. (2000). *The Vision Web: op reis naar's werelds spannendste ondernemingsvorm.* Schiedam: Scriptum.

Deryagina, M. A., & Butovskaya, M. L. (1992). *Primate Ethology.* Moscow: Moscow State University.

Deryagina, M. A., & Butovskaya, M. L. (2004). *Primate Systematics and Behavior.* Moscow: Russian Village Academy.

Detrain, C. & Deneubourg, J.-L. (2006). Self-organized structures in a superorganism: do ants "behave" like molecules? *Physics of Life Reviews 3,* 162–187.

Devigne, C., Renon, A., & Detrain, C. (2004). Modulation of recruitment according to home range marking in ants. *Animal Behaviour, 67,* 1023–1029.

Devreotes, P. (1989). *Dictyostelium discoideum*: a model system for cell-cell interactions in development. *Science, 245,* 1054-1058.

Dewey, J. (1916). *Democracy and Education. An Introduction to the Philosophy of Education.* New York: Macmillan.

Dewsbury, D. A. (1978). *Comparative Animal Behaviour.* New York: McGraw-Hill.

Dictionary.com. (2014). *21st Century Lexicon.* http://dictionary.reference.com.

Dillon, P., Craft, A., Best, P., with Rigby, A., & Simms, K. (2007). *Turning Peases West Inside Out: Flexible Educational Environments for Developing Possibilities and Pedagogies.* Creative Partnerships Durham Sunderland.

Dlussky, G. M. (1981). Communication principles in ants. In: *Readings in Commemoration of N. A. Kholodny* (pp.3-33). Leningrad: Nauka.

Dlussky, G. M. (1984). Principles of organization of families in social insects. In: B. P. Manteufel & A. A. Zakharov (Eds.), *The Behavior of Insects* (pp.2-25). Moscow: Nauka.

DNAX (2014). *DNAX Research Institute Overview.* http://venturebeatprofiles.com/company/profile/dnax-research-institute.

Dubey, G. P. & Ben-Yehuda, S. (2011). Intercellular nanotubes mediate bacterial communication. *Cell, 144,* 590-600.

Dubynin, V. A., Kamensky, A. A., Sapin, M. P., & Sivoglazov, V. N. (2003). *Regulatory Systems of the Human Organism.* Moscow: Drofa.

Duda, V. I., Pronin, S. V., El'-Registan, G. I., Kapreliants, A. S. & Mityushin, L. L. (1982). The formation of dormant refractile cells in *Bacillus cereus* under the influence of an autoregulatory factor. *Microbiology (Moscow), 51(1),* 77-81.

Duda, V. I., Vypov, M. G., Sorokin, V. V., Mityushina, L. L. & Lebedinsky, A. V. (1995). Formation of hemoprotein-containing extracellular structures by bacteria. *Microbiology (Moscow), 64,* 69-73.

Dunbar, R. I. M. (1992). Neocortex size as a constraint on group size in primates. *Journal of Human Evolution, 22,* 469–493.

Dunbar, R. I. M. (1998). The social brain hypothesis. *Evolutionary Anthropology, 6,* 178-190.

Dunn, C. W. (2005). Complex colony-level organization of the deep-sea siphonophore *Bargmannia elongata (Cnidaria, Hydrozoa)* is directionally asymmetric and arises by the subdivision of pro-buds. *Developmental Dynamics, 234,* 835–845.

Dunn, C. W. (2007). Siphonophores. *Current Biology, 19(6),* R233-R234.

Dunn, C. W. & Wagner, G. P. (2006). The evolution of colony-level development in the *Siphonophora (Cnidaria:Hydrozoa). Development Genes & Evolution, 216,* 743-754.

Dunn, C. W., Pugh, P. R., & Haddock, S. H. D. (2005). Molecular phylogenetics of the *Siphonophora (Cnidaria),* with implications for the evolution of functional specialization. *Systematic Biology, 54(6),* 916–935.

Dwyer, D. J., Kohanski, M. A., & Collins, J. J. (2008). Network opportunities for bacteria. *Cell, 135,* 1153-1156.

Easton, D. (1965). *A Framework for Political Analysis.* Englewood Cliffs, NJ: Prentice-Hall.

Economic Theory. Microeconomics (1, 2). (2005). G. P. Zhuravleva (Ed.). Moscow: Dashkov & Co.

Eibl-Eibesfeldt, I. (1997). Human ethology: origin and prospects of a new discipline. In: A. Schmitt, K. Altzwanger, K. Grammer, & K. Schäfer (Eds.), *New Aspects of Human Ethology* (pp.1-23). New York: Plenum Press.

Eibl-Eibesfeldt, I. (1998). Us and the others: the familial roots of ethnonationalism. In: I. Eibl-Eibesfeldt & F. K. Salter. (Eds.), *Indoctrinability, Ideology and Warfare* (pp.21–53). New York & Oxford: Berghahn books.

Eigen, M. & Schuster, P. (1977). The hypercycle. A principle of natural self-organization. Part A: Emergence of the hypercycle *Naturwissenschaften, 64,* 541-565.

Eigen, M. & Schuster, P. (1979). *The Hypercycle. A Principle of Natural Self-organization.* Berlin, Heidelberg, & New York: Springer-Verlag.

Ekbia, H. R. & Kling, R. (2005). Network organization: symmetric cooperation or multivalent negotiation? *Information Society, 21(3),* 155-168.

El'-Registan, G. I., Khokhlova, Yu. M., Duzha, M. V. & et al. (1979). Identification and characterization of specific autoregulatory factors synthesized by pro- and eukaryotic microorganisms. In: *Izvestiya AN SSSR (Reports of the Academy of Sciences of the USSR), 48(6),* 869-877.

Elenkin, A. A. (1936). *The Blue-Green Algae of the USSR. General Characteristics.* Moscow & Leningrad: USSR Academy of Sciences Publishing Company.

Ellermeier, C. D., Hobbs, E. C., Gonzalez-Pastor, J. E., & Losick, R. (2006). A three-protein signaling pathway governing immunity to a bacterial cannibalism toxin. *Cell, 124,* 549-559.

Elsner, W., Hocker, G. & Schwardt, H. (2008). Simple and complex organizational forms: markets, hierarchies, and networks in an 'organizational triangle'. *Discussion-Papers Series No.010-2008.* Bremen: Faculty 7. Business Studies & Economics. Department of Economics. Universität Bremen.

Engdahl, F. W. (2011). *Egypt's Revolution: Creative Destruction for a Greater Middle East?* http://www.engdahl.oilgeopolitics.net.

Erdös, P. & Rényi, A. (1959). On random graphs. *Publicationes Mathematicae, 6,* 290-297.

Erdös, P. & Rényi, A. (1960). On the evolution of random graphs. *Publications of the Mathematics Institute of the Hungarian Academy of Sciences, 5,* 17–60.

Esau, A. (2004). *The Courts and the Colonies.* Vancouver, BC: UBC Press.

Es'kov, K. Yu. (2004). *The History of the Earth and Life on it: From the Chaos to Humankind.* Moscow: ENAS Novation Center Publishers.

Evans, H. E. (1958). The evolution of social life in wasps. *Proceedings of the Tenth International Congress on Entomology, 2,* 449-457.

Evans, T. (2004). Complex networks. *Contemporary Physics, 45,* 455-474.

Feiser, J. (2004). Evolution of the al-Qaeda brand name. *Asia Times Online,* August 13.

Filatova, O. A., Fedutin, I. D., Ivkovich, T. V., Nagaylik, M. W., Burdin, A. M., & Hoyt, E. (2008). The function of multi-pod aggregations of fish-eating killer whales (*Orcinus orca*) in Kamchatka, Far East Russia. *Journal of Ethology, 27,* 33-41.

Flack, J. C., Girvan, M., de Waal, F. B. M., & Krakauer, D. C. (2006). Policing stabilizes construction of social niches in primates. *Nature, 439(7075),* 426-430.

Flanagan, J. G. & Rayner, S. (eds.). (1988). *Rules, Decisions, and Inequality in Egalitarian Societies.* Brookfield, Hongkong, Singapore, Sydney: Avebury.

Fomin, O. N., Mamonov, M. V., Bryantsev, I. I., & Zabbarov, A. G. (2009). *Network Structures of the Third Sector in Relation to Power–Civil Society Interactions.* Saratov: Volga Region Academy of State Service.

Fossey, D. (1983). *Gorillas in the Mist,* Boston: Houghton Mifflin Company.

Foucault, M, (1978). *The History of Sexuality. Volume 1. An Introduction.* New York: Random House.

Foucault, M. (2003 [1976]). *Society Must Be Defended.* London: Penguin.

Frances, J., Levačić, R., Mitchell, J, & Thompson, G. (1991). Introduction. In: G. Thompson, J. Frances, R. Levačić, & J. Mitchell (Eds.), *Markets, Hierarchies, and Networks: The Coordination of Social Life* (pp.1-20). London, Thousand Oaks (CA), & New Delhi: SAGE Publications.

Franks, N. R. (1986). Teams in social insects: group retrieval of prey by army ants (*Eciton burchelli*: *Hymenoptera: Formicidae*). *Behavioral Ecology & Sociobiology, 18(6),* 425-429.

Fraser, O. N. & Aureli, F. (2008). Reconciliation, consolation and postconflict behavioral specificity in chimpanzees. *American Journal of Primatology, 70,* 1–10.

Freeman, L. C. (1979 [1978]). Centrality in social networks. Conceptual clarification. *Social Networks, 1,* 215-239.

Fridman, V. S. (2007). Delimitating human and animal sociality: where is the boundary? In: A. V. Oleskin (Ed.), *Biopolitics. An Open Interdisciplinary Seminar at the Biology Faculty of Lomonosov Moscow State University. Issue 3* (pp.32-38). Moscow: Biology Faculty of Moscow State University.

Fridman, V. S. (2009). The social structure of vertebrates in terms of network organization (in the example of the territoriality of some bird species). In: E. S. Lobakova et al. (Eds.), *The Life Sciences and Education. Fundamental Integration Problems* (pp.336-340). Moscow: MAX Press.

Fried, M. (1967). *The Evolution of Political Society: An Essay in Political Anthropology.* New York: Random House.

Fukuyama, F. (2002). *Our Posthuman Future: Consequences of the Biotechnology Revolution.* New York: Farrar, Straus, and Giroux.

Fuqua, W. C., Winans, S. C., & Greenberg, E. P. (1994). Quorum sensing in bacteria: the LuxR-LuxI family of cell density-responsive transcriptional regulators. *Journal of Bacteriology, 176(2),* 269-275.

Gall, M. & Breeze, N. (2008). Music and eJay: An opportunity for creative collaborations in the classroom. *International Journal of Educational Research 47,* 27–40.

Gavron, D. (2000). *The Kibbutz Awakening from Utopia.* Lanham: Rowman & Littlefield.

Gerasimov, V. V. & Darkov, A. A. (1984). Organizational principles in fish shoals. In: *Systems Principles and Ethological Approaches in Population Research* (pp.143-158). Puschino: NCBI.

Ghiglieri, M. P. (1987). Sociobiology of the great apes and the hominid ancestor. *Journal of Human Evolution, 16,* 319-357.

GMO Directive. (2001). Directive 2001/18/EC of the European Parliament and of the Council of 12 March 2001 on the deliberate release into the environment of genetically modified organisms and repealing Council Directive 90/220/EE. *Official Journal of the European Communities, L106/*2.

GOA. (2014). *Good Old Age. Official Website.* http:/starikam.org.

Goldberg, E. (2001). *The Executive Brain: Frontal Lobes and the Civilized Mind.* Oxford, New York etc.: Oxford University Press.

Golovlev, E. L. (1999). Academician Nikolai Dmitrievich Yerusalimsky (1901-1967). *Microbiology (Moscow), 68(6),* 800-808.

González-Pastor, J. E. (2011). Cannibalism; a social behavior in sporulating *Bacillus subtilis. FEMS Microbiological Reviews, 35(3),* 415-424.

Goodall, J. (1994). Gombe chimpanzee politics. In: Schubert, G. & Masters, R. D. (Eds.), *Primate Politics* (pp.105-137). New York & London: Lanham.

Goodall, J. (1996). Order without law. In: *Law, Biology and Culture* (pp.242-252). Gruter Institute, Dartmouth: McGraw-Hill.

Gorny, M. B. (2006). Think tanks and Centers of Public Policy: research activities in the example of the Strategy Center. An introduction. In: M. B. Gorny (Ed.), *Think Tanks and Centers of Public Policy: Their Research Activities* (pp.5-50). Saint-Petersburg: Norma.

Gottfredson, L. S. (2001). Biological basis of differences in human intelligence. In: M. Butovskaya (Ed.), *Evolution, Behaviour and Society: Human Ethology* (pp.31-33). Moscow: Russian State University for Humanities.

Granovetter, M. (1973). The strength of weak ties. *American Journal of Sociology, 78(6),* 1360-1380.

Granovetter, M. (1985). Economic action and social structure: The problem of embeddedness. *American Journal of Sociology, 91(3),* 481–510.

Grassé, P.-P. (1959). La reconstruction du nid et les coordinations interindividuelles. La théorie de la stigmergie. *Insectes Sociaux,* 6, 41-84.

Greiner, L. L., Edwards, J. L., Shao, J., Rabinak, C., Entz, D., & Apicella, M. A. (2005). Biofilm formation by *Neisseria gonorrhoeae. Infection & Immunity, 73,* 1964-1970.

Grinshaw, D., Marchington, M., Rubey, J., & Willmott, H. (2005). Introduction: fragmenting work across organizational boundaries. In: M. Marchington et al., (Eds.), *Fragmenting Work: Blurring Organizatiuonal Boundaries and Disordering Hierarchies* (pp.1-31). Oxford, New York, Auckland, & Cape Town: Oxford University Press.

Gromov, A. V. & Kuzin, O. S. (1990). *"Neformaly" (Informal Groups): Who is Who?* Moscow: Mysl'.

Grove, M. (2009). Hunter-gatherer movement patterns: causes and constraints. *Journal of Anthropological Archaeology, 28,* 222-233.

Gruby, R. L. & Basurto, X. (2014). Multi-level governance for large marine commons: Politics and polycentricity in Palau's protected area networks. *Environmental Science & Policy, 36,* 48-60.

Gruter Institute. (2014). *Gruter Institute for Law and Behavioral Research. Official Website.* http://www.gruterinstitute.org.

Guimerà, R., Danon, L., Díaz-Guilera, A., Giralt, F., & Arenas, A. (2006). The real communication network behind the formal chart: community structure in organizations. *Journal of Economic Behavior & Organization, 61,* 653-667.

Guimerà, R. & Nunes Amaral, L. A. (2006). Functional cartography of complex metabolic networks. *Nature, 433,* 895–900.

Gurwitsch, A. (1944). *Theory of the Biological Field.* Moscow: Sovietskaya Nauka.

Gusev, B. A. & Bobrova, N. I. (1989). On intercellular interactions in microbial populations. In: *Microbiological Research in West Siberia* (pp.55-59). Novosibirsk: Nauka Publ. Co. (Siberian Subdivision).

Habermas, J. (1990). *Moral Consciousness and Communicative Action.* Cambridge, MA: MIT Press.

Hahn, M. W. & Kern, A. D. (2005). Comparative genomics of centrality and essentiality in three eukaryotic protein-interaction networks. *Molecular Biology & Evolution, 22(4),* 803-806.

Håkansson, H. & Ford, D. (2002). How should companies interact in business networks? *Journal of Business Research, 55,* 133-139.

Hall-Stoodley, L., Costerton, J. W. & Stoodley, P. (2004). Bacterial biofilms: from the natural environment to infectious diseases. *Nature Reviews Microbiology, 2,* 95-108.

Hamilton, W. D. (1964). The genetical evolution of social behavior. *Journal of Theoretical Biology, 7,* 1-52.

Harcourt, A. H. (1992). Coalitions and alliances: are primates more complex than non-primates? In: A. H. Harcourt & F. B. M. de Waal (Eds.), *Coalitions and Alliances in Humans and Other Animals* (pp. 445–471). Oxford: Oxford University Press.

Harcourt, J. L., Sweetman, G., Johnstone, R. A., & Manica, A. (2009). Personality counts: the effect of boldness on shoal choice in three-spined sticklebacks. *Animal Behaviour, 77,* 1501–1505.

Hardt, M., & Negri, A. (2001). *Empire.* Cambridge, MA: Harvard University Press.

Harshey, R. M. (1994). Bees aren't the only ones: swarming in Gram-negative bacteria. *Molecular Microbiology, 16(3),* 389-394.

Hayden, B. (1995). Pathways to power: principles for creating socioeconomic inequalities. In: T. D. Price & G. M. Feinman (Eds.), *Foundations of Social Inequality* (pp. 15–86). New York: Plenum.

Heatwole, H. F. & Taylor, J. (1987). *Ecology of Reptiles.* Chipping Norton: Surrey Beatly & Sons.

Helbing, D., Farkas, I., & Vicsek, T. (2000). Simulating dynamical features of escape panic. *Nature, 407*, 487–490.

Hemelrijk, C. K. (2002). Despotic societies, sexual attraction, and the emergence of male 'tolerance': an agent based model. *Behaviour, 139,* 729–747.

Hickson, J., Diane Yamada, S., Berger, J., Alverdy, J., O'Keefe, J., Bassler, B, & Rinker-Schaeffer, C. (2009). Societal interactions in ovarian cancer metastasis: a quorum-sensing hypothesis. *Clinical & Experimental Metastasis,26(1),* 67-76.

Higashi, S., Ito, F., Sugiura, N., & Ohkawara, K. (1994). Worker's age regulates the linear dominance hierarchy in the queenless ponerine ant, *Pachycondyla sublaevis (Hymenoptera: Formicidiae). Animal Behaviour, 47(1),* 179-184.

Hill, R. A., Bentley, R. A., & Dunbar R. I. M. (2008). Network scaling reveals consistent fractal pattern in hierarchical mammalian societies. *Biology Letters, 4(6),*748–751.

Hochberg, M. E., Rankin, D. J., & Taborsky, M. (2008). The coevolution of cooperation and dispersal in social groups and its implications for the emergence of multicellularity. *BMC Evolutionary Biology*, 8, 238.

Hölldobler, B. & Wilson, E. O. (1990). *The Ants.* Cambridge, MA: Harvard University Press.

Hölldobler, B. & Wilson, E. O. (2009). *The Superorganism: The Beauty, Elegance, and Strangeness of Insect Societies.* New York: W.W. Norton.

Hölldobler, B. & Wilson, E. O. (2010). *The Leafcutter Ants: Civilization by Instinct.* New York: W.W. Norton.

Hooper, P. L., Kaplan, H. S., & Boone, J. L. (2010). A theory of leadership in human cooperative groups. *Journal of Theoretical Biology, 265,* 633-646.

Hopfield, J. J. (1982). Neural networks and physical systems with emergent collective computational abilities. *Proceedings of the National Academy of Sciences of the USA, 79,* 2554-2558.

Hsu, C.-L. & Park, H. W. (2012). Mapping online social networks of Korean politicians. *Government International Quarterly, 29,* 169-181.

ICBL. (2014). *International Campaign to Ban Landmines.* Official website. http://www.icbl.org/index.php.

Immelmann, K. & Beer, C. (1989). *A Dictionary of Ethology.* Cambridge (MA) & London: Harvard University Press.

Isaeva, V. V. (2012). Self-organization of biological systems. *RAS Herald. Series Biology, No.2,* 144-153.

ISCS. (2014). *Institute for the Study of Complex Systems. Official Website.* http://complexsystems.org.

Ito, F. & Higashi, S. (1991). A linear dominance hierarchy regulating reproduction and polyethism of the queenless ant *Pachycondyla sublaevis. Naturwissenschaften, 78(2),* 80-82.

IZEW. (2014). *Welcome to the International Centre for Ethics in the Sciences and Humanities (IZEW).* http://www.uni-tuebingen.de/en/facilities/zentrale-einrichtungen /international-centre-for-ethics-in-the-sciences-and-humanities.html.

Jansson, H., Johanson, M., & Ramström, J. (2007). Institutions and business networks: A comparative analysis of the Chinese, Russian, and West European markets. *Industrial Marketing Management, 36,* 955-967.

Jantsch, E. (1992). *Die Selbstorganisation des Universums. Vom Urknall zum menschlichen Geist.* München & Wien: Hanser.

Jennings, H. (1943). *Leadership and Isolation: A Study of Personality in Interpersonal Relations.* New York: Longmans.

Jeong, H., Mason, S. P., Barabási, A.-L., & Oltvai, Z. N. (2001). Lethality and centrality in protein networks. *Nature, 411,* 41-42.

Jindal, R. (2011). Reducing the size of internal hierarchy: The case of multi-unit franchising. *Journal of Retailing, 87(4),* 549-562.

Johnson, B. R. (2010). Division of labor in honeybees: form, function, and possible mechanisms. *Behavioral Ecology & Sociobiology, 64(3),* 305–316.

Johnson, L. R. (2008). Microcolony and biofilm formation as a survival strategy for bacteria. *Journal of Theoretical Biology, 251,* 24-34.

Jung, D. F. & Lake, D. A. (2011). Markets, hierarchies, and networks: an agent-based organizational ecology. *American Journal of Political Science, 55(4),* 972-990.

Jurcizek, J. A. & Bakaletz, L. O. (2007). Biofilms formed by nontypeable *Haemophilus influenzae* in vitro contain both double-stranded DNA and type IV pilin protein. *Journal of Bacteriology, 189(10),* 3868-3875.

Kahler, M. (2009a). Networked policies: agencies, power and governance. In: M. Kahler (Ed.), *Networked Politics: Agency, Power, and Governance* (pp.1-20). San Diego: University of California.

Kahler M. (2009b). Collective action and clandestine networks: the case of Al Qaeda. In: M. Kahler (Ed.), *Networked Politics: Agency, Power, and Governance* (pp.103-124). San Diego: University of California.

Kaiser, D. (2007). Bacterial swarming: re-examination of cell-movement patterns. *Current Biology, 17,* R561-570.

Kaiser, D. & Dworkin, M. (2008). From glycerol to the genome. In D. E. Whitworth (Ed.), *Myxobacteria. Multicellularity and Differentiation* (pp.6-7). Washington, DC: ASM Press.

Kalan, A. & Rainey, H. (2009). Hand-clapping as a communicative gesture by wild female swamp gorillas. *Primates, 50,* 273–275.

Kamnev, A. N. (1997). Creating a system of permanent active biology education. In: *International Conference on Biology, the Humanities, and Education* (pp.68-75). Moscow: Moscow State University, Biology Faculty.

Kandel, E. R., Schwartz, J. H., & Jessell, T. M. (2000). [1981]. *Principles of Neural Science.* New York: McGraw-Hill.

Kaplan, H. B. & Plamann, L. (1996). A *Myxococcus xanthus* cell density-sensing system required for multicellular development. *FEMS Microbiology Letters, 139,* 89-95.

Kappeler, P. M., Silk, J. S., Burkart, J. M., & van Schaik, C. P. (2010). Primate behavior and human universals: exploring the gap. In: P. M. Kappeler & J. B. Silk (Eds.), *Mind the Gap* (pp.3-15). Berlin & Heidelberg: Springer-Verlag.

Karatan, E. & Watnick, P. (2009). Signals, regulatory networks, and materials that build and break bacterial biofilms. *Microbiology & Molecular Biology Reviews, 73,* 310-347.

Karpukhin, O. I. (1996). Managing organizational structures in the realm of culture. *Sotsial'no-politicheskiy Zhurnal (Socio-political Journal), No.6,* 165-174.

Kavtaradze, D. N. (1997). Interactive teaching methods in biology and ecology. In: *International Conference on Biology, the Humanities, and Education* (pp.30-40). Moscow: Moscow State University, Biology Faculty.

Kavtaradze, D. N. (1998). *Teaching and Playing. Introduction into Active Teaching Methods.* Moscow: Flinta.

Kelley, J. L., Morrell, L. J., Inskip, C., Krause, J., Croft, D. P. (2011). Predation risk shapes social networks in fission-fusion populations. *PLoS One, 6(8),*e24280.

Kenis, P. & Schneider, V. (1991). Policy networks and policy analysis: scrutinizing a new analytical toolbox. In: B. Marin & R. Mayntz (Eds.), *Policy Network: Empirical Evidence and Theoretical Considerations* (pp.25-59), Frankfurt am Main: Campus Verlag.

Kenney, M., Horgan, J., Horne, C., Vining, P., Carley, C. M., Bigrigg, W., Bloom, M., & Braddock, K. (2013). Organizational adaptation in an activist network. Social networks, leadership, and change in al-Muhajiroun. *Applied Ergonomics, 44,* 739-747.

Khokhlov, A. S. (1988). *Microbial Low Molecular Weight Autoregulators.* Moscow: Nauka Publ. Co.

Kilpatrick, W. H. (1918). The project method. *Teacher College Record, 19(4),* 319-335.

Kipyatkov, V. E. (1991). *The World of Social Insects.* Leningrad: Leningrad University Publ. Co.

Kirovskaya, T. A. (2004). The phenomenon of "anticipatory reflection" in distributed networks structures: its application to microbiology in Russia in the 20th century. In: *Proceedings of the RAS Institute of History of Science and Technology. Vol. 3.* (pp.77-101). Moscow: Sputnik Publ. Co.

Kirovskaya, T. A., & Oleskin, A. V. (2003). The population- and communication-centered paradigm in Russian microbiology. In: A. V. Oleskin (Ed.), *Biopolitics. An Open Interdisciplinary Seminar at the Biology Faculty of Lomonosov Moscow State University* (pp.8-13). Moscow: Biology Faculty of Moscow State University.

Kirovskaya, T. A., & Oleskin, A. V. (2004). The population- and communication-centered paradigm in Russian microbiology in the second half of the 20th century. An example of decentralized network organization in the scientific community. In: *Proceedings of the Institute of History of Science and Technology, Russian Academy of Sciences. Vol. 2.* (pp.42-99). Moscow: Sputnik Publ. Co.

Kiwanis. (2013). Kiwanis International Website. http://www.kiwanis.org.

Klausner, A. (1988). DNAX keeps a scholar's focus. *Bio-technology, 6(4),* 373–378.

Knauft, B. (1991). Violence and sociality in human evolution. *Current Anthropology, 32,* 391-428.

Knauft, B. (1994). Culture and cooperation in human evolution. In: L. Sponsel & T. Gregor (Eds.) *The Anthropology of Peace and Nonviolence* (pp. 37–67). Boulder, CO: Lynne Rienner.

Knoke, D. (1994). *Political Networks.* Cambridge (UK), New York, & Melbourne: Cambridge University Press.

Knyazeva, E. N. & Kurdyumov, S. P. (2007). *Synergetics: Time Non-Linearity and Evolution Landscapes.* Moscow: KomKniga Publishers.

Kobyakov, A., Averyanov, V., & Kucherenko, V. (2005). *The Russian Doctrine.* http//:www.rusdoctrina.ru.

Kohonen, T. (2001 [1989]). *Self-Organizing Maps.* Berlin & New York: Springer-Verlag.

Kolmer, K. & Heinze, J. (2000). Rank order and division of labour among unrelated cofounding ant queens. *Proceedings of the Royal Society of London B, 267,* 1729-1734.

Kooiman, J. (1993). Social-political government introduction. In: J. Kooiman (Ed.), *Modern Governance. New Government-Society Interactions* (pp.1-60). London: Sage.

Kosevich, I. A. (2006). Changes in the patterning of a hydroid colony. *Zoology, 109,* 244-259.

Kosslyn, S. M., & Koenig, O. (1992). *Wet Mind: The New Cognitive Science.* New York: Free Press.

Kovalev, V., Malinetsky, G., & Matvienko, Yu. (2013). "Network-centric" warfare concept with respect to the Russian army: "Strength potentiation" or a mental trap? *Economic strategies, No.5,* 40-51.

Kozma, R. & Puljic, M. (2013). Hierarchical random cellular neural networks for system-level brain-like signal processing. *Neural Networks, 45,* 101-110.

Krause, J. & Ruxton, G. D. (2002). *Living in Groups.* Oxford, New York et al.: Oxford University Press.

Krause, J., Lusseau, D., & James, R. (2009). Animal social networks: an introduction. *Behavioral Ecology & Sociobiology, 63,* 967-973.

Krebs, V. E. (2002). Mapping networks of terrorist cells. *Connections, 24(3),* 43-52.

Krebs, V. E. (2008). *Connecting the Dots. Tracking Two Identified Terrorists.* http://orgnet.com/tnet.html.

Kropotkin, P. (1972 [1902]). *Mutual Aid: A Factor of Evolution.* New York: New York University Press.

Kroos, L., Kuspa, A., & Kaiser, D. (1990). Defects in fruiting body development caused by Tn5 *lac* mutation in *Myxococcus xanthus. Journal of Bacteriology, 172,* 484-487.

Kuramoto, Y. (1975). Self-entrainment of a population of coupled nonlinear oscillators. *Lecture Notes in Physics, 39,* 420-422.

Kuramoto, Y. (1984). *Chemical Oscillations, Waves, and Turbulence.* New York: Springer-Verlag.

Laland, K. N. & Hoppitt, W. (2003). Do animals have culture? *Evolutionary Anthropology, 12,* 150–159.

Lamprecht, J. (1996). What makes an individual the leader of its group? An evolutionary concept of distance regulation and leadership. *Social Science Information, 35(4),* 595–617.

Lazerson, M. (1988). Organizational growth of small firms: An outcome of markets and hierarchies? *American Sociological Review, 55(3),* 330-342.

Le Hellaye, Y., Goossens, B., Jamart, A., & Curtis, D. J. (2010). Acquisition of fission-fusion social organization in a chimpanzee (*Pan troglodytes troglodytes*) community released into the wild. *Behavioral Ecology & Sociobiology, 64(3),*349-360.

Lebedev, D. A. (2007). *A Syncretic Theory of Natural Social Evolution.* Moscow: LKI Publishers.

Ledeneva, A. (1998). *Russia's Economy of Favours. Blat, Networking and Informal Exchange.* Cambridge: Cambridge University Press.

Lee, H. (2002). Aligning supply chain strategies with product uncertainties. *California Management Review, 44,* 105–119.

Lehmann, J. & Boesch, C. (2004). To fission or to fusion: effects of community size on wild chimpanzee (*Pan troglodytes verus*) social organisation. *Behavioral Ecology & Sociobiology, 56,* 207–16.

Lehmann, J. & Boesch, C. (2009). Sociality of the dispersing sex: the nature of social bonds in West African female chimpanzees, *Pan troglodytes. Animal Behaviour, 77,* 377–387.

Lemon, K. R., Earl, A. M., Vlamakis, H. C., Aguilar, C., & Kolter R. (2008). Biofilm development with an emphasis on *Bacillus subtilis. Current Topics in Microbiology, 322,* 1-16.

Lengeler, J. W. (2004). Signal transduction and global control systems: the basis of Systems Biology. In: B. Holand et al. (Eds.), *Bacterial Neural Networks. Book of Abstracts* (p.5). San-Feliu-de-Guichols (Spain): EURESCO Conference.

Leontovich, A. A. & Marfenin, N. N. (1990). Relationships among the main intracolony processes during branching in colonial hydroids. *Zhurnal Obschei Biologii (Journal of General Biology), 51(3),* 353-362.

Leviatan, U. (1982). Counterbalancing the ill effects of hierarchy: the case of the kibbutz industrial organization. *Journal of Social & Biological Structures, 5,* 141-159.

Lewis, T. G. (2013). Cognitive stigmergy: A study of emergence in small-group social networks. *Cognitive Systems Research, 21,* 7-21.

Lijphart, A. (1968). *The Politics of Accomodation.* Berkeley: University of California Press.

Lijphart, A. (1977). *Democracy in Plural Societies.* New Haven: Yale University Press.

Lindelauf, R., Borm, P., & Hamars, H. (2009). The influence of secrecy on the communication structure of covert networks. *Social Networks, 31,* 126-137.

Liseev, I. K. (2004). Role of modern biology in shaping new cultural norms. In: I. K. Liseev (Ed.), *Biology and Culture.* Moscow: Kanon.

Living Lakes (2014). *Living Lakes – The International Network.* http://www.globalnature.org.

Lobel, T. E. & Agami-Rozenblat, O. (1993). Self-esteem, need for achievement, and sex-role orientation among kibbutz and urban men and women, *Personal & Individual Differences, 15(5),* 523-529.

Long, T., Tu, K. C., Wang, Y., Mehta, P., Ong, N. P., Bassler, B. L., & Wingreen, N. S. (2009). Quantifying the integration of quorum-sensing signals with single-cell resolution. *PLoS Biology, 7,* e1000068.

Longman Dictionary of Contemporary English (2003). Edinburgh Gate, Harlow (UK): Pearson Education Ltd.

Lorenz, K. (1966). *On Agression.* London: Methuen.

Losco, J. (2001). Evolutionary theory and political thought. In: A. Somit & S. A. Peterson (Eds.), *Research in Biopolitics. Vol.8. Evolutionary Approaches in the Behavioral Sciences: Toward a Better Understanding of Human Nature* (pp.161-180). Amsterdam, London & New York: JAI Imprint. An Imprint of Elsevier Science.

Low, B. S. (2000). *Why Sex Matters? A Darwinian Look at Human Behavior.* Princeton, NJ: Princeton University Press.

Lumsden, C. J., & Wilson, E. O. (1981). *Genes, Mind and Culture: The Coevolutionary Process.* Harvard: Harvard University Press.

Lusseau, D. (2003). The emergent properties of a dolphin social network. *Proceedings of the Royal Society London B (Biology Letters), 270,* S186-S188.

Lusseau, D. & Newman, M. E. J. (2004). Identifying the role that animals play in their social networks. *Proceedings of the Royal Society London B (Biology Letters), 271,* S477-S481.

Madden, J. R., Drewe, J. A., Pearce, G. P., & Clutton-Block, T. H. (2009). The social network structure of a wild meerkat population. 2. Intragroup interactions. *Behavioral Ecology and Sociobiology, 64,* 81-95.

Makris, N. C., Ratilal, P., Jagannathan, S., Gong, Z., Andrews, M., Bertsatos, I., Godø, O. R., Nero, R. W., & Jech, J. M. (2009). Critical population density triggers rapid formation of vast oceanic fish shoals. *Science, 323,* 1734–1737.

Mandeville, B. (1988 [1732]). *The Fable of The Bees: or, Private Vices, Public Benefits. Vol. 1.* Indianapolis: Liberty Funds.

Mangelsdorf, D. J., Thummel, C., Beato, M., & et al. (1995). The nuclear receptor superfamily: The second decade. *Cell, 83,* 835–839.

Marfenin, N. N. (1993). *Coloniality Phenomenon.* Moscow: Moscow State University Publisher.

Marfenin, N. N. (1999). Modular organization concept in developmental terms. *Zhurnal Obschei Biologii (Journal of General Biology), 60(1),* 6-17.

Marfenin, N. N. (2002). Decentralized self-regulation of integrity of colonial polyps. *Zhurnal Obschei Biologii (Journal of General Biology), 63(1),* 26–39.

Marfenin, N. N. (2009). Principles of organization and functioning withinhydroid colonies. In: E. S. Lobakova et al. (Eds.). The Life Sciences and Education. Fundamental Integration Problems (pp.317-321). Moscow: MAX Press.

Marfenin, N. N. (2011). *Principles of Organization and Functioning of the Network Structure within a Hydroid Colony.* Presentation at a Meeting of the Club of Biopolitics, a Subdivision of the Moscow Society of Natural Scientists.

Markov, A. (2009). "Quorum sensing": making collective decisions in the micro- and macroworld. http://elementy.ru/news/431043.

Marshall (1920 [1890]). *Principles of Economics.* London: Macmillan and Co. Ltd.

Maryanski, A., & Turner, J. H. (1992). *The Social Cage. Human Nature and the Evolution of Society.* Stanford (Calif.).: Stanford University Press.

Maslow, A. H. (1954). *Motivation and Personality.* New York: Harper & Row.

Masters, R. D. (1989). *The Nature of Politics.* New Haven & London: Yale University Press.

Masters, R. D. (1991). Conclusion: primate politics and political theory. In: G. Schubert & R. D. Masters (Eds.), *Primate Politics* (pp. 221–250). Carbondale: Southern Illinois University Press.

Masters, R.D. (1994) Why study serotonin, social behavior and the law? In: *The Neurotransmitter Revolution. Serotonin, Social Behavior and the Law* (pp.3-16). Carbondale; Edwardsville: Southern Illinois University Press.

Masters, R. D. (2001). Biology and politics: linking nature and nurture. *Annual Review of Political Science, 4,* 345-369.

Matessi, G., Matos, R. J., Peake, T. M., McGregor, P. K., & Dabelsteen, T. (2010). Effects of social environment and personality on communication in male Siamese fighting fish in an artificial network. *Animal Behaviour, 79,* 43–49.

Matsuhashi, M., Shindo, A., Ohshina, H., Tobi, M., Endo, S., Watanabe H., Watanabe, H., Pankrushina, A. N. (1996a) Cellular signals regulating antibiotic sensitivities of bacteria. *Microbial Drug Resistance, V. 2(1),* 91–93.

Matsuhashi, M., Pankrushina, A. N., Endoh, K., Watanabe, H., Ohshima, H., Tobi, M., Endo, S., Mano, Y., Hyodo, M., Kaneko, T., Otani, S., & Yoshimura. S. (1996b). *Bacillus carbonifillus* cells respond to growth-promoting physical signals from cells of homologous and heterologous bacteria. *Journal of General & Applied Microbiology, 42,* 315–323.

Maynard Smith, J. (1978). The evolution of behavior. *Scientific American, 239,* 176-192.

McBride, M. & Hewitt, D. (2013). The enemy you can't see: an investigation of the disruption of dark networks. *Journal of Economic Behavior & Organization, 93,* 32-50.

McClurg, S. D. & Lazer, D. (2014). Political networks. *Social Networks, 36,* 1-4..

McCowan, B., Beisner, A. B., Capitanio, J. P., Jackson, M. E., Cameron, A. N., Seil, S., Atwill, E. R., & Fushing, H. (2011). Network stability is a balancing act of personality, power, and conflict dynamics in rhesus macaque societies. *PLoS One, 6(8)*, e22350.

McCulloch, W. S. & Pitts, W. H. (1943). A logical calculus of the ideas immanent in nervous activity. *Bulletin of Mathematical Biophysics, 5,* 115-133.

McFadden, C. (1986). Colony fission increases particle capture rates of a soft coral: advantages of being a small colony. *Journal of Experimental Marine Biology & Ecology, 103*, 1-20.

McLellan, R. & Nicholl, B. (2013). Creativity in crisis in Design & Technology: Are classroom climates conducive for creativity in English secondary schools? *Thinking Skills & Creativity, 9,* 165-185.

Mehra, A., Smith, B. R., Dixon, A. L. & Robertson, B. (2006). Distributed leadership in teams: The network of leadership perceptions and team performance. *The Leadership Quarterly 17,* 232–245.

Meir, I. A. (1990). Monitoring two kibbutz houses in the Negev desert. *Building & Environment, 25(2),* 189-194.

Merriam-Webster Dictionary. (2014). http://www.merriam-webster. com/dictionary.

Mescon, M. H., Albert, M., & Khedouri, F. (1986). *Management.* New York: Harper & Row Publishers.

Meulemann, L. (2008). *Public Management and the Metagovernance of Hierarchies, Networks and Markets.* Heidelberg: Physica-Verlag. 2008. 399 p.

Meyer, P. (1996). Grundformen menschlicher Sozialsysteme aus der Perspektive der EE. In: R. Riedel & M. Delpos (Hrsg.), *Die Evolutionäre Erkenntnistheorie im Spiegel der Wissenschaften* (S. 200–224). Wien: Wiener Universitätsverlag.

Meyer-Emerick, N. (2007). Public administration and the life sciences: revisiting biopolitics. *Administration and Society, 38(6)*, 689-708.

Mezhuev, V. M. (2008). Civil society and modern Russia. In: *Man and Culture from the Perspective of the Developing Civil Society of Russia* (p.6). Moscow: Insitute of Philosophy, Russian Academy of Sciences.

Michels, R. (1949). *Political Parties.* New York: Free Press.

Mikheev, V. N. (2006). *Environment Heterogeneity and Trophic Interactions in Fish.* Moscow: Nauka.

Mikkelsen, T. S, Hillier, L. W, Eichler, E. E, Zody, M. C., & et al. (2005). Initial sequence of the chimpanzee genome and comparison with the human genome. *Nature, 437,* 69–87.

Milgram, S. (1967). The small world problem. *Psychology Today, 2,* 60–67.

Millner, B. Z. (2006). *Organization Theory. A Guidebook.* Moscow: Infra-M.

Miralles Canals, J. J. (2009). Fourth-generation warfare: jihadist methods and percolation. *Mathematical & Computer Modeling, 50*, 896-909.

Mitani J. C., Gregory F. Grether, G. F., & Rodman, P. S. (1991). Associations among wild orang-utans: sociality, passive aggregations or chance? *Animal Behaviour, 42,* 33-36.

Mitchell, M. (2006). Complex systems: Network thinking. *Artificial Intelligence, 170,* 1194–1212.

Mittal, N., Budrene, E. O., Brenner, M. P. & Van Oudenaarden, A. (2003). Motility of *Escherichia coli* cells in clusters formed by chemotactic aggregation. *Proceedings of the National Academy of Sciences of the U.S.A., 100,* 13259-13263.

Moiseev, N. N. (1987). *Algorithms of Development.* Moscow: Nauka Publ.

Mondragón Corporation (2014). *Mondragón Official Website* http://www.mondragon-corporation.com/eng/about-us.

Monroe, D. (2007). Looking for chinks in the armor of bacterial biofilms. *Plos Biology, 5,* 1-4.

Moreno J. L. (1932). *Applications of the Group Method to Classification.* N.Y.: National Committee on Prisons and Prison Labor.

Morrell, L. J., Hunt, K. L., Croft, D. P., & Krause, J. (2007). Diet, familiarity and shoaling decisions in guppies. *Animal Behaviour, 74,* 311–319.

Morrison-Shetlar, A. & Marwitz, M. (2001). *Teaching Creatively: Ideas in Action.* Eden Prairie: Outernet.

Motroshilova, N. V. (2009). On the modern concept of civil society. *Voprosy Filosofii (Philosophical Issues), No.6,* 12-32.

Mourrier, J., Vercelloni, J., & Planes, S. (2012). Evidence of social communities in a spatially structured network of a free-ranging shark species. *Animal Behaviour, 83,* 389-401.

Mourtzis, D., Doukas, M, & Psarommatis, F. (2012). Design and planning of decentralised production networks under high product variety demand. *Procedia CIRP, 3,* 293 – 298.

MSNS (2014). *Moscow Society of Natural Scientists Official Website.* http://moipros.ru.

Mulyukin, A.L., Suzina, N. E., El'-Registan, G. I., Danilevich, V. N. (2013). Effective PCR detection of vegetative and dormant bacterial cells due to a unified method for preparation of template DNA encased within cell envelopes. *Microbiology (Moscow), 82(3),* 300-311.

Mutzel, R. (1995). Introduction. Molecular biology, growth and development of the cellular slime mold *Dictyostelium discoideum. Experientia, 51(12),* 1103-1110.

Myers, D. G. (2010). *Social Psychology.* New York: McGraw-Hill.

Nakamura, M. (2003). "Gatherings" of social grooming among wild chimpanzees: implications for evolution of sociality. *Evolution, 44,* 59-71.

Naug, D. (2008). Structure of the social network and its influence on transmission dynamics in a honeybee colony. *Behavioral Ecology & Sociobiology, 62,* 1719–1725.

Networks. (2003). *We Are Everywhere. The Irresistible Rise of Global Anticapitalism.* K. Arier, G. Chesters, T. Credland, J. Jordan, A. Stern, & J. Whity (Eds.). http://artactivism.gn.apc.org.

Neu, D., Everett, J., Rahaman, A. S., & Martinez, D. (2013). Accounting and networks of corruption. *Accounting, Organization, & Society,38,* 505-524.

Neuman, L. H., Pardue, K. T., Grady, J. L., Gray, M. T., Hobbins, B., Edelstein, J., Herrman, J. W. (2009). What does an innovative teaching assignment strategy mean to nursing students? *Nursing Education Perspectives, 30 (3),* 159–163.

Newman, M. E. J. (2003). The structure and function of complex networks. *SIAM Review, 45(2),* 167–256.

Newman, M. E. J. (2012). *Networks. An Introduction.* Oxford & New York: Oxford University Press.

Newman, M. E. J., Barabási, A.-L. & Watts, D. J. (2006). *The Structure and Dynamics of Networks.* Princeton, NJ: Princeton University Press.

Nezdyurov, A. L. & Sungurov, A. Yu. (2008). Interatctions between power bodies and civil society structrures: possible scenarios and their implementation in the social and political life of Russia. In: L. I. Yablokov (Ed.), *Factors Involved in the Development of Civil Society and the Mechanisms of its Interaction with the State* (pp.209-236). Moscow: Vershina.

Nicholls, J. G., Martin, A. R., Fuchs, P. A., Brown, D. A., Diamond, M. E., & Weisblat, D. (2012). *From Neuron to Brain.* Sunderland (MA): Sinauer Associates, Inc.

Nicolis, G. & Prigogine, I. (1977). *Self-organization in Nonequilibrium Systems.* New York: Wiley.

Nikolaev, Yu. A. (1992). Distant interactions between microbial cells. *Microbiology (Moscow), 61(6),* 1066-1071.

Nikolaev, Yu. A. (2000). Distant informational interactions in bacteria. *Microbiology (Moscow), 69(5),* 597-605.

Nikolaev, Yu. A. & Prosser, J. (2000). Extracellular factors influencing *Pseudomonas fluorescens* to glass. *Microbiology (Moscow), 69(2),* 231-236.

Nikolaev, Yu. A. & Plakunov, V. K. (2007). A biofilm: a "city of microbes" or an analog of a multicellular organism? *Microbiology (Moscow), 76,* 149-163.

Norris, K. S. & Schilt, C. R. (1988). Cooperative societies in three-dimensional space: on the origins of aggregations, flocks, and schools, with special reference to dolphins and fish. *Ethology & Sociobiology, 9,* 149-179.

Norris, V., Molina, F., & Gewirtz, A.T. (2013). Hypothesis: bacteria control host appetites. *Journal of Microbiology, 195(3),* 411-416.

Novgorodova, T. A. (2003). Intraspecies diversity of behavior models in the ant *Formica cunicularia glauca* during trophobiosis. *Uspekhi Sovremennoi Biologii (Advances in Modern Biology), 123(3),* 229-233.

Oleinik, A. (2003). A model of network capitalism. *Voprosy Ekonomiki (Issues in Economics), No.8,* 132-149.

Oleskin, A. V. (1994a). Biopolitics (Pt. 1–3). A series of articles. *Moscow State University Herald (Vestnik). Biology Series, 2-4.*

Oleskin, A. V. (1994b). Social behaviour of microbial populations. *Journal of Basic Microbiology, 34(6),* 425–439.

Oleskin, A. V. (1996). Making a new case for voluntary cooperation-based fission-fusion structures in human society. *Social Science Information,* 35(4), 619-627.

Oleskin, A. V. (1998). Social network structures from the biopolitical viewpoint. *Polis, No. 1,* 68-86.

Oleskin, A. V. (2007a). Network structures as a biopolitical project. *Herald of the Russian Academy of Sciences, 77(6),* 598–601.

Oleskin, A. V. (2007b). *Biopolitics. Political Potential of Modern Life Sciences.* Moscow: Nauchnyi Mir.

Oleskin, A. V. (2008). Prospects of biopolitics. *Gosudarstvennaya Sluzhba (State Service), No.4,* 88-95.

Oleskin, A. V. (2009). Biosociality of the unicellulars (in the example of prokaryotes). *Zhurnal Obschei Biologii (Journal of General Biology), 70(3),* 225-238.

Oleskin, A. V. (2012). *Biopolitics. The Political Potential of the Life Sciences.* New York: Nova Science Publishers.

Oleskin, A. V. (2013). Networks as non-hierarchical and non-market structures: their implementation in biological and social systems. *Economic Strategies, 5,* 2-7.

Oleskin, A. V. (2014). Network structures in biological systems. *Biology Bulletin Reviews, 74(1),* 47-70.

Oleskin, A. V. & Kirovskaya, T. A. (2005). The network self-identification issue. In: E. G. Zakharchenko et al. (Eds.), *The Logos of Life and the Hermeneutics of Corporality* (pp.600-622). Moscow: Russian Institute of Culture Studies.

Oleskin, A. V. & Kirovskaya, T. A. (2006). Research on population organization and communication in microorganisms. *Microbiology (Moscow), 75,* 1-6.

Oleskin, A. V. & Kirovskaya, T. A. (2007). Network structure in microbiology. *Herald of the Russian Academy of Sciences, 77(2),* 139-148.

Oleskin, A. V., & Masters, R. D. (1997). Biopolitics in Russia: history and prospects for the future. *Research in Biopolitics, 5,* 279—299.

Oleskin, A. V., Kirovskaya, T. A., Botvinko, I. V. & Lysak, L. V. (1998). Effects of serotonin (5-hydroxytryptamine) on the growth and differentiation of microorganisms. *Microbiology (Moscow), 67,* 251-257.

Oleskin, A. V., Botvinko, I. V., & Tsavkelova, E. A. (2000). Colony organization and intercellular communication in microorganisms. *Microbiology (Moscow), 69(3),* 309–327.

Oleskin, A. V., Pivovarova, L. V., Kartashova, E. R., & Gusev, M. V. (2001). Teaching biopolitics in terms of school biology syllabi. *Moscow University Herald. Biology Series, No. 3,* 3-13.

Oleskin, A. V., Shishov, V. A., & Malikina, K. D. (2011). Symbiotic biofilms and brain neurochemistry. In: W. C. Bayley (Ed.) *Biofilms: Formation, Development and Properties* (pp.163-193). New York: Nova Science Publishers.

Olson, J. (2005). Maybe it is time to rediscover bureaucracy. *Working Paper, No.10.* http://www.sv.uio.no/arena/english/research/publications/arena-publications/workingpapers/working-papers2005/wp05_10.pdf.

Ostrom, E. (2005). *Understanding Institutional Diversity.* Princeton: Princeton University Press.

Ostrom, E. (2010). Polycentric systems for coping with collective action and global environmental change. *Global Environmental Change, 20,* 550–557.

Ostrom, E., Parks, R. B. & Whitake, G. P. (1978). *Patterns of Metropolitan Policing.* Cambridge, MA: Ballinger.

Ostrom, V. (1962). The political economy of water development. *American Economic Review, 52,* 450–458.

Ostrom, V., Tiebout, C., & Warren, R. (1961). The organization of government in metropolitan areas: a theoretical inquiry. *The American Political Science Review, 55,* 831–842.

Otto, K. (2011). Signalling processes implicated in *Escherichia coli* biofilm formation. In: W. C. Bayley (Ed.) *Biofilms: Formation, Development and Properties* (pp.195-225). New York: Nova Science Publishers.

Panov, E. N. (2001). *Escape from Solitude. The Individual and the Collective in Nature and in Human Society*. Moscow: Lazur'.

Parish, A. R. (1993). Sex and food control in the "uncommon chimpanzee": How bonobo females overcome a phylogenetic legacy of male dominance. *Ethology & Sociobiology, 15,* 157-179.

Parker, S. C. (2008). The economics of formal business networks. *Journal of business Venturing, 23,* 627-640.

Partridge, B. L. (1982). Structure and function of fish shoals. *Scientific American, 245,* 114–123.

Patriquin, K. J., Leonard, M. L., Broders, H. G., & Garroway, C. J. (2010). Do social networks of female long-eared bats vary with reproductive period and age? *Behavioral Ecology & Sociobiology, 64,* 899–913.

Pavlov, D. S., & Kasumyan, A. O. (2002). *Fish Diversity in Terms of Nutrition Types and Strategies (A Trophic Classification of Fish).* Moscow: Moscow University Press.

Pavlov, D. S., & Kasumyan, A. O. (2003). *Fish Schooling Behavior.* Moscow: Moscow University Press.

Pavlova, I. B., Kulikovsky, A. V. Botvinko, I. V., Dzhentemirova, K. M., & Drozdova, T. I. (1990). Electron microscopic studies on the development of bacteria in colonies. The morphology of bacterial colonies. *Journal of Microbiology, Epidemiology & Immunology, No. 12,* 15–20.

Pavlova, I. B., Levchenko, K. M. & Bannikova, D. A. (2007). *Atlas of the Morphology of Populations of Pathogenic Bacteria*. Moscow: Kolos.

Pearson, H. T. & Shelton, D. E. (2010). A large-brained social animal. In: B. Würsig. & M. Würsig (Eds.), *Dusky Dolphins: Master Acrobats off Different Shores. Chapter 16.* (pp.333-353). London: Academic Press.

Pelvig, D. P, Pakkenberg, H, Stark, A. K & Pakkenberg, B (2008). Neocortical glial cell numbers in human brains. *Neurobiology of Aging, 29(11),* 1754–1762.

Pertseva, E. (2013). *Funday* has come to *Sportmaster. Kommersant, 125(5156).* http://www.kommersant.ru/doc/2235941.

Pivovarova, L. V., Korzhenevskaya, T. G., & Gusev, M. V. (2002). An integrative biology-teaching scenario. *Biology at School, No.6,* 17-25.

Plakunov, V. K. & El'-Registan, G. I. (2004). Research on the physiology and biochemistry of microorganisms at the Institute of Microbiology of the Russian Academy of sciences. *Microbiology (Moscow), 73(5),* 659-667.

Podolny, J. M. & Page, K. L. (1998). Network forms of organization. *Annual Review of Sociology, 24,* 57–76.

Polanyi, M. (1966). *The Tacit Dimension*. London: Routledge.

Poldervaart, S. W. (2009). The utopian or DiY-strategy of social movements. In: D. M. Bondarenko & O. I. Kavykin (Eds.), *Hierarchy and Power in the History of Civilization. Fifth International Conference*. Moscow: Center for Civilizational and Regional Studies of the RAS, Institute for African Studies of the RAS.

Political Dictionary of the State of Israel. (1993). New York: McMillan Publ. Co.

Popov, L. (2009). *Ape Society Demonstrates Why Instincts Play Such an Important Role in Human Society*. http://membrana.ru/particle/787.

Porshnev, B. F. (1974). *The Beginnings of Humankind's History: Paleopsychological Issues*. Moscow: Mysl'.

Povinelli, D.J., Nelson, K.E., & Boysen, S.T. (1992). Comprehension of role reversal in chimpanzees: evidence of empathy? *Animal Behaviour, 43,* 633–640.

Powell, S. & Tschinkel, W. R. (1999). Ritualized conflict in *Odontomachus brunneus* and the generation of interaction-based task allocation: a new organizational mechanism in ants. *Animal Behaviour, 58(5)*, 965-972.

Powell, W. W. (1990). Neither market nor hierarchy: network forms of organization. *Research in Organizational Behavior, 12,* 295-336.

Powell, W. W. (1996). Interorganizational collaboration in the biotechnology industry. *Journal of Institutional & Theoretical Economics, 152,* 197-215.

Pradhan, G. P. Tennie, C., & van Schaik, C. P. (2012). Social organization and the evolution of cumulative technology in apes and hominins. *Journal of Human Evolution, 63,* 180-190.

Pulsifer, L. (2013). *Interactive Teaching & Learning*. http://www.ehow.com /about_5552709_interactive-teaching-learning.html.

Pusey, A. E. (2001). Of genes and apes: chimpanzee social organization and reproduction. In: F. B. M. de Waal (Ed.), *Tree of Origin: What Primate Behavior Can Tell Us about Human Social Evolution* (pp.11-37). Cambridge, MA: Harvard University Press.

Putnam, R. D., (2000). *Bowling Alone: The Collapse and Revival of American Community*. New York: Simon & Schuster.

Putnam, R. D., Leonardi, & R., Nanetti, R. Y. (1993). *Making Democracy Work: Civic Traditions in Modern Italy*. Princeton, NJ: Princeton University Press.

Raab, J. H. & Milward, B. (2003). Dark networks as problems. *Journal of Public Administration Research & Theory, 13,* 413–439.

Radaev, V. V. (2002a). The main trends in the development of modern economic sociology. In: V. V. Fadeev (Ed.), *Economic Sociology. New Approaches to Institutional and Network Analysis* (pp.3-18). Moscow: ROSSPEN.

Radaev, V. V. (2002b). Institutional approach and deformalization of rules in the Russian economy. In: V. V. Fadeev (Ed.). *Economic Sociology. New Approaches to Institutional and Network Analysis* (pp.157-209). Moscow: ROSSPEN.

Radakov, D. V. (1972). *Fish Shoaling as an Ecological Phenomenon*. Moscow: Nauka.

Radcliffe-Brown, A. R. (1931). *The Social Organization of Australian Tribes*. Melbourne: Oceania Monographs. No. 1.

Ramos-Fernández, G., Boyer, D., Aureli, F., & Vick, L. G. (2009). Association networks in spider monkeys (*Ateles geoffroyi*). *Behavioral Ecology & Sociobiology, 63,*999–1013.

Rani, S. A., Pitts, B., Velucharny, R. A., Levandovski, Z., Davison, W. M., Buckingham-Meyer, K., & Stewart, P. S. (2007). Spatial patterns of DNA replication, protein synthesis, and oxygen concentration within bacterial biofilms reveal diverse physiological states. *Journal of Bacteriology, 189,* 4223-4233.

Rasse, P. & Deneubourg, J. L. (2001). Dynamics of nest excavation and nest size regulation of *Lasius niger* (*Hymenoptera: Formicidae*). *Journal of Insect Behavior, 14,* 433–449.

Rathbone, A. & Rowley, C. K. (2002). Terrorism. *Public Choice, 111*, 1–10.

Ratnieks, F. L. W. & Reeve, H. K. (1992). Conflict in single-queen hymenopteran societies: the structure of conflict and processes that reduce conflict in advanced eusocial species. *Journal of Theoretical Biology, 158(1),* 33-65.

Rauprich, O., Matsushita, M., Weijer, C. J., Siegert, F., Esipor, S. E., & Shapiro, J. A. (1996). Periodic phenomena in *Proteus mirabilis* swarm colony development. *Journal of Bacteriology, 178(22),* 6525-6538.

Reeve, H. K. & Hölldobler, B. (2007). The emergence of a superorganism through intergroup competition. *Proceedings of the National Academy of Sciences of the USA, 104(23),* 9736-9740.

Remedios, R., Logothetis, N. K., & Kayser, C. (2009). Monkey drumming reveals common networks for perceiving vocal and nonvocal communication sounds. *Proceedings of the National Academy of Sciences of the U. S. A., 106,* 18010–18015.

Ren, B., Au, K. Y., & Birtch, T. A. (2009). China's business network structure during institutional transitions. *Asia Pacific Journal of Management, 26,* 219-240.

Répertoire officiel des communes de Suisse. Mutations. (2013). Eidgenössisches Departement des Innern EDI. Bundesamt für Statistik BFS.

Reznikova, Z. I. (2001). *Populations and Species on the Scales of War and Peace. Ecological and Evolutionary Aspects of Interspecies Relationships in Animals (Competition, Parasitism, and Symbiosis).* Moscow: Logos.

Reznikova, Z. I. (2005). *Intelligence and Language in Animals and Humans,* Moscow: Akademkniga.

Reznikova, Zh. I. & Novgorodova, T. A. (1998). Role distribution and information exchange in worker ant groups. *Uspekhi Sovremennoi Biologii (Advances in Modern Biology), 118(3),* 345-356.

Rhodes, R. A. W. (1997). *Understanding Governance. Policy Networks, Governance, Reflexivity, and Accountability.* Buchingham & Philadelphia: Open University Press.

Rice, S.A. (1927). The identification of blocs in small political bodies. *American Political Science Review, 21,* 619–627.

Rich, A. (1999). *Think Tanks, Public Policy, and the Politics of Expertise. Ph. D. Dissertation.* Yale University.

Richardson, A. J. & Kilfoyle, E. (2009). Accounting in markets, hierarchies, and networks: The role of accounting in the transnational governance of postal transactions *Accounting, Organization, & Society, 34,* 939-956.

Richerson, P. J. & Boyd, R. (1998). The evolution of human ultrasociality. In: I. Eibl-Eibesfeldt & F. K. Salter (Eds.), *Indoctrinability, Ideology and Warfare* (pp.71-95). New York & Oxford: Berghahn Books.

Richerson, P. J., & Boyd, R. (2001). The evolution of subjective commitment of groups: a tribal instincts hypothesis. In: R. M. Nesse (Ed.), *The Evolution of Commitment* (pp. 186–220). New York: Russel Sage Foundation.

Richerson, P. J. & Boyd, R. (2004). Culture is part of human biology: why the superorganic concept serves the human sciences badly. In: N. K. Udumyan (Ed.), *Challenge to Cognition. Strategies of Science Development in the Modern World* (pp.344-373). Moscow: Nauka.

Rimsky, V. l. & Sungurov, A. Yu. (2002). "Think tanks", democracy centers, and Centers of Public Policy. In: A. Yu. Sungurov (Ed.), *"Think Tanks" and Centers of Public Policy: International and Initial Russian Experience* (pp.6-29). Saint-Petersburg: Norma.

Römmele, A. (2003). Political parties, party communication and new information and communication technologies. *Party Politics, 9(1),* 7-20.

Ronfeldt, D. &. Arquilla, J. (2001). Networks, netwars, and the fight for the future. *First Monday, 6(10).* http://www.firstmonday.org/ojs/index.php/fm/article/ viewArticle/889.

Rosenberg, M. (2009). Proximate causation, group selection, and the evolution of hierarchical human societies: system, process, and pattern. In: A. M. Prentiss et al. (Eds.), *Macroevolution in Human Prehistory* (pp.23-49). Berlin & Heidelberg: Springer-Verlag + Business-Media.

Rosenblatt, F. (1957). The Perceptron – a perceiving and recognizing automaton. *Report* 85-460-1, Cornell Aeronautical Laboratory.

Rosner, M. (1971). *Hierarchy and Democracy in Kibbutz Industry.* Givat Haviva, Israel: Institute for Social Research on the Kibbutz (In Hebrew).

Rosner, M. (1975). Principles, types, and problems of direct democracy in the kibbutz. In: B. Horvat, M. Marković, & R. Supek (Eds.), *Self-Governing Socialism. Vol. 1* (pp.287-298). New York: International Arts and Sciences Press.

Rountree, R. A. & Sedberry. G. R. (2009). A theoretical model of shoaling behavior based on a consideration of patterns of overlap among the visual fields of individual members. *Acta Ethologica, 12(2),* 61–70.

Ruffle, B. J. & Sosis, R. (2006). A field test on Israeli kibbutz members and city residents. *Journal of Economic Behavior & Organization, 60,* 147-163.

Ryabchuk, E. (2005). Present-day genetic engineering. *Energia, No.6,* 55–59.

Sageman, M. (2004). *Understanding Terror Networks.* Philadelphia: University of Pennsylvania.

Sageman, M. (2008). *Leaderless Jihad. Terror Networks in the Twenty-first Century.* Philadelphia: University of Pennsylvania.

Sakai, M., Hishii, T., Takeda, S., & Kohshima, S. (2006). Flipper rubbing behavior in wild bottlenose dolphins (*Tursiops aduncus*). *Marine Mammal Science, 22(4),* 966–978.

Sakyte, E. & Ragulskis, M. (2011). Self-calming of a random network of dendritic neurons. *Neurocomputing, 74,* 3912–3920.

Samoilov, V. I. & Vasiliev, Yu. M. (2009). Mechanisms of the social behavior of vertebrate tissue cells; culture models. *Zhurnal Obschei Biologii (Journal of General Biology), 70(3),* 239–244.

Sardanyés, J. & Solé, R. V. (2007). Spatio-temporal dynamics in simple asymmetric hypercycles under weak parasitic coupling. *Physica D, 231,* 116–129.

Sardanyés, J., Duarte, J., Januário, C., & Martins, N. (2012). Topological entropy of catalytic sets: Hypercycles revisited. *Communications in Nonlinear Science and Numerical Simulation, 17,* 795-803.

Satt, E. & Ginzburg, H. (1992). On the dynamic effects of using hired labor in the kibbutz – theory and case studies. *Journal of Comparative Economics, 16,* 688-700.

Scarino, A. & Liddicoat, A. J. (2009). *Teaching and Learning Languages. A Guide.* Carlton (Australia): Curriculum Corporation.

Schlegel, H. G. (2006). *Allgemeine Mikrobiologie.* Stuttgart: Georg Thieme Verlag.

Schmidt-Hempel, P. (1990). Reproductive competition and the evolution of workload in social insects. *American Naturalist, 135(4),* 501-526.

Schneirla, T. C. (1971). *Army Ants.* San Fransisco: Freeman.

Scott, J. (2000). *Social Network Analysis: A Handbook.* Second Edition London: Sage.

Segel, L. A. & Cohen, J. R. (Eds.). (2001). *Design Principles for the Immune System and Other Distributed Autonomous Systems.* New York: Oxford University Press.

Sekowska, A., Masson, J. B., Celani, A., Danchin, A., & Vergassola, M. (2009). Repulsion and metabolic switches in the collective behavior of bacterial colonies. *Biophysical Journal, 97,* 688-698.

SEMCO. (2014). *Semco Partners.*http://www.semco.com.br/en.

Semyonov, D. V. (2001). *Reptiles.* Moscow: AST.

Sergeev, V. M. & Sergeev, K. V. (2003). Mechanisms of evolution of the political system of society: social hierarchies and social networks. *Political Studies (Polis), No.3,* 6-13.

Setrov, M. I. (1971). *Organization of Biosystems. Methodological Overview of the Principles of Organization of Living Systems.* Leningrad: Nauka Publ. Leningrad Subdivision.

Sewertzowa, L. B. (1929). Über den Einfluβ der mitogenetischen Strahlen auf die Vermehrung der Bacterien. *Biologisches Zentralblatt, 49,* 212–225.

Shapiro, J. A. (1988). Bacteria as multicellular organisms. *Scientific American, 256(6),* 82-89.

Shapiro, J. A. (1995). The significances of bacterial colony patterns. *BioEssays, 17(7),* 597-607.

Shapiro, J A. (1998). Thinking about bacterial populations as multicellular organisms. *Annual Review of Microbiology, 52,* 81-104.

Shishov, V. A., Kirovskaya, T. A., Kudrin, V. S., & Oleskin, A. V. (2009). Amine neuromediators, their precursors, and oxidation products in the culture of *Escherichia coli* K-12. *Applied Biochemistry & Microbiology, 45,* 494-497.

Shleeva, M. O., Salina, E. G., & Kapreliants, A. S. (2010). The dormant forms of myxobacteria. *Microbiology (Moscow), 73(1),* 3-15.

Shpakov, A. O. (2009). QS-type non-peptide signal molecules of bacteria. *Microbiology (Moscow), 78,* 163-175.

Shupatova, N. N. (1999). Some peculiarities of the behavior and the formation of directed water currents in bryozoan colonies *(Bryozoa: Gymnolaemata). Zhurnal Obschei Biologii (Journal of General Biology), 60(1),* 118-128.

Sih, A., Hanser, S. F., McHugh, K. A. (2009). Social network theory: new insights and issues for behavioral ecologists. *Behavioral Ethology & Sociobiology, 63,* 975-988.

Silberman, M. (1996). *Active Learning: 101 Strategies to Teach Any Subject.* Boston: Allyn & Bacon.

Simmel, G. (1908). *Soziologie.* Leipzig: Duncker & Humblot.

Simmons, R. & Birchall, J. (2008). The role of co-operatives in poverty reduction: network perspectives. *Journal of Socio-Economics, 37,* 2131-2140.

Simpson, P. K. (1990). *Artificial Neural Systems.* Elmsford, NY: Pergamon Press.

Smirnov, K. K., Smirnov, S. G., & Smirnova, V. G. (1982). The lag phase, the phase of anticipatory reflection of the bacterial culture. In: *Physico-chemical Research on Pathogenic Enterobacteria During the Cultivation Process* (p.42). Ivanovo: Ivanovo State Medical Institute.

Smirnov, S. G. (2004). *Bacterial Ethology.* Ivanovo: Ivanovo State Medical Academy.

Smirnova, T. A., Didenko, L. V., Azizbekyan, R. R., & Romanova, Yu. M. (2010). *Microbiology (Moscow), 79(4),* 435-446.

Smith, E. A. (2010). Communication and collective action: language and the evolution of human cooperation. *Evolution and Human Behavior, 31,* 231–245.

Smorgunov, L. V. & Timofeeva, L. N. (Eds.). (2012). *GR-Links at the State Level: Theory, Practical Experience, and the Mechanism of Interaction of Businesses and of Civil Society with the State. A Guide-book.* Moscow: Russian Political Encyclopedia.

Snarey, J., Reimer, J., & Kohlberg, L. (1985). The kibbutz as a model for moral education; a longitudinal cross-cultural study. *Journal of Applied Developmental Psychology, 6,* 151-172.

Snow, C. C., Miles, R. E., & Coleman, H. J. (1992). Managing 21st century network organizations. *Organizational Dynamics, 20(3),* 5-20.

Soengas, B. (2009). Power relations in a BaKoya Pygmy community (Gabon). In: D. M. Bondarenko & O. I. Kavykin (Eds.), *Hierarchy and Power in the History of Civilization. Fifth International Conference* (pp.102-103). Moscow: Center for Civilizational and Regional Studies of the RAS, Institute for African Studies of the RAS.

Søgaard-Andersen, L. (2004). Cell polarity, intercellular signalling and morphogenetic cell movements in *Myxococcus xanthus*. *Current Opinion in Microbiology, 7,* 587–593.

Soloviev, E. (2004). *Network Structures of Transnational Terrorism.* httm://www.intertrends.ru/five/006.htm.

Somit, A. (1968). Toward a more biologically oriented political science. *Midwest Journal of Political Science, 12,* 550-567.

Somit, A. (1972). Biopolitics. *British Journal of Political Science,* 2, 209–238.

Somit, A., & Peterson, S. A. (1997). *Darwinism, Dominance, and Democracy.* Westport (CA): Praeger Publ.

Somit, A. & Peterson, S. A. (1998). Biopolitics after three decades: a balance sheet. *British Journal of Political Science, 28,* 555–571.

Somit, A. & Peterson, S. A. (2001a). Biopolitics in the year 2000. In: S. A. Peterson & A. Somit (Eds.), *Research in Biopolitics. Vol.8. Evolutionary Approaches in the Behavioral Sciences: Toward a Better Understanding of Human Nature* (pp.181-201). Amsterdam, London & New York: JAI Imprint. An Imprint of Elsevier Science.

Somit, A. & Peterson, S. A. (2001b). Darwinism, dominance, and democracy: a reaffirmation. *Politics and the Life Sciences, 20(2),* 227–231.

Somit, A. & Peterson, S. A. (2011). Introduction. In: S. A. Peterson & A. Somit (Eds.), *Research in Biopolitics. Vol.9. Biology and Politics. The Cutting Edge* (pp.3-10). UK, North America, Japan, etc.: Emerald Group Publ. Ltd.

Sparrow, M. (1991). The application of network analysis to criminal intelligence. An assessment of the prospects. *Social Networks, 13,* 251-274.

Stark, D. (2002). Heterarchy: asset ambiguity and diversity management in postsocialist countries. In: V. V. Radaev (Ed.), *Economic Sociology: New Approaches to Institutional and Network Analysis* (pp. 47-95). Moscow: Russian Economics Encyclopedia.

Startsev, Yu. N. (2011). *Networks: The State-of-the-Art and Potentialities.* Chelyabinsk: Encyclopedia.

Stehfest, K. M., Patterson, T. A., Dagorn, L., Holland, K. N., Itano, D., & Semmens, J. M. (2013). Network analysis of acoustic tracking data reveals the structure and stability of fish aggregations in the ocean. *Animal Behaviour, 85,* 839-848.

Steinberg, J. (1976). *Why Switzerland?* Cambridge et al.: Cambridge University Press.

Strauss, L. (1964). *The City and Man.* Chicago: Rand McNally.

Stumpf, M. P. H., Kelly, W. P., Thorne, T., & Wiuf, C. (2007). Evolution at the system level: the natural history of protein interaction networks. *Trends in Ecology & Evolution, 22(7),* 366-373.

Suchak, M. & de Waal, F. B. M. (2012). Monkeys benefit from reciprocity without the cognitive burden. *Proceedings of the National Academy of Sciences of the USA, 109,* 15191-15196.

Sumina, E. L. (2006). The behavior of filamentous cyanobacteria in laboratory culture. *Microbiology (Moscow), 75(4),* 532-537.

Summers, K. (2005). The evolutionary ecology of despotism. *Evolution & Human Behavior, 26,* 106-135.

Sumpter, D. J. T. (2006). The principles of collective animal behavior. *Philosophical Transactions of the Royal Society B., 361,* 5-22.

Sungurov, A. Yu. (2006). *The Strategy Center: Brief Information.* http://www.strategy-spb.ru/index.php?do=about&doc=17.

Sungurov, A. Yu. (2007). *Civil Society and its Development in Russia.* Saint-Petersburg: Saint-Petersburg Subdivision, Higher School of Economics.

Sungurov, A. Yu., Zakharova, O. S., Petrova, L. A., & Raspopov, N. P. (2012). Mediator institutions and their development in contemporary Russia. 1. Public chambers and consultancy commissions: experience at the federal and regional levels. *Political Studies (Polis), No. 1,* 165-178.

Sutherland, I. W. (2001). Biofilm polysaccharides: a strong and sticky framework. *Microbiology, 147,* 3-9.

Teknologirådet (Danish Board of Technology). (2007). *EPTA - Genetically Modified Plants and Food.* http://www.tekno.dk.

Terekhin, A. T. & Budilova, E. V. (1995). Network mechanisms of physiological regulation. *Uspekhi Fisiologicheskikh Nauk (Advances in Physiological Sciences), 26(1),* 75-97.

Tetz, V. V., Rybalchenko, O. V., & Savkova,G. A. (1990). Ultrastructural features of microbial colony organization. *Journal of Basic Microbiology, 30,* 597–607.

Tetz, V. V., Rybalchenko, O. V. & Savkova, G. A. (1993). Ultrastructure of the surface film. *Journal of General Microbiology, 139,* 855-858.

Tetz, V. V., Korobov, V. P., Artemenko, N. K., Lemkina, L. M, Panjkova, N. V. & Tetz, G. V. (2004). Extracellular phospholipids of isolated bacterial communities. *Biofilms., 1,* 149-155.

Thierry, B. (2007). Unity in diversity: lessons from macaque societies. *Evolutionary Anthropology, 16,* 224–238.

Thompson, d'Arcy W. (1945). *On Growth and Form.* Cambridge: Cambridge University Press & New York: Macmillan.

Thorelli, H. B. (1986). Networks: between markets and hierarchies. *Strategic Management Journal, 7,* 37–51.

Tinbergen, N. (1968). On war and peace in animals and man. *Science, 160,* 1411–1418.

Tönnies, F. (1988). *Community and Society (Gemeinschaft und Gesellschaft).* New Brunswick & Oxford: Transaction Book.

Trivers, R. L. (1972). Parent-offspring conflict. *American Zoologist, 14,* 249-264.

Trkman, P. & Desouza, K. C. (2012). Knowledge risks in organizational networks: an exploratory framework. *Journal of Strategic Information Systems, 21,* 1-17.

Tumasjan, A., Sprenger, T. O., Sandner, P. G., & Welpe, I. M. (2010). Predicting elections with Twitter: What 140 characters reveal about political sentiment. *Proceedings of the Fourth International AAAI Conference on Weblogs and Social Media* (pp. 178–195). Menlo Park, CA: AAAI Press.

Twin Oaks. (2014). *Twin Oaks Intentional Community.* http://www.twinoaks.org.

Uher, J., Asendorpf, J. B., & Call, J. (2008). Personality in the behaviour of great apes: temporal stability, cross-situational consistency and coherence in response. *Animal Behaviour, 75,* 99-112.

United Nations Millennium Declaration. (2000) *United Nations General Assembly Resolution* 2. Session 55, 18 September 2000.

van Alstyne, M. (1997). The state of network organizations: a survey in three frameworks. http://ccs.mit.edu/papers/CCSWP192/CCSWP192.html.

van der Dennen, J. M. G. (2011). The biopolitics of primates. In: S. A. Peterson & A. Somit (Eds.), *Research in Biopolitics. Vol.9. Biology and Politics. The Cutting Edge* (pp.53-96). UK, North America, Japan, etc.: Emerald Group Publ. Ltd.

van der Wusten, H. (2004). The distribution of political centrality in the European state system. *Political Geography, 23,* 677-700.

van Lawick-Goodall, J. (1968). The behaviour of free-living chimpanzees in the Gombe Stream Reserve. *Animal Behaviour Monographs, 1,* 161–311.

van Lindert, P. (2009). Transnational linking of local governments: The consolidation of the Utrecht–León municipal partnership. *Habitat International, 93,* 173-180.

Vanhaeren, M. & d'Ērrico, F. (2005). Grave goods from the Saint-Germain-la-Riviere burial: evidence for social inequality in the Upper Paleolithic. *Journal of Anthropological Archaeology, 24,* 117-134.

Vasil'ev, Yu. M. (2000). Normal and tumor cells. In: *Modern Natural Science Encyclopedia. Vol. 2. General Biology* (pp.189-194). Moscow: Magister-Press.

Viscido, S. V., Parish, J. K., & Grunbaum, D. (2007). Factors influencing the structure and maintenance of fish shoals. *Ecological Modeling, 206,* 153-165.

Vlavianos-Arvanitis, A. (1985). *Biopolitics – Dimensions of Biology.* Athens: Biopolitics International Organisation.

Vlavianos-Arvanitis A. (Ed.). (2003). *Bio-Syllabus for European Environmental Education. A Textbook for the Better Understanding and Appreciation of the Bio-Environment.* Athens: Biopolitics International Organisation.

von Bertalanffy, L. (1968). *General System Theory.* New York: George Braziller.

von Rueden, C., Gurven, M., & Kaplan, H. (2008). The multiple dimensions of male social status in an Amazonian society. *Evolution & Human Behavior, 29,* 402-415.

Vygotsky, L. S. (1997). The history of the development of higher mental functions. In: R. W. Rieber (Ed.). *The Collected Works of L. S. Vygotsky. Vol. 4* (pp.1–251). New York: Plenum Press.

Vysotsky, V. V., Zaslavskaya, P. L., Mashkovtseva, A. V., & Baulina, O. I. (1991). Polimorphism as a developmental pattern in populations of prokaryote organisms. *Biological Sciences, No.12,* 5-17.

Wander, A. (2008). A history of terror: Al-Qaeda 1988-2008. *The Observer,* 13 July.

Wang, Y. & Tanaka, A. (2011). From hierarchy to hybrid: The evolving nature of inter-firm governance in China's automobile groups. *Journal of Business Research, 64(1),* 74-80.

Wasserman, F. (1992). *Neural Computing. Theory and Practice.* Moscow: Mir Publishers.

Waters, C. M. & Bassler, B. L. (2005). Quorum sensing: cell-to-cell communication in bacteria. *Annual Review of Cell & Developmental Biology, 21,* 319-346.

Watkins, R. (2005). *75 e-Learning Activities: Making Online Learning Interactive.* San Fransisco: Pfeiffer.

Watnick, P. I. & Kolter, R., (2000). Biofilm, city of microbes. *J. Bacteriol., 182,* 2675-2679.

Watts, D. J. (2003). *Six Degrees: The Science of a Connected Age.* New York: Gardner's Books.

Watts, D. J. & Strogatz, S. H. (1998). Collective dynamics of 'small-world' networks. *Nature, 393,* 440-442.

Watts, D. P. (2010). Dominance, power, and politics in nonhuman and human primates. In: P. M. Kappeler & J. B. Silk (Eds.), *Mind the Gap* (pp.109-138). Berlin & Heidelberg: Springer-Verlag.

Weber, A. W. & Goriunova, O. I. (2013). Hunter-gatherer migration, mobility and social relations: a case study from the Early Bronze Age Baikal Region, Siberia. *Journal of Anthropological Archaeology, 32,* 330-346.

Weber, M. (1956 [1907]). *Staatssoziologie.* Berlin: Duncker & Humblot.

Wey, T., Blumstein, D. T., Shen, W., & Jordán, F. (2008). Social network analysis of animal behaviour: a promising tool for the study of sociality. *Animal Behaviour, 75,* 333–344.

Whallon, R. (2006). Social networks and information: «non-utilitarian» mobility among hunter-gatherers. *Journal of Anthropological Archaeology, 25,* 259-270.

Wheeler, W. M. (1911). The ant-colony as an organism. *Journal of Morphology, 22,* 307-325.

Wheeler, W. M. (1928). *The Social Insects.* New York: Harcourt, Brace & Co..

Wetherington, J., Serrano, G., & Dingledine, R. (2008). Astrocytes in the epileptic brain. *Neuron, 58,* 168-178.

Whimpenny, J., Manz, W. & Szewzyk, U. (2000). Heterogeneity in biofilms. *FEMS Microbiol. Reviews, 24,* 661-671.

White Book of the Baikal Region.(2004). *White Book of the Baikal Region.* Buryat Regional Association Specializing in Lake Baikal.

White, H. (2002). Markets and firms: deliberations on the prospects of economic sociology. In: V. V. Fadeev (Ed.), *Economic Sociology. New Approaches to Institutional and Network Analysis* (pp.96-118). Moscow: ROSSPEN.

Whitehead, H. (2008). *Analyzing Animal Societies: Quantitative Methods for Vertebrate Social Analysis.* Chicago: Chicago University Press.

Whiten, A. (2010). Ape behavior and the origins of human culture. In: P. M. Kappler & J. B. Silk (Eds.), *Mind the Gap* (pp.429-450). Berlin & Heidelberg: Springer-Verlag.

Whiten, A, & Mesoudi, A. (2008). Establishing an experimental science of culture: animal social diffusion experiments. *Philosophical Transactions of the Royal Society B, 363,* 3477–3488.

Whyte, W. F. & Whyte, K. K. (1988). *Making Monragon: the Growth and Dynamics of the Worker Cooperative Complex.* Ithaca (N. Y.): ILR Press.

Wiener, N. (1948). *Cybernetics.* New York: Wiley.

Wigoder, G. (1994). *New Encyclopedia of Zionism and Israel.* Madison, N. J., L.& Cranbury, N. J.: Farleigh Dickinson University Press and Associated University Presses.

Wilson, D. S. & Sober, E. (1989). Reviving the superorganism. *Journal of Theoretical Biology, 136,* 337-356.

Wilson, D. S. & Sober, E. (1994). Reintroducing group selection to the human behavioral sciences. *Behavioral & Brain Sciences, 17,* 585-654.

Wilson E. O.(1971). *The Insect Societies.* Cambridge, MA, London: Harvard University Press.

Wilson, E. O. (1975). *Sociobiology: The New Synthesis.* Cambridge, MA, London: Harvard University Press.

Woodburn, J. (1982). Egalitarian societies. *Man, 17,* 431-451.

Wray, M. K., Mattila, H. R., & Seeley, T. D. (2011). Collective personalities in honeybee colonies are linked to colony fitness. *Animal Behaviour, 81,* 559-568.

Yerusalimsky, N. D. (1952). *The Physiology of the Development of Pure Bacterial Cultures.* D. Sci. (Biol.) Dissertation. Moscow: Institute of Microbiology. USSR Academy of Sciences.

Zaitsev, V. A. (1992). Behavior synchronization and individual distances in wild boar (*Sus scrofa* L.) groups. *Zhurnal Obschei Biologii (Journal of General Biology*), 53(2), 243-257.

Zakharov, A. A. (1987). Ant societies. *Science in the USSR, No. 1,* 118—127.

Zakharov, A. A. (1991). *Organization of Communities in Ants.* Moscow: Nauka.

Zakharov, A. A. (1995). Alliances of workers in the family of the ants of the genus *Formica (Hymenoptera, Formicidae). Uspekhi Sovremennoi Biologii (Advances in Modern Biology), 115(6),* 459–469.

Zakharov, A. A. (2005). Functional differentiation and dominance in the development of biosociality. *Zoologicheskiy Zhurnal (Zoological Journal), 84(1),* 38-53.

Zakharov, A. A. (2009). Network and hierarchical structures in ants. In: E. S. Lobakova et al. (Eds.), *The Life Sciences and Education. Fundamental Integration Problems* (pp.305-309). Moscow: MAX Press.

Zakharov, A. A. & Orlova, T. A. (1984). Behavioral structure and individual development in ants. In: B. P. Manteufel & A. A. Zakharov (Eds.), *The Behavior of Insects* (pp. 43-63). Moscow: Nauka.

Zaleski, P. S. (2006). Global non-governmental administrative system. In: D. Gawin & P. Glinski (Eds.), *Civil Society in the Making.* Warsaw: IFiS Publishers.

Zara (2013). *Zara Company.* Official Website. http://www.zara.com/ru/en.

Zelený, M. (1988). Tectology. *International Journal of General Systems, 14,* 331-343.

Zhuzhikov, D. P. & Shatov, K. S. (1984).The structure of the family of the termite *Anacanthotermes ahngerianus.* In: B. P. Manteufel & A. A. Zakharov (Eds.), *The Behavior of Insects* (pp.26-45). Moscow: Nauka.

Glossary

AFFILIATION: In ethology, affiliation is social *behavior* involving an individual animal's tending to approach and remain near conspecifics (Dewsbury, 1978), particularly those belonging to the same family or social group. In humans, affiliation is associated with feelings of physical comfort, safety, and belonging.

AGGRESSION: In ethology, the term means approaching an opponent and inflicting damage on it or at least generating stimuli that cause the opponent to submit (Tinbergen, 1968). Similarly, in human society, aggression is physical or verbal behavior aimed at causing damage, according to Myers (2010).

AGONISTIC BEHAVIOR comprises all conflict-related forms of social behavior and includes: (a) aggression and relevant signals accompanying aggression or preceding it, i.e., aggressive/threatening displays; (b) conciliatory (buffer) behavior; and (c) avoidance/isolation.

AGONISTIC HIERARCHY: The highest rank in such a *hierarchy* is acquired via agonistic interactions involving aggression or threatening displays demonstrating forcefulness and assertiveness.

ASSORTATIVITY: A network is considered asortative if nodes with similar properties (e.g., similar *degree* values) are preferentially linked to one another; if, conversely, dissimilar nodes prefer to attach, a network is called *disassortative*.

ANTICIPATORY REFLECTION means that the initial stage of the development of a system includes second-order stages (substages) that resemble the later stages of its development (Smirnov et al., 1982); the system's development is a temporarily self-similar (fractal) process.

BETWEENNESS of a node is based on the number of paths between pairs of nodes that pass through a particular node (Croft et al., 2008, p.174). A criterion of network hierarchization can be obtained by calculating the sum of differences between B_{max}, the maximum betweenness value among the nodes in a given network, and B_i, the betweenness of any node i in the network (Freeman, 1979).

BIOPOLITICS: Defined in my recent works as the totality of all kinds of interactions between the life sciences and politics, including both the political potential of biology and the biological implications of politics (Oleskin, 2008, 2012). Most biopoliticians prefer less general interpretations of biopolitics including: (1) interpretations emphasizing the political potential of biology, exemplified by the following: biopolitics is "... used to describe the approach of those political scientists who use biological concepts, with neo-Darwinian evolutionary theory at the center, and biological research techniques to study, explain, predict, and sometimes even to prescribe political behavior" (Somit & Peterson, 2011, p. 3) and (2) the effects produced by the political system on the biology of its citizens/subjects (Foucault, 2003 [1976]) or on the whole biosphere ("bios", Vlavianos-Arvanitis, 1985, 2003). In Foucault's works, the term "biopolitics" is used as a synonym for "biopower," but some of his followers stress that these two words have different meanings: biopolitics deals with political protest against political system-exercized biopower.

BIOSOCIAL SYSTEMS: Systems composed of biological individuals or their groups that are characterized by communication, affiliation, and cooperation among them. Biosocial systems consist of individuals/groups of the same species (homotypic systems) or, alternatively, belonging to several species (heterotypic systems, or associations; Oleskin, 2012, p.226).

BUREAUCRACY: A scenario of social/political organization based upon centralized hierarchy, a narrow specialization of all members of a bureaucratic organization, and the dependence of relations between them on their official status and formal regulations.

CELLULAR STRUCTURES (and the **cellular paradigm**): Most structures made up of free-living cells (microorganisms) or cultivated animal or plant cells represent almost completely flat networks. Hierarchization can be considered a temporary or exceptional phenomenon in cell network structures, except for special cell types (see *neural (neuronal) structures*). In some of these structures, cells and their groups (clusters, microcolonies, rafts) can be functionally differentiated. The dominant role in regulating the activities of the entire network (colony or biofilm) is typically played by chemical communication involving diffusing signal molecules that are exemplified by the pheromones of quorum-sensing systems. Behavioral coordination in cell structures is facilitated by the matrix that performs the structural, protective, and communication-promoting function.

CENTRALITY: "The extent to which a given node occupies a position that is important to the structure of the whole network" (Croft et al., 2008, p.173). *Node degree* (degree centrality), *betweenness*, and *closeness* (see these items) are centrality measures. Each centrality measure can characterize not only a single node but a whole network as well, representing a criterion of its *hierarchization (centralization).* It can be calculated as the sum of the differences between the largest value of a node's centrality in the network (in terms of one of these measures) and those of all n nodes in this network.

CLOSENESS of a network node is calculated as the reciprocal of the averaged path length between a given node and all other nodes of the network: $Б = 1/(\Sigma l/n)$, where l is path length and n the number of nodes in the network. The hierarchization of a network on the basis of closeness centrality of its nodes is also defined in the literature, similar to degree and betweenness centrality, as the sum of the differences between the maximum closeness value and the closeness values of other network nodes.

CLUSTERING COEFFICIENT (*C*) of a single network node corresponds to the ratio between actual links among its neighbors and those potentially possible (which would make the part of the network around the node in question a complete graph; Croft et al., 2008; Newman, 2012). The clustering coefficient can be averaged for any part of a network or for the network as a whole, becoming its integral feature (Watts & Strogatz, 1998). Another definition of the term "clustering coefficient" at the level of a whole network (Newman, 2003, 2012) is the probability that two neighbores of a given network node are also directly linked to one another, thereby forming a triangle (a closed path).

COMMUNICATION: Information exchange between individuals and/or their groups. Communication implies that the individuals receiving the message can choose whether and how they will respond to the message; it is based upon the *stimulus → receiving individual → variable (chosen) response* pattern.

CONTROLLING ACTION (CONTROL): In contrast to *communication,* this term means that those under control have no choice. Controlling implies the *stimulus → preprogrammed response* pattern of interactions.

COMMUNITY STRUCTURE: If a network structure includes parts with dense ties between nodes (i.e., a high *clustering coefficient* value) that are sparsely connected to one another, the network is characterized by *community structure*, or *cliquishness.*

COOPERATION: Cooperation denotes interactions between two or more individuals for the purpose of solving a problem or carrying out a task. Alternatively, cooperation is defined from the viewpoint of a whole group (biosocial system): cooperators contributing to the collective good are contrasted with cheaters (free riders) exploiting it (Hochberg et al., 2008, modified).

COORDINATION: Concerted behavior of individuals, e.g., animals in a biosocial system or human individuals in society, often involving differentiated social roles.

CUT SET: A group of nodes whose removal disconnects a given pair of nodes in a network (Newman, 2012, p.147).

DENSITY of a network: The proportion of actual links between nodes out of the total number of possible links (Croft et al., 2008, p.173).

DOMINANCE: In ethological terms, it is "a social relationship when one individual can monopolize resources at the other's expense or usurp them from the others by using force or threatening to do so" (Watts, 2010, p. 110). Dominance is widespread in various species including primates, and it is obviously related to human society-specific phenomena such as political power.

EGALITARIAN STRUCTURES (and the **egalitarian paradigm**): Egalitarian structures are defined in this work as three-dimensional network structures formed by some primate species including such great apes as the chimpanzee and the bonobo. The same term is applied to primitive human hunter-gatherers and its modern analogs. The egalitarian paradigm is based on the following principles: (1) Respect for individual freedoms (particularly the freedom of choice); (2) Partial hierarchization of the structure associated with respect for high-ranking network members and a significant contribution of dominance-submission relationships; however, no network member can *permanently* exercise total control over the whole structure; (3) Loose links between network members (typified by chimpanzee *fission-fusion* groups); (4) Loyal (cooperative) interactions among individuals.

EQUIPOTENTIAL STRUCTURES (and the **equipotential paradigm**): Equipotential structures are completely *flat* (leaderless) *networks* exemplified by decentralized fish structures (although formed by some other animals as well). Such fish shoals or schools "can be defined as temporary groups of individuals that mostly belong to the same species and have a similar size and physiological status; they lack permanent leadership and intragroup dominance, and they are characterized by intense interrelationships between individuals that tend to aggregate and display coordinate uniform responses to external stimuli" (Pavlov & Kasumyan, 2003, p.143).

ETHOLOGY: The field of biology dealing with animal behavior. Classical ethology emphasized research on innate species-specific behavior forms and mechanisms that were studied under natural conditions (during field studies). Human ethology applied methods originally developed by classical ethology to research on human behavior, including that in political situations (political ethology, an area very closely related to *biopolitics*) (Caton, 1998).

EUSOCIAL STRUCTURES (and the **eusocial paradigm**): Eusocial structures are three-dimensional networks that possess the following properties (Wilson, 1971; Hölldobler & Wilson, 1990, 2009): (1) Coexistence of at least two generations (the parents and the offspring) in the same system. (2) Cooperation in nurturing the young as well as in other collective activities. (3) Classification of individuals into several castes. Typical eusocial structures are formed by many social insects as well as by such mammals as mole-rats.

FLAT NETWORKS are characterized by an equal importance of all its members (network nodes) and a lack of partial leaders. An almost ideally horizontal structure occurs in shoals of fish or seals. The individual that happens to be the first in a moving leaderless *(equipotential)* shoal migrates towards the back of the shoal after a short period of time and is replaced by another individuals who does not "lead the way" for a long time either.

FRACTAL (SELF-SIMILAR) PROPERTIES: If a structure possesses fractal properties, this means that it is characterized by similar or even identical structural patterns at different levels/scales. An army typically represents a fractal structure. Each of its divisions organizationally is a reduced copy of the whole army; each regiment in a division organizationally is a reduced copy of the whole division, and so on.

GLOCALISM (glocal = global + local): In philosophical terms, a worldview based upon the dialectic unity of the global and the local, the macro- and the micro-level. In political terms, this is the current trend that weakens state- (nation-) level political structures by activating two mutually complementary processes: (1) the strengthening of local-level social and political structures; (2) the establishment of international structures including politically influential global organizations.

HEDONIC HIERARCHY: The top of a hedonic hierarchy is occupied by the most interesting, capable, intelligent, sociable, or knowledgeable individual to whom the whole group pays special attention (Chance, 1967).

HETERARCHY: A social system that has no ranks or can be ranked in several different ways. Heterarchy means that the system involved lacks a centralized *hierarchy*; it may have a spilt hierarchy with partial leaders (a *three-dimensional network*) or no hierarchy at all (a *flat network*). Heterarchy is contrasted with *homoarchy* that is characterized by a centralized hierarchy (Crumley, 1995; Bondarenko, 2004).

HETEROGENEOUS NETWORKS: The parts (nodes, subnetworks) of such network structures are characterized by different properties; accordingly, they can perform different functions, as is the case with partial leaders in *hiramas*.

HETERONOMOUS SYSTEMS: These systems are characterized by different organizational structures at different structural levels. Some of them, nonetheless, may possess *fractal/self-similar* properties. This is exemplified by eusocial systems formed by ants. Their *networks consist of hierarchies* that, in their turn, are made up of networks composed of hierarchies. Overall, there is a repetitive interlevel pattern that makes the entire system self-similar.

HIERARCHY: A structure with a fixed single center (leader, pacemaker, dominant individual, boss, president, etc.). The other elements in the structure are subordinate to the central element. These subordinate elements often differ among themselves in terms of rank, forming a typical hierarchical step-ladder. "Applied to social organization, hierarchy is synonymous with rank order and involves the concept of social dominance. Individuals in a group yield to others in contention for something, such as food and mate, according to a more or less linear order" (Immelmann & Beer, 1989, p. 131).

HIRAMA (High-Intensity Research and Management Association): A network structure with *partial creative leaders* whose functions are limited to specific subproblems within the framework of the general interdisciplinary task dealt with by the whole network (Oleskin, 1994a, 1996, 2007a, b; 2012). A hirama-type network structure also may have a *psychological* leader. The psychological leader provides support, advice, and psychological help; the leader creates an atmosphere that promotes efficient work on all subproblems. A network of this kind typically includes an *"external affairs"* leader. The individual with this role is responsible for propagandizing hirama-promoted ideas, establishing contacts with other organizations, and shaping the pastime and leisure activities, thus contributing to the development of informal loyal relationships among members. Modified *hiramas* can also have other kinds of partial leaders.

HOMOGENEOUS NETWORKS: The parts (nodes, subnetworks) of such network structures possess similar or identical properties, which predisposes them to perform coinciding functions and increases the stability and adaptability of the whole structure.

HOMONOMOUS SYSTEMS are systems with structurally similar (isomorphic) levels. Apart from pure structural types (networks of networks, hierarchies made up of hierarchies, or markets consisting of markets), homonomous sytems include mixed structures, provided that the same ratio between different structure types and the same pattern of interactions between them is characteristic of all their structural levels. Homonomous systems possess manifest fractal properties: parts of the system are reduced copies of the whole system in structural terms. Cf. *heteronomous systems.*

INTERACTIVE NETWORKS engage in intense interaction with the social milieu. Their boundaries are dynamic, fuzzy, and often unclear. These features are typical of many global networked organizations.

ISOLATIONIST NETWORKS are characterized by strong "external skeletons"—closed borders—and unlimited communication *only* inside these borders. In human society, they are exemplified by various closed clandestine decentralized societies.

KIN ALTRUISM: Sacrificial behavior for the benefit of close relatives such as siblings, parents, or offspring that have a large number of common genes (Hamilton, 1964). Helping a relative at the expense of one's own reproduction or survival chances actually promotes the spreading of the helper's genes within the population.

LEADERSHIP: Guiding a group's movements—leading the group literarily—and controlling other collective behaviors (Lamprecht, 1996).

LOYAL BEHAVIOR: All kinds of friendly interactions among individuals that cement a biosocial system. Of particular importance for both humans and animals are the forms of loyal behavior denoted as *affiliation* and *cooperation.*

MARKET STRUCTURES: Social structures where competition dominates over cooperation. They include autonomous agents that enter into competitive relations with one another, while cooperation is only limited to contracts between the independent agents (Powell, 1990; Meulemann, 2008; Kahler, 2009a). Analogs of markets, i.e., competition-based structures existing in human societies lacking true market relationships as well as in communities formed by various biological species, can be called *quasi-market* structures (Oleskin, 2012). The term "quasi-market" is used in a different sense by some economists: it denotes interactions between former departments of a firm that have acquired some degree of autonomy and entered into contract-based relations.

(THE) MATRIX: This term denotes the totality of network-consolidating factors and mechanisms including (1) a material structural basis, e.g., the extracellular biopolymers that bind cells together, and (2) immaterial regulatory factors such as social norms and restrictions. For example, the behavior of social insects, such as ants or bees, is influenced by a complex system of social norms and rituals (Zakharov, 1991; Hölldobler & Wilson, 2009).

MODULES (and the **modular paradigm**): Modules are interconnected repetitive structural units (Marfenin, 1993, 1999) exemplified by zooids (polyps and medusas) in colonial cnidarians. Modules form coherent higher-order systems, often referred to as decentralized *modular organisms*. With respect to the neural networks of the brain, the term "module" denotes a network dealing with a specific function. It may represent a local structure (or a group of adjacent brain structures) or a delocalized brain network that is responsible for a certain activity such as, e.g., memorizing information.

NETWORK PARADIGMS (paradigms of network organization): The main organizational patterns (paradigms) of network structures in biological systems. In this work, the *cellular, modular, equipotential, eusocial, neural,* and *egalitarian* paradigms are singled out; the author admits that other paradigms may exist that are not considered in this work. Each paradigm is not necessarily confined to a specific group of biological species. Some paradigms apply to a relatively wide variety of species. In addition, social structures of many species can be described in terms of more than one paradigm. This is exemplified by social insects: although predominantly considered from the viewpoint of the eusocial paradigm they, to an extent, fit the modular paradigm emphasizing the quasiorganismic features of their social structures.

NETWORK STRUCTURES (NETWORKS): The term has been defined in at least two different ways in the literature: (1) in the broad sense, denoted in this work as *sensu lato* networks: networks are any sets of items, which are called vertices or nodes, with connections between them, called edges or links (Newman, 2003, 2012) and (2) in the more specific meaning called *sensu stricto networks* herein: networks are systems that lack a central pacemaker (leader, dominant element), and their activities and collective behaviors result from cooperation among their members often involving a number of partial leaders with limited power and competence. This interpretation is used in the social sciences (Powell, 1990; Castells, 1996, 2004; Meulemann, 2008; Kahler, 2009a, b). Such *sensu stricto networks* are contrasted with structures that contain a single center (*hierarchies*) as well as with those whose elements predominantly compete with one another (*markets* and their analogs, *quasi-market structures*).

NEURAL (NEURONAL) STRUCTURES and **the neural paradigm:** Neural networks arestructures formed by nervous cells (neurons) or their social or artificial analogs. These networks typically exhibit the following features: (1) collective information processing and decision-making; (2) an associative mode of operation; (3) adaptivity, i.e., the ability to change their structure in order to optimally solve problems; and (4) an enhanced reliability.

NODE DEGREE (degree centrality of a node): The degree of a node in a network is defined as the number of its neighbors with whom it is directly linked. This measure provides one of the criteria for estimating the *hierarchization (centralization)* of a network. It is defined as the sum of the differences between the largest value of a node's degree centrality in the network (K_{max}) and those of all n nodes in this network (Freeman, 1979).

PATH LENGTH (geodesic distance): The minimum number of nodes one has to traverse to get from one node to another (Newman, 2012, p.10).

RANDOM NETWORKS are created as follows: "we start with n nodes and connect every pair of nodes with probability p" (Almaas et al., 2007, p.9). Random networks include typical nodes whose *node degrees* have the maximum probability. The node degree distribution is approximately that of a Gaussian curve with short tails

RECIPROCAL ALTRUISM: Sacrificial behavior that benefits those who are ready to perform an analogous act of altruism benefitting the altruist (Trivers, 1971). Indirect reciprocal altruism means that altruism is rewarded via third persons in whose eyes the altruist has a good reputation (Alexander, 1979).

SCALABILITY: The capacity of a structure (e.g., a network) to change its size without altering the organizational pattern (Kahler, 2009a).

SCALE-FREE NETWORKS comply with the following two principles: "First, networks grow through the addition of new nodes linking to nodes already present in the system. Second, there is a higher probability to link to a node with a larger number of connections, a property called preferential attachment" (Almaas et al., 2007, p.9). Such Barabási-Albert (BA) networks tend to follow the *power law* in terms of their node degree distribution.

SMALL-WORLD EFFECT: Most pairs of nodes in most networks seem to be connected by a disproportionately short path through the network (Newman, 2003, p.9), compared to the numbers of nodes in the network. For instance, within the living cell, "paths of only three to four reactions can link most pairs of metabolites" (Barabási & Oltvai, 2004, p.106).

SOCIAL COGNITIVE FUNCTION (SCF): This function enables humans and, to an extent, higher animals (e.g., apes, monkeys, and cetaceans), to obey social behavioral norms and to respond adequately to the behavior of others.

THEORY OF MIND (TOM): The capacity to understand others and to know what they can and what they cannot know. TOM is linked to the *social cognitive function.*

THREE-DIMENSIONAL NETWORK includes *partial leaders* whose functions are limited to a particular situation/period of time/area of activity. Such a network may have a large number of leaders. Despite the partial leaders, the structure lacks the pyramidal shape that is characteristic of hierarchies. Nonetheless, the partial leaders typically represent nodes with a relatively high *node degree* (K), i.e., hubs.

Author's Contact Information

Prof. Alexander V. Oleskin,
General Ecology Department,
Biology Faculty,
Moscow State University,
Moscow, Russia
Email: AOleskin@Rambler.ru

Acknowledgments

I wish to thank most sincerely my wife, Julia P. Oleskina, for her constant support during the whole period of my work on the manuscript. I am very grateful to my mother, Elena P. Feofilova, and my daughter, Liza A. Oleskina. Dr. Andrey Postnov's technical assistance is gratefully acknowledged. The work on this book was also stimulated by the thought-provoking meetings and conferences of the *Club of Biopolitics (Moscow Society of Natural Scientists).*

Index

A

B

C

D

E

F

G

H

S

T